TURBOMACHINERY ROTORDYNAMICS

TURBOMACHINERY ROTORDYNAMICS

Phenomena, Modeling, and Analysis

DARA CHILDS
Texas A & M University
College Station, Texas

A Wiley-Interscience Publication

JOHN WILEY & SONS, INC.

New York • Chichester • Brisbane • Toronto • Singapore

Library of Congress Cataloging in Publication Data:
Childs, Dara W.
    Turbomachinery rotordynamics: phenomena, modeling, and analysis/
Dara Childs.
        p.   cm.
    Includes index.
    ISBN 0-471-53840-X
    1. Turbomachines—Dynamics.   2. Rotors—Dynamics.   I. Title.
TJ267.C46 1993
621.406—dc20   92-35452

Printed in the United States of America

10 9 8 7 6 5 4 3 2 1

This book is dedicated to my wife Susan,
whose cheerful support made it possible.

CONTENTS

PREFACE

This book has been prepared to introduce engineers with a background in mechanics to a collection of phenomena and related analysis techniques associated with dynamics of rotating machinery, i.e., turbines, compressors, pumps, power-transmission shafting, etc. It is intended primarily for practicing engineers who are involved in the design, analysis, or operation of turbomachinery, but will also serve as a text for a separate course in rotordynamics or as a supplementary reference for machine-design courses.

I began working on this book in the late 1970s while I was at the University of Louisville and substantially completed the first three chapters. My efforts on the book ended in 1980 when I moved to Texas A&M University (TAMU) and became heavily involved in research related to the influence of liquid and gas seals on rotordynamics. A sabbatical at the ETH in Zurich, Switzerland (kindly provided by TAMU) gave me the time necessary to complete most of this reference.

In my view, literature related to rotordynamics covers several general stages. In the beginning, rotordynamics was the study of vibrations related to the rotor's structural dynamics, without concern for the bearings. Books by Stodola (1927) and Biezeno and Grammel (1959), for example, examine critical-speed calculations for flexible rotors. Then, beginning in the early 1960s, most of the attention focused on hydrodynamic bearings, this was largely stimulated by Lund and Sterlicht (1962) and Lund (1964). Gunter's work (1966) related to rotordynamic stability problems, combined with Lund's (1974) method for calculating damped critical speeds, stimulated a great deal of interest in rotor-bearing stability problems. In the mid 1970s, rotordynamic instability experiences with various high-pressure compressors and the high-pressure fuel turbopump of the Space Shuttle main engine focused a great deal of attention on the influence of fluid-structure-interaction forces, particularly forces due to liquid and gas seals, impellers, and turbines.

Today, adequate rotordynamic models for high performance turbomachinery must account for the structure, the bearings, and all of the remaining forces. The present text is concerned with the development of adequate, computationally oriented, component and system models for the analysis of rotors. No attempt is made to provide exhaustive lists of references or to outline the historical development of rotordynamics. To assist the reader in developing an appreciation for rotordynamics, the text material has been organized into the following three basic parts:

(a)   *Part 1.* An introduction to rotordynamics phenomena is provided by Chapter 1, based on comparatively simple idealized models. An appreciation of the material provided in this chapter is necessary for an understanding of the types of problems to be anticipated in the development and operation of rotating machinery; however, the material provided in this chapter is of little direct value in the development of models for "real" rotors.

(b)   *Part 2.* Chapters 2–7 remedy this deficiency by first discussing the development of structural-dynamic models, and eigenanalysis for undamped flexible rotors Chapter 2. The initial focus of Chapter 2 is in the development of lumped-parameter representations for flexible rotors. This objective is met by first deriving appropriate component equations of motion and then explaining the general matrix procedure used in developing a rotor-stiffness matrix. Eigenanalysis procedures to be used with a general matrix formulation of the rotor vibration equation are then reviewed. The Myklestad-Prohl transfer-matrix approach for modeling and eigenanalysis of rotors is also reviewed for lumped-parameter rotor representation. The concluding flexible-rotor representation presented in this chapter is based on finite elements. The finite-element approach is introduced for the slender, Euler-beam model, and then extended to a slender beam with rotary inertia and gyroscopic coupling.

The completion of a structural dynamics model for a rotor based on the results of Chapter 2 generally constitutes only the initial step in the development of an adequate rotordynamic model, since the principal analysis complication which arises in rotordynamics, as compared to other areas of vibration analysis, is the requirement of modeling forces arising due to hydrodynamic bearings, squeeze-film dampers, seals, etc. Chapters 3–6 provide derivations of models for these forces which are suitable for rotordynamic analysis. Specifically, analytic component models are derived which define the external fluid forces acting on a rotor as a consequence of its motion.

A rotordynamic system model is obtained by adding the appropriate fluid-force component models to the structural dynamic model of Chapter 2. Methods for assembling and analyzing completed rotordynamics models are reviewed in Chapter 7.

(c)   *Part 3.* To demonstrate the lessons of the book, the alternate technology development (ATD) high-pressure fuel turbopump (HPFTP) of the

Space Shuttle main engine (SSME) is analyzed in Chapter 8. The model for the ATD-HPFTP includes liquid and gas seals, impellers, and turbines. The structural dynamics of both the rotor and the housing are accounted for. Linear analysis demonstrates critical speeds, synchronous response, and stability calculations. Transient solutions demonstrate the influence of bearing "dead bands" on the solution response.

When I began this book, I believed (naively) that a book could be assembled which covered "everything" about rotordynamics. In preparing the present book, a large part of my time was spent deciding on material to reject. To arrive at a book of reasonable proportions, some important material had to be omitted. Because of the emphasis on modeling and analysis, balancing is not covered. Torsional and coupled lateral-torsional vibrations are not covered. Also, there is no coverage for actively controlled bearings and dampers, because I have simply not had occasion to work closely on these topics. Similarly, space (and limited direct knowledge) have restricted the coverage of relevant computational fluid dynamics work related to impellers, seals, and turbines.

Having provided the above disclaimers, I feel that mastery of the contents of this book will allow the reader to model, analyze, and understand the rotordynamics of most operating turbomachinery. There is clearly enough material and alternative viewpoints left over to encourage other would-be authors in the area of rotordynamics.

ACKNOWLEDGEMENT

I am indebted to several people and organizations who have helped materially in the development of this book. First, my thanks to the succession of typists and wordprocessors who waded through my handwritten drafts and successive revisions. This includes Donna Greenwell at the University of Louisville and then Anne Owens, Suzan Taylor, and Melinda Price at Texas A&M University (TAMU).

Several colleagues were kind enough to review chapters from my book. Harold Nelson at Texas Christian University and Chang-Ho Kim at KIST in South Korea reviewed and made helpful corrections on the first three chapters. Luis San Andres of TAMU provided very helpful suggestions for Chapter Three. Joe Scharrer of RSR, Inc., made several useful suggestions for Chapter Five. Jørgen Nikolajsen, formerly of TAMU, reviewed Chapter Seven and identified several points which needed clarification. To all these gentlemen, my warmest thanks.

As noted in the Preface, much of the book was completed during my sabbatical at the ETH in Zurich. My thanks are extended to the TAMU Association of Former Students who funded the sabbatical and Gerhard Schwitzer who provided a very congenial working environment at the ETH.

My thanks are also extended to a range of technical societies and private companies for permission to print many of the figures in this book. Specifically, thanks are extended to: the American Society of Mechanical Engineers (ASME), the Institute of Mechanical Engineers (IMechE), the Society of Tribologist and Lubrication Engineers (STLE), the Turbomachinery Laboratory at TAMU, and to Pratt & Whitney at West Palm Beach, Florida and Rocketdyne division of Rockwell International, Conoga Park, California.

TURBOMACHINERY ROTORDYNAMICS

1

INTRODUCTION TO
ROTORDYNAMIC PHENOMENA

1.1 INTRODUCTION

The material presented in this chapter was selected to introduce and explain the nature of rotordynamic phenomena from comparatively simple analytic models. As would be expected, the phenomena demonstrated by flexible rotors and the techniques employed for their analysis are basically similar to other areas of vibrations and structural dynamics. Specifically, one is generally concerned with linear resonance phenomena (natural frequency and frequency-response calculations), linear instabilities, parametric instabilities, and forced steady-state and transient nonlinear response.

Considering these topics separately, the following sections consider the frequency-response characteristics of flexible rotors due to imbalance and product-of-inertia disturbances, gravity loading, and shaft distortion or bow. Rotor forward and backward critical speeds are defined, and their relationships to rotor natural frequencies are examined. The phenomenon of rotor subsynchronous whirl (linear instability) is introduced using internal rotor damping as a representative destabilizing mechanism. The potential for improving rotor stability by adding damping at bearings or by introducing orthotropic stiffness characteristics at bearing supports is examined. Rotor shaft-stiffness orthotropy is used as an example of mechanisms which have the potential for parametric excitation of rotor instabilities. Nonsymmetric clearance effects at bearings or rubbing due to a "jammed" or off-centered seal are also shown to have the potential for rotor parametric excitation. Further, these mechanisms are also shown to support fractional-frequency subharmonic rotor resonances. Additional nonlinear mechanisms examined here are (a) symmetric synchronous response of rotors in bearing clearances, (b) interaction of a rotor and stator across a clearance, and (c) a transient solution of a critical-speed transition.

A thorough understanding of material presented in this chapter is necessary (and generally sufficient) to *understand* the *desired* design behavior of rotating machinery and to recognize the distinctive characteristics of *undesirable* dynamic behavior.

1.2 THE JEFFCOTT MODEL: CRITICAL SPEEDS AND SYNCHRONOUS IMBALANCE RESPONSE

The Jeffcott* (1919) flexible motor model is illustrated in Figure 1.1, and consists of a flat disk supported by a uniform, massless, flexible shaft, which is supported at its ends by rigid frictionless bearings. The X, Y, Z coordinate system is inertial, with Z the nominal axis of rotor rotation. The x, y, z system is fixed to the disk, and its origin is defined by the vector **R** relative to the X, Y, Z system. The mass center of the disk does not lie on the elastic axis of the shaft, and its position relative to the origin of the x, y, z system is defined by the vector **a**. The (only) rotation of the x, y, z system relative to the X, Y, Z system is defined by the angle ϕ. The equations of motion for the system illustrated in Figure 1.1 can be stated from Newton's laws of motion as

$$m\ddot{R}_X + kR_X = f_X + ma_X\dot{\phi}^2 + ma_Y\ddot{\phi},$$

$$m\ddot{R}_Y + kR_Y = f_Y + ma_Y\dot{\phi}^2 - ma_X\ddot{\phi},$$

$$J_z\ddot{\phi} = M_z + ma_Y\ddot{R}_X - ma_X\ddot{R}_Y, \tag{1.1}$$

where

$$a_X = a_x \cos\phi - a_y \sin\phi,$$

$$a_Y = a_x \sin\phi + a_y \cos\phi. \tag{1.2}$$

Further, k is the shaft-stiffness coefficient, m is the rotor (disk) mass, and J_z is its moment of inertia about the z axis. The components of the external force vector are denoted by f_X, f_Y, and the component of the external moment vector along the z axis is M_z.

---

*Crandall (1983) suggests that, by right of precedence, this model might properly be called the Föppl (1895) model, since he was the first investigator to use the model and demonstrate that a rotor could operate supercritically. On the European continent this model is normally called the Laval rotor in honor of Carl Laval (1845–1913) who first built a single-stage steam turbine which actually looks like the model of Figure 1.1.

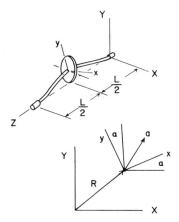

1.1 The Jeffcott flexible-rotor model.

An inspection of Eqs. (1.1) shows that transverse motion of the rotor (as defined by R_X, R_Y) can be readily excited by the components of the external force vector f_X, f_Y. Typically, however, the dominant source of rotor excitation is the vector **a**, the rotor imbalance vector. We are initially concerned with the response of the rotor displacement vector **R** to the imbalance vector **a** for constant running speed; i.e., $\ddot{\phi} = 0$, $\dot{\phi} = \omega$, $\phi = \omega t$. The dimensionality of the governing equations (1.1) can be reduced by introducing the following complex variables:

$$R = R_X + jR_Y, \qquad A = a_X + ja_Y, \qquad a = a_x + ja_y \qquad (1.3)$$

to obtain

$$\ddot{R} + \lambda^2 R = \omega^2 a e^{j\omega t} = \omega^2 A, \qquad \lambda^2 = k/m. \qquad (1.4)$$

The steady-state solution for $\omega \neq \lambda$ is

$$R = B(\omega) a e^{j\omega t} = B(\omega) A, \qquad (1.5)$$

where

$$B(\omega) = \frac{\omega^2}{\lambda^2 - \omega^2} = \frac{1}{(\lambda/\omega)^2 - 1}. \qquad (1.6)$$

For $\omega = \lambda$, the following particular solution results:

$$R = \frac{t\lambda}{2} a e^{j(\omega t - \pi/2)} = \frac{t\lambda}{2} A e^{-j\pi/2}. \qquad (1.7)$$

In words, these solutions show that the displacement vector **R** and imbalance vector **a** are in phase for a running speed ω which is much less than the natural frequency λ. For ω much greater than λ, the two vectors are 180° out of phase. For $\omega = \lambda$, the amplitude of **R** grows linearly with time, and its phase is 90° behind **a**. Figure 1.2 illustrates these three conditions.

A physical interpretation of the above results is that a rotor's motion is harmonic when viewed from the side and a circular orbit when viewed axially. The frequency of this motion coincides with the running speed ω and is said to be "synchronous." If the rotor's running speed is increased slowly from zero, a running speed will eventually be approached for which the rotor's amplitude diverges linearly with time, and this speed is customarily called the "critical speed" of the rotor. The steady-state response of the rotor above its critical speed is also synchronous. For the Jeffcott model, the rotor's critical speed is indistinguishable from its natural frequency; however, as we shall see in Section 1.4, this is not generally the case.

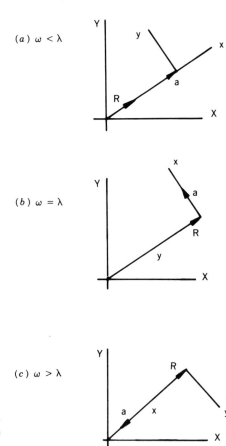

$(a)\ \omega < \lambda$

$(b)\ \omega = \lambda$

$(c)\ \omega > \lambda$

Figure 1.2   Synchronous response of the Jeffcott-rotor model for (a) $\omega \ll \lambda$, (b) $\omega \gg \lambda$, and (c) $\omega = \lambda$.

The effect of external (viscous) damping on the Jeffcott model of Eq. (1.1) will now be considered. The force due to external viscous damping is defined by

$$\mathbf{F}_{de} = -c_e\left(\mathbf{I}\dot{R}_X + \mathbf{J}\dot{R}_y\right) = -c_e\dot{\mathbf{R}}, \tag{1.8}$$

and is seen to be proportional to the velocity of the rotor in the X, Y, Z system $\dot{\mathbf{R}}$, but with opposite direction. The addition of external viscous damping to Eq. (1.1) yields

$$m\ddot{R}_X + c_e\dot{R}_X + kR_X = f_X + ma_X\dot{\phi}^2 + \ddot{\phi}ma_Y,$$

$$m\ddot{R}_Y + c_e\dot{R}_Y + kR_Y = f_Y + ma_Y\dot{\phi}^2 - \ddot{\phi}ma_X,$$

$$J_z\ddot{\phi} = M_z + ma_Y\ddot{R}_X - ma_X\ddot{R}_Y. \tag{1.9}$$

The constant-running-speed version of this equation, comparable to Eq. (1.4), is

$$\ddot{R} + 2\zeta_e\lambda\dot{R} + \lambda^2 R = \omega^2 ae^{j\omega t} = \omega^2 A, \tag{1.10}$$

where

$$2\zeta_e\lambda = c_e/m. \tag{1.11}$$

The steady-state solution is

$$R = C(\omega)ae^{j\omega t} = |C(\omega)|Ae^{j\psi}, \tag{1.12}$$

where

$$|C(\omega)| = \omega^2 \Big/ \left[(\lambda^2 - \omega^2)^2 + 4\zeta_e^2\lambda^2\omega^2\right]^{1/2},$$

$$\tan\psi = -2\zeta_e\lambda\omega/(\lambda^2 - \omega^2). \tag{1.13}$$

In physical terms, the rotor is in phase with the imbalance vector at low running speeds ($\omega/\lambda \ll 1$), is 180° out of phase at high running speeds ($\omega/\lambda \gg 1$), and is 90° behind the imbalance vector at the rotor's undamped critical speed $\omega = \lambda$. A comparison of Eqs. (1.6) and (1.13) shows that external damping causes the rotor's motion to be bounded at the critical speed. Figure 1.3 illustrates the solution of Eq. (1.12) for various damping factors ζ_e.

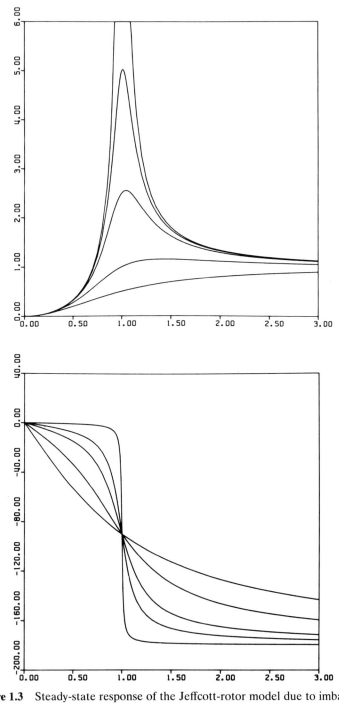

Figure 1.3 Steady-state response of the Jeffcott-rotor model due to imbalance.

The solution for the homogeneous version of Eq. (1.10) can be stated:

$$R_h = A_1 e^{\lambda\{-\zeta_e + j(1 - \zeta_e^2)^{1/2}\}t} + A_2 e^{\lambda\{-\zeta_e - j(1 - \zeta_e^2)^{1/2}\}t}, \tag{1.14}$$

and consists of two damped, precessionally spiraling orbits with the common period $T = 2\pi/\lambda(1 - \zeta_e^2)^{1/2}$. The precessional rotation of the first term in this solution is in the same direction as the rotor's rotation (forward precession), while the direction of the second is opposite (backward precession). In mathematical terms, the Jeffcott model has complex-conjugate roots at $-\zeta_e \lambda \pm j\lambda(1 - \zeta_e^2)^{1/2}$ and both a forward and backward critical speed at $\omega = \lambda(1 - \zeta_e^2)^{1/2}$. However, from Eq. (1.12) only the forward precessional motion is excited by rotor imbalance. We will return to a discussion of rotor forward and backward critical speeds in Sections 1.4 and 1.5.

The steady-state solutions of Eqs. (1.5) and (1.12) explicitly require constant running speed; i.e., $\dot\phi = \omega$; $\ddot\phi = 0$. From the last of Eq. (1.1), this condition is met by requiring

$$M_z = m\left(a_X \ddot{R}_Y - a_Y \ddot{R}_Y\right) = m(\mathbf{A} \times \ddot{\mathbf{R}})_z.$$

The term on the right is the z component of the cross-product of

$$\mathbf{A} = \mathbf{I}a_X + \mathbf{J}a_Y, \qquad \ddot{\mathbf{R}} = \mathbf{I}\ddot{R}_X + \mathbf{J}\ddot{R}_Y.$$

For zero external damping, Figures 1.2(a) and 1.2(b) show that the cross-product $(\mathbf{A} \times \ddot{\mathbf{R}})$ is zero for $\omega \neq \lambda$. From Figure 1.2(c) and Eq. (1.5), the torque

$$M_z = mt\left(\frac{\lambda^3}{2}\right)a^2 \tag{1.15}$$

is required at $\omega = \lambda$. This result clearly shows that the spin-axis torque must supply the energy for the unbounded lateral deflections which are predicted at the undamped critical speed. When external damping is provided, Eqs. (1.12) and (1.13) show that the torque

$$M_z = -ma^2\omega^2|C(\omega)|\sin\psi \tag{1.16}$$

is required to maintain the constant running speed ω, implying that external damping will cause the rotor to decelerate if the torque of Eq. (1.16) is not applied.

The coordinate transformation

$$V_x = V_X \cos\phi + V_Y \sin\phi,$$
$$V_y = -V_X \sin\phi + V_Y \cos\phi \tag{1.17}$$

can be used to restate Eq. (1.9) as

$$
\begin{aligned}
m\ddot{R}_x &+ c_e\dot{R}_x - 2\dot{\phi}m\dot{R}_y - c_e\dot{\phi}R_y + \left(k - \dot{\phi}^2m\right)R_x \\
&= f_x + \dot{\phi}^2ma_x + \ddot{\phi}\left(a_y + R_y\right), \\
m\ddot{R}_y &+ c_e\dot{R}_y + 2\dot{\phi}m\dot{R}_x + c_e\dot{\phi}R_x + \left(k - \dot{\phi}^2m\right)R_y \\
&= f_y + \dot{\phi}^2ma_y - \ddot{\phi}m\left(a_x + R_x\right), \\
J_z\ddot{\phi} &= M_z + ma_y\left(\ddot{R}_x - 2\dot{\phi}\dot{R}_y - \dot{\phi}^2R_x - \ddot{\phi}R_y\right) \\
&\quad - ma_X\left(\ddot{R}_y + 2\dot{\phi}\dot{R}_x - \dot{\phi}^2R_y + \ddot{\phi}R_x\right).
\end{aligned}
\tag{1.18}
$$

These equations define the components of **R** in the rotor-fixed x, y, z coordinate system. A steady-running-speed version of these equations, comparable to Eq. (1.10), follows:

$$
\ddot{r} + 2\zeta_e\lambda\dot{r} + j2\omega\dot{r} + \left(\lambda^2 - \omega^2\right)r + j2\zeta_e\lambda\omega r = \omega^2 a,
\tag{1.19}
$$

where

$$
r = R_x + jR_y.
\tag{1.20}
$$

The solution is

$$
r = a\left|C(\omega)\right|e^{j\psi},
\tag{1.21}
$$

which shows that synchronous shaft motion is nonoscillatory to an observer moving with the rotor-fixed x, y, z coordinate system. In physical terms, synchronous rotor motion does not cause alternating stress in the shaft.

1.3   THE JEFFCOTT MODEL: RESPONSE TO GRAVITY LOADING AND SHAFT "BOW"

We are concerned in this section with the following two additional types of common rotor forcing functions:

(a) gravity loading of horizontal rotors, and
(b) static rotor misalignment due to shaft "bow."

Gravity loading is the simpler of the two, and is accounted for in the following restatement of Eq. (1.9):

$$
\begin{aligned}
m\ddot{R}_X + c_e\dot{R}_X + kR_X &= ma_X\dot{\phi}^2 + \ddot{\phi}ma_Y, \\
m\ddot{R}_Y + c_e\dot{R}_Y + kR_Y &= ma_Y\dot{\phi}^2 - \ddot{\phi}ma_X - mg.
\end{aligned}
\tag{1.22}
$$

From Eq. (1.12), the steady-speed solution to this equation is

$$\boldsymbol{R} = R_X + jR_Y = a|C(\omega)|e^{j(\omega t + \psi)} - jg/\lambda^2. \tag{1.23}$$

In the rotor-fixed x, y, z coordinate system, this solution has the form

$$r = R_x + jR_y = a|C(\omega)|e^{j\psi} - j(g/\lambda^2)e^{-j\omega t}. \tag{1.24}$$

Equation (1.24) shows that gravity loading (as opposed to imbalance loading) yields harmonic motion to an observer fixed to the rotor. Hence gravity loading of a horizontal rotor (or any other nonsynchronous rotor loading) will yield alternating rotor bending stresses, which in turn give rise to internal damping forces. The significance of internal damping forces on rotor stability is discussed in Section 1.5. Gravity excitation of rotors that have stiffness orthotropy (see Section 1.7) causes a resonance at a running speed that is approximately one-half the nominal critical speed.

Figure 1.4 illustrates a bowed shaft, specifically, a shaft whose static equilibrium curve does not coincide with the Z axis of rotation. From Eq. (1.18), the steady-speed, rotor-fixed model for this situation is

$$m\begin{Bmatrix} \ddot{R}_x \\ \ddot{R}_y \end{Bmatrix} + \begin{bmatrix} c_e & -2\omega m \\ 2\omega m & c_e \end{bmatrix} \begin{Bmatrix} \dot{R}_x \\ \dot{R}_y \end{Bmatrix} + (k - \omega^2 m) \begin{Bmatrix} R_x \\ R_y \end{Bmatrix}$$

$$= k\begin{Bmatrix} R_{0x} \\ R_{0y} \end{Bmatrix} + \omega^2 \begin{Bmatrix} ma_x \\ ma_y \end{Bmatrix}, \tag{1.25}$$

where R_{0x}, R_{0y} are the x, y, z components of the static equilibrium vector

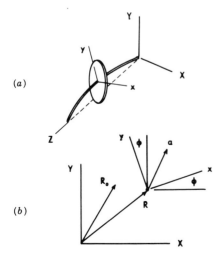

(a)

(b)

Figure 1.4   A bowed shaft.

$\mathbf{R}_0$. The x, y, z force components kR_{0x}, kR_{0y} are analogous to the X, Y, Z gravity force components in Eq. (1.22) in that their magnitudes are independent of both running speed and time. The stationary X, Y, Z version of Eq. (1.25) is

$$\ddot{R}_X + 2\zeta_e\lambda\dot{R}_X + \lambda^2 R_X = \lambda^2 R_{0X} + \omega^2 a_X,$$
$$\ddot{R}_Y + 2\zeta_e\lambda\dot{R}_Y + \lambda^2 R_Y = \lambda^2 R_{0Y} + \omega^2 a_Y, \tag{1.26}$$

where

$$R_{0X} = R_{0x}\cos\phi - R_{0y}\sin\phi,$$
$$R_{0Y} = R_{0x}\sin\phi + R_{0y}\cos\phi. \tag{1.27}$$

From Eq. (1.3), the complex-variable statement of these equations is

$$\ddot{R} + 2\zeta_e\lambda\dot{R} + \lambda^2 R = \omega^2 a + \lambda^2 R_0 = \left(\omega^2 a + \lambda^2 R_0 e^{j\alpha_0}\right)e^{j\omega t}, \tag{1.28}$$
$$R_0 = R_{0X} + jR_{0Y} = R_0 e^{j(\omega t + \alpha_0)}. \tag{1.29}$$

From Eq. (1.28), the nonzero static equilibrium vector $\mathbf{R}_0$ yields a synchronous excitation vector in the stationary X, Y, Z system in a manner which is analogous to the imbalance vector $\mathbf{a}$, except that its magnitude is constant, and independent of ω.

The characteristics of the steady-state solution to Eq. (1.26) principally depend on the relative magnitude and phasing of the imbalance $\mathbf{a}$ and static equilibrium $\mathbf{R}_0$ vectors, and will be examined by the following restatement of Eq. (1.28):

$$\ddot{R} + 2\zeta_e\lambda^2\dot{R} + \lambda^2 R = a\left(\omega^2 + r_0\lambda^2 e^{j\alpha_0}\right)e^{j\omega t},$$
$$r_0 = R_0/a. \tag{1.30}$$

The steady-state solution is

$$R = ac(\omega)e^{j\omega t}, \tag{1.31}$$

where $c(\omega)$ is the complex function

$$c(\omega) = |c(\omega)|e^{j\alpha} = \left(\omega^2 + r_0\lambda^2 e^{j\alpha_0}\right)e^{j\psi} \Big/ \left\{\left(\lambda^2 - \omega^2\right)^2 + \left(2\zeta_e\lambda\omega\right)^2\right\}^{1/2},$$
$$\tan\psi = -2\zeta_e\lambda\omega/(\lambda^2 - \omega^2). \tag{1.32}$$

For a zero-magnitude, static-equilibrium vector, $R_0 = r_0 = 0$, this solution reduces to the steady-state imbalance solution of Eq. (1.12). An examination of Eq. (1.32) shows that the solution including shaft bow can yield solutions which differ markedly from solutions due to imbalance alone. Figures 1.5(a)

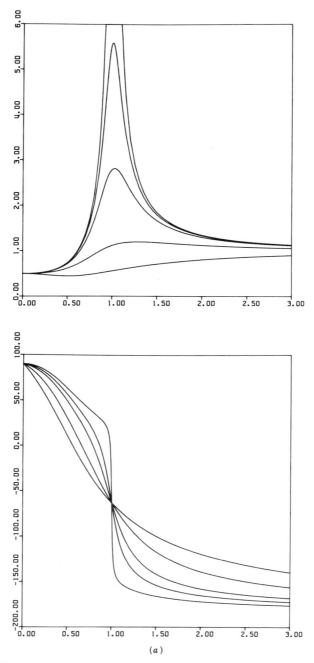

Figure 1.5(a)   Steady-state response to the Jeffcott-rotor model due to imbalance and shaft bow as defined by Eq. (1.32) for $r_0 = 0.5$, $\alpha_0 = \pi$.

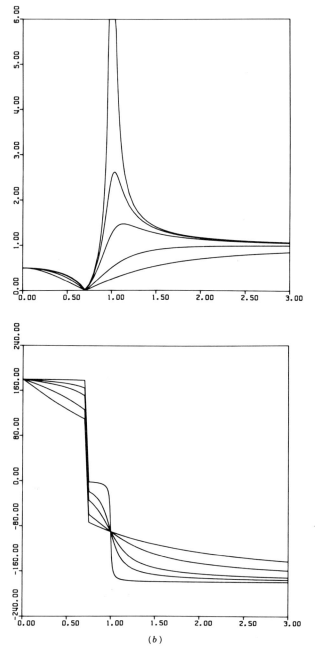

Figure 1.5(b)   Steady-state response to the Jeffcott-rotor model due to imbalance and shaft bow as defined by Eq. (1.32) for $r_0 = 0.5$, $\alpha_0 = \pi/2$.

and (b) illustrate the solution of Eq. (1.32) for $r_0 = 0.5$ with α_0 equal to $\pi/2$ and π, respectively. These results are generally self-explanatory in terms of Eq. (1.32), but they differ markedly from the solution of Figure 1.3.

The principal practical consequence of the results of Figure 1.5 is that shaft bow is as likely to be present in slender shafts as shaft unbalance. Furthermore, a shaft which is initially unbowed can become bowed due to (a) thermal stresses caused by rubbing or (b) gravity sag during shutdown, etc. Hence, an analysis of the experimental steady-state dynamics characteristics of a rotor may make no sense when viewed from the imbalance-only results of Figure 1.3, but be perfectly reasonable in terms of the combined imbalance/shaft-bow model results of Figure 1.5.

1.4 THE STODOLA-GREEN (RIGID-BODY) MODEL FOR ROTORDYNAMICS: EFFECTS OF ROTARY INERTIA AND GYROSCOPIC COUPLING

The analysis of the preceding section uses the simplicity of the Jeffcott model to advantage in introducing the concepts of rotor critical speeds and synchronous motion. However, the Jeffcott model is basically a particle or point-mass representation and is inadequate to explain phenomena that arise due to the rigid-body character of flexible rotating equipment. The Stodola-(1927) Green (1948) model of Figure 1.6 consists of a rigid disk (not necessarily thin) used here to examine the influence of rigid-body parameters on the rotor's natural frequencies, critical speeds, and synchronous response. From a practical viewpoint, the Stodola-Green model can be used to explain many of the dynamic consequences of an over-hung turbine wheel design for rotating equipment. The development of this section follows Den Hartog (1952).

Figure 1.7 illustrates the Euler angles β_X, β_Y, ϕ, which are used to define the disk's orientation. The x, y, z coordinate system of this figure is fixed to the disk, and the X, Y, Z system is an inertial frame with Z the nominal axis of rotation. The angles β_X, β_Y define the *elastic* rotations of the disk and are assumed to be small. From Section 2.2, the appropriate governing vibration

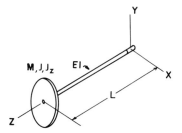

Figure 1.6 The Stodola-Green flexible rotor model.

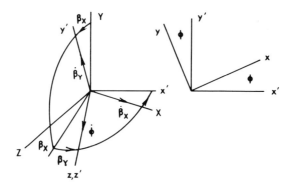

Figure 1.7   Euler-angle definitions for the Stodola-Green flexible-rotor model.

equations for the X, Z and Y, Z planes are

$$m\ddot{R}_X = f_X + \bar{f}_X + ma_X\dot{\phi}^2 + ma_Y\ddot{\phi},$$

$$J\ddot{\beta}_Y = M_Y + \overline{M}_Y + \dot{\phi}J_z\dot{\beta}_X + \dot{\phi}^2J_{XZ} + \ddot{\phi}J_{YZ},$$

$$m\ddot{R}_Y = f_Y + \bar{f}_Y + ma_Y\dot{\phi}^2 - ma_X\ddot{\phi},$$

$$J\ddot{\beta}_X = M_X + \overline{M}_X - \dot{\phi}J_z\dot{\beta}_Y + \dot{\phi}^2J_{YZ} + \ddot{\phi}J_{XZ}, \tag{1.33}$$

while the running speed is defined by the moment-of-momentum equation

$$J_z\ddot{\phi} = M_Z + \ddot{\beta}_X J_{XZ} + \ddot{\beta}_Y J_{YZ} - \ddot{R}_Y ma_X + \ddot{R}_X ma_Y. \tag{1.34}$$

In these equations, J is the diametral moment of inertia (i.e., $J = J_{XX} = J_{YY}$), and J_z is the polar moment of inertia. The products of inertia (J_{XY}, J_{YZ}), are defined in terms of their rotor-fixed counterparts J_{xz}, J_{yz} by

$$J_{XZ} = J_{xz}\cos\phi - J_{yz}\sin\phi,$$

$$J_{YZ} = J_{xz}\sin\phi + J_{yz}\cos\phi. \tag{1.35}$$

The elastic reaction forces ($\bar{f}_X, \bar{f}_Y$) and moments ($\overline{M}_X, \overline{M}_Y$) are defined by the following stiffness matrices:

$$\begin{Bmatrix} \bar{f}_X \\ \overline{M}_Y \end{Bmatrix} = -\frac{12EI}{L^3}\begin{bmatrix} 1 & -L/2 \\ -L/2 & L^2/3 \end{bmatrix}\begin{Bmatrix} R_X \\ \beta_Y \end{Bmatrix} = -[K_{XZ}]\begin{Bmatrix} R_X \\ \beta_Y \end{Bmatrix},$$

$$\begin{Bmatrix} \bar{f}_Y \\ \overline{M}_X \end{Bmatrix} = -\frac{12EI}{L^3}\begin{bmatrix} 1 & L/2 \\ L/2 & L^2/3 \end{bmatrix}\begin{Bmatrix} R_Y \\ \beta_X \end{Bmatrix} = -[K_{YZ}]\begin{Bmatrix} R_Y \\ \beta_X \end{Bmatrix}. \tag{1.36}$$

The remaining terms in Eqs. (1.33)–(1.34) were introduced in the preceding section in connection with Eq. (1.1).

The Stodola-Green model is obviously more complicated than the Jeffcott model of the preceding section. The new parameters and phenomena introduced by this model will be considered in the following sequential steps:

(a) The effect of the diametral disk inertia (rotary inertia) on the zero-speed rotor natural frequencies is considered.

(b) The effect of the gyroscopic coupling terms $(\dot{\phi}J_z\dot{\beta}_X, \dot{\phi}J_z\dot{\beta}_Y)$ in causing the model's natural frequencies to depend on running speed is examined, with the consequent necessary distinction between rotor natural frequencies and critical speeds. The concept of asynchronous (backward) critical speeds is introduced and compared to synchronous (forward) critical speeds.

(c) The products of inertia are shown to be analogous to the imbalance components in providing a source of synchronous rotor excitation, and the nature of the synchronous response of the Stodola-Green model is examined.

Disk Diametral (Rotary) Inertia Effect on Zero-Running-Speed Natural Frequencies

The appropriate model to examine the zero-speed natural frequencies of the Stodola-Green model is

$$\begin{bmatrix} m & 0 \\ 0 & J \end{bmatrix} \begin{Bmatrix} \ddot{R}_X \\ \ddot{\beta}_Y \end{Bmatrix} + [K_{XZ}] \begin{Bmatrix} R_X \\ \beta_Y \end{Bmatrix} = 0,$$

$$\begin{bmatrix} m & 0 \\ 0 & J \end{bmatrix} \begin{Bmatrix} \ddot{R}_Y \\ \ddot{\beta}_X \end{Bmatrix} + [K_{YZ}] \begin{Bmatrix} R_Y \\ \beta_X \end{Bmatrix} = 0.$$

The motion in the X, Z and Y, Z planes is seen to be uncoupled, and, because of the stiffness and inertial symmetry of the model, either of the planar matrix equations above can be used to obtain the characteristic equation

$$\hat{\lambda}^4 - 4(1 + d^{-1})\hat{\lambda}^2 + 4d^{-1} = 0, \tag{1.37}$$

where

$$d = 3J/mL^2, \qquad \lambda_0^2 = 3EI/mL^3, \qquad \hat{\lambda} = \lambda/\lambda_0.$$

The parameter d is a measure of the "rotary-inertia effect" of the disk. For a thin disk of radius r, it reduces to $d = 3r^2/4L^2$. The eigenvalue λ_0^2 is

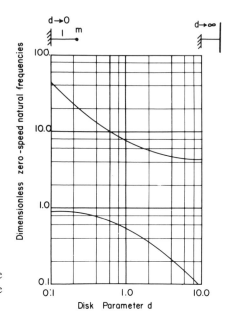

Figure 1.8 Zero-speed eigenvalues for the Stodola-Green flexible-rotor model. The effect of disk or rotary inertia.

obtained for a disk with zero diametral moment of inertia, i.e., a point mass. The general eigenvalue solution to Eq. (1.37) can be stated:

$$\hat{\lambda}_1^2 = 2(1 + d^{-1}) - 2\left[(1 + d^{-1})^2 - d^{-1}\right]^{1/2},$$

$$\hat{\lambda}_2^2 = 2(1 + d^{-1}) + 2\left[(1 + d^{-1})^2 - d^{-1}\right]^{1/2}, \qquad (1.38)$$

with the following two limiting-case solutions: (a) $d = 0$; $\lambda_1^2 = \lambda_0^2$, $\lambda_2^2 = \infty$, and (b) $d^{-1} = 0$, $\lambda_1^2 = 0$, $\lambda_2^2 = 4\lambda_0^2$. The general solution to Eq. (1.37) is illustrated in Figure 1.8 with physical models for the two limiting-case solutions. The form of the rotor mode shapes (eigenvectors) for the first and second eigenvalues are illustrated in Figure 1.9.

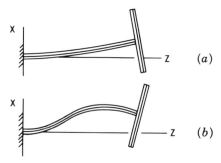

Figure 1.9 Zero-speed eigenvectors (mode shapes) for Stodola-Green flexible-rotor model.

The principal effect of disk diametral (or rotary) inertia at zero running speed is seen to be an additional eigenvalue associated with the additional rotational degree of freedom. The principal practical significance of disk inertia is the reduction of the first (lowest) natural frequency of the rotor.

The Effect of Gyroscopic Coupling on Rotor Natural Frequencies. Synchronous and Asynchronous Critical Speeds

For the general constant-running-speed case, the system characteristic equation can be obtained by substituting the assumed solution

$$\begin{Bmatrix} R_X \\ \beta_Y \end{Bmatrix} = \begin{Bmatrix} c_1 \\ c_2 \end{Bmatrix} \cos pt, \qquad \begin{Bmatrix} R_Y \\ \beta_X \end{Bmatrix} = \begin{Bmatrix} c_1 \\ -c_2 \end{Bmatrix} \sin pt \qquad (1.39)$$

into either of the homogeneous equations,

$$[L]\begin{Bmatrix} \ddot{R}_X \\ \ddot{\beta}_Y \end{Bmatrix} - \omega \begin{bmatrix} 0 & 0 \\ 0 & J_z \end{bmatrix} \begin{Bmatrix} \dot{R}_Y \\ \dot{\beta}_X \end{Bmatrix} + [K_{XZ}]\begin{Bmatrix} R_X \\ \beta_Y \end{Bmatrix} = 0,$$

$$[L]\begin{Bmatrix} \ddot{R}_Y \\ \ddot{\beta}_X \end{Bmatrix} + \omega \begin{bmatrix} 0 & 0 \\ 0 & J_z \end{bmatrix} \begin{Bmatrix} \dot{R}_X \\ \dot{\beta}_Y \end{Bmatrix} + [K_{YZ}]\begin{Bmatrix} R_Y \\ \beta_X \end{Bmatrix} = 0,$$

to yield

$$\hat{p}^4 - b\hat{\omega}\hat{p}^3 - 4(1 + d^{-1})\hat{p}^2 + 4b\hat{\omega}\hat{p} + 4d^{-1} = 0;$$

$$\hat{\omega} = \omega/\lambda_0, \qquad \hat{p} = p/\lambda_0, \qquad b = J_Z/J. \qquad (1.40)$$

For a given running speed this equation defines four real roots, two positive and two negative, which are denoted, respectively, by $\hat{p}_1, \hat{p}_2$ and $\hat{p}_{-1}, \hat{p}_{-2}$. Figure 1.10 illustrates the nature of the solutions to Eq. (1.40) for a range of running speeds. At $\omega = 0$, the solution reduces to $\hat{p}_1 = -\hat{p}_{-1} = \hat{\lambda}_1, \hat{p}_2 = -\hat{p}_{-2} = \hat{\lambda}_2$, with $\hat{\lambda}_1$ and $\hat{\lambda}_2$ defined by Eq. (1.37). The physical significance of the positive and negative roots can be better appreciated by a review of assumed solution forms in Eq. (1.39). As illustrated in Figure 1.11, the positive and negative roots correspond, respectively, to counterclockwise and clockwise rotor precession. The direction of the rotor precession and rotor rotation are the same for positive roots $\hat{p}_i$, and opposite for negative roots $\hat{p}_{-i}$. The intersections of the line $\hat{p} = \hat{\omega}$ with the solution curves of Figure 1.10 define the model's (only) forward or synchronous critical speed $\hat{p}_{1c}$, while the intersections of the line $\hat{p} = -\hat{\omega}$ with these curves define two backwards or asynchronous critical speeds $\hat{p}_{-1c}, \hat{p}_{-2c}$. Note from this figure

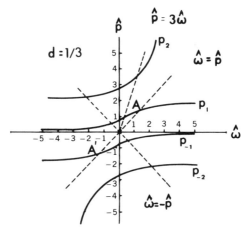

Figure 1.10 Natural frequencies for the Stodola-Green flexible-rotor model ($d = \frac{1}{3}$).

that

$$\hat{p}_{-1c} < \hat{\lambda}_1 < \hat{p}_{1c},$$

which is the customary result for rotors; i.e., the gyroscopic coupling terms yield forward and backward critical speeds which are, respectively, above and below their associated zero-running-speed natural frequencies. The influence of the gyroscopic coupling terms in elevating the roots $\hat{p}_1(\hat{\omega})$, $\hat{p}_2(\hat{\omega})$ with increasing running speed is frequently described as a "gyroscopic stiffening" effect. In Figure 1.10, the $\hat{p}_2$ root increases so rapidly with increasing ω that there is no intersection between this curve and the $\hat{p} = \hat{\omega}$ line. Dimentberg

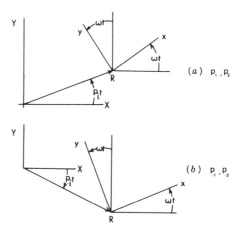

Figure 1.11 Motion corresponding to (a) forward precession associated with positive p_1, p_2 roots to Eq. (1.40), and (b) reverse precession for negative roots p_{-1}, p_{-2}.

(1961) has demonstrated that including the additional flexibility due to shear deflections will lower the second rotor natural frequency sufficiently to yield a second forward critical speed.

In rotating equipment, hydrodynamic forces or support nonlinearities frequently give rise to harmonic rotor loadings that are integer multiples of the running speed. Hence forcing functions arise, which cause a precessional motion of the rotor in the direction of rotation, but at a multiple of the running-speed frequency ω, and a resonant condition exists if one of the excitation frequencies coincides with the positive $\hat{p}_1, \hat{p}_2$ roots of Eq. (1.40). The intersection of line $\hat{p} = 3\hat{\omega}$ with the $\hat{p}_2$ curve in Figure 1.10 illustrates this possibility.

The defining relationship for forward critical speeds can be obtained directly by substituting $\hat{\omega} = \hat{p}$ into Eq. (1.40) with the result

$$\hat{p}^4(b - 1) - \hat{p}^2 4(b - 1 - d^{-1}) - 4d^{-1} = 0. \qquad (1.41)$$

The two real roots to this equation are denoted by A and A' in Figure 1.10. The defining equation for the backward critical speeds is obtained by substituting $\hat{\omega} = -\hat{p}$ into Eq. (1.40) to obtain

$$\hat{p}^4(1 + b) - 4(b + 1 + d^{-1})\hat{p}^2 + 4d^{-1} = 0. \qquad (1.42)$$

The forward critical speed of the Stodola-Green model has precisely the same practical significance as the critical speed of the Jeffcott model; viz., operation of a rotor at or near a forward critical speed causes large and potentially damaging deflections. Further, as we shall see in Section 1.5, the modes associated with forward critical speeds are potentially subject to unstable whirling motion. Conversely, the backward critical speeds are (generally) not excited by rotor imbalance, and the stability of modes associated with backward critical speeds tends to be enhanced by the factors which destabilize synchronous modes. The discussion which follows on synchronous rotor response should clarify the differences in response of modes associated with forward and backward critical speeds.

Steady-State Rotor Response

A comparison of the Stodola-Green model as defined in Eqs. (1.33)–(1.35) with the Jeffcott model of Eqs. (1.1) and (1.2) reveals a similarity between terms involving products of inertia J_{xz}, J_{yz} and those involving imbalance components a_x, a_y. Figure 1.12 illustrates a physical situation which gives rise to products of inertia. The disk in this figure is not perpendicular to the shaft; hence, there is an angular misalignment between the shaft axis and the principal axis of the disk. This misalignment causes a dynamic load on the rotor which is analogous in all respects to the dynamic loads induced by rotor

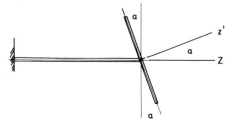

Figure 1.12 Static angular misalignment of disk and shaft which causes a synchronous product-of-inertia disturbance.

imbalance. From Eq. (1.33) the appropriate steady-speed ($\phi = \omega t$, $\dot{\phi} = \omega$, $\ddot{\phi} = 0$) equations are

$$[L]\begin{Bmatrix} \ddot{R}_X \\ \ddot{\beta}_Y \end{Bmatrix} - \omega \begin{bmatrix} 0 & 0 \\ 0 & J_z \end{bmatrix}\begin{Bmatrix} \dot{R}_Y \\ \dot{\beta}_X \end{Bmatrix} + [K_{XZ}]\begin{Bmatrix} R_X \\ \beta_Y \end{Bmatrix} = \omega^2 \begin{Bmatrix} ma_X \\ J_{XZ} \end{Bmatrix},$$

$$[L]\begin{Bmatrix} \ddot{R}_Y \\ \ddot{\beta}_X \end{Bmatrix} + \omega \begin{bmatrix} 0 & 0 \\ 0 & J_z \end{bmatrix}\begin{Bmatrix} \dot{R}_X \\ \dot{\beta}_Y \end{Bmatrix} + [K_{YZ}]\begin{Bmatrix} R_Y \\ \beta_X \end{Bmatrix} = \omega^2 \begin{Bmatrix} ma_Y \\ -J_{YZ} \end{Bmatrix}, \quad (1.43)$$

where

$$\begin{Bmatrix} ma_X \\ J_{XZ} \end{Bmatrix} = \begin{Bmatrix} ma_x \\ J_{xz} \end{Bmatrix} \cos \omega t - \begin{Bmatrix} ma_y \\ J_{yz} \end{Bmatrix} \sin \omega t,$$

$$\begin{Bmatrix} ma_Y \\ -J_{YZ} \end{Bmatrix} = \begin{Bmatrix} ma_x \\ -J_{xz} \end{Bmatrix} \sin \omega t + \begin{Bmatrix} ma_y \\ -J_{yz} \end{Bmatrix} \cos \omega t. \quad (1.44)$$

The steady-state solution to these equations is obtained by substituting the assumed solution,

$$\begin{Bmatrix} R_X \\ \beta_Y \end{Bmatrix} = \begin{Bmatrix} a_1 \\ a_2 \end{Bmatrix} \cos \omega t - \begin{Bmatrix} b_1 \\ b_2 \end{Bmatrix} \sin \omega t,$$

$$\begin{Bmatrix} R_Y \\ \beta_X \end{Bmatrix} = \begin{Bmatrix} a_1 \\ -a_2 \end{Bmatrix} \sin \omega t + \begin{Bmatrix} b_1 \\ -b_2 \end{Bmatrix} \cos \omega t, \quad (1.45)$$

into either of Eq. (1.43).

The assumed solution of Eq. (1.45) is clearly synchronous, as can be seen from the complex-variable definition

$$\boldsymbol{R} = R_X + jR_Y = (a_1 + jb_1)e^{j\omega t}, \quad (1.46)$$

which shows that the disk center is precessing in a circular orbit at the same frequency and in the same direction as the imbalance vector $A = ae^{j\omega t}$. In addition, the result shows that the backward critical speeds of Eq. (1.43) are not excited by imbalance or product-of-inertia disturbances. However, Den Hartog (1952) and other investigators noted that backwards or asynchronous critical speeds are occasionally encountered in the operation of rotating machinery. Dimentberg (1961) demonstrated that orthotropy in the (stationary) support structure stiffness (not the shaft stiffness) provides an explanation for this phenomenon. For example, suppose that a spring is added to the disk end of the shaft in Figure 1.6, resulting in an increase in stiffness in the X direction such that the $[K_{XZ}]$ matrix in Eq. (1.36) becomes

$$[K_{XZ}] = \frac{-3EI}{L^3} \begin{bmatrix} (4+q) & -2L \\ -2L & 4L^2/3 \end{bmatrix}, \tag{1.47}$$

and $[K_{YZ}]$ is unchanged. The appropriate steady-state solution for Eq. (1.43) is now

$$\begin{Bmatrix} R_X \\ \beta_Y \\ R_Y \\ \beta_X \end{Bmatrix} = \begin{Bmatrix} a_1 \\ a_2 \\ a_3 \\ a_4 \end{Bmatrix} \cos \omega t + \begin{Bmatrix} b_1 \\ b_2 \\ b_3 \\ b_4 \end{Bmatrix} \sin \omega t, \tag{1.48}$$

which leads to the following complex-variable solution:

$$R = \tfrac{1}{2}[(a_1 + b_3) + j(a_3 - b_1)]e^{j\omega t} + \tfrac{1}{2}[(a_1 - b_3) + j(a_3 + b_1)]e^{-j\omega t}$$

$$= Fe^{j\omega t} + Ge^{-j\omega t}.$$

The complex coefficients F, G are the amplitudes associated, respectively, with forward and backward precessional motion. Figures 1.13–1.15 demonstrate the nature of solutions to Eq. (1.43) for $a_x = a$, $a_y = J_{xz} = J_{yz} = 0$. The variables illustrated in Figure 1.13 are $f = |F|/a$ and $g = |G|/a$ for $q = 0.4$, and demonstrate that the backward critical speeds of Figure 1.10 are excited by imbalance if support stiffness orthotropy is provided, and that comparatively sharp peaks are associated with these resonances. Figure 1.14 illustrates $(R/a)_{max}$ versus $\hat{\omega}$ for $q = 0$ (symmetric stiffness, synchronous response) and $q = 0.4$ (orthotropic support stiffness, mixed synchronous and asynchronous response). The asymmetry provided by $q = 0.4$ provided some additional stiffness and slightly elevates the forward-critical-speed location. Figure 1.15 demonstrates the elliptical* nature of the solutions when support-stiffness orthotropy is provided.

---

*Consult Appendix A on the definition of elliptical orbits from the solution of Eq. (1.48).

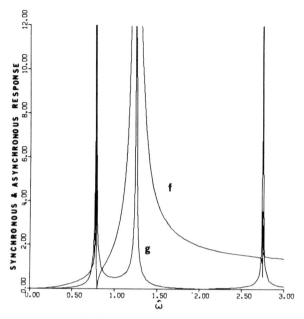

Figure 1.13 Synchronous $f = |F|/a$ and asynchronous $g = |G|/a$ response amplitudes for Eq. (1.43) with $a_x = a$; $a_y = J_{xz} = J_{yz} = 0$; $q = 0.4$.

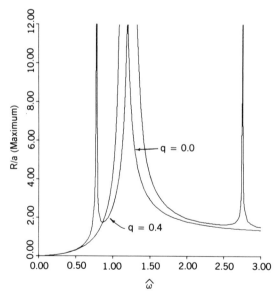

Figure 1.14 Maximum values for R/a solutions to Eq. (1.43) for $q = 0.0$ and $q = 0.4$.

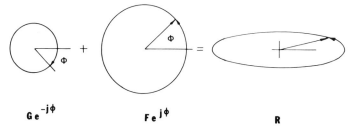

Figure 1.15   "Elliptical" orbit solution.

1.5   ROTOR STABILITY: EFFECTS OF INTERNAL AND EXTERNAL DAMPING

In the preceding sections, we have considered synchronous rotor motion, which is characterized by equal precession and rotation rates for the rotor. We found that large rotor deflections could be anticipated at rotor critical speeds, conditions that arise when the running speed coincides with a rotor natural frequency. At an undamped rotor critical speed, the rotor amplitudes are predicted to grow linearly with time. Flexible rotors are typically subject to a second type of potentially destructive motion which has the following characteristics:

(a) Below a given operating speed, "the onset speed of instability," denoted by ω_s, the rotor's motion is stable and synchronous. Above this speed, there is a subsynchronous component to the rotor's motion. The onset speed of instability always exceeds the rotor's first critical speed.

(b) For running speeds above the onset speed of instability, the subsynchronous component diverges exponentially with time. The precessional motion associated with the subsynchronous component is in the same direction as the rotor's rotation.

(c) The occurrence (or absence) of rotor instability is largely independent of the state of the rotor balance.

This type of motion is referred to here as "whirling."

Figure 1.16 is taken from Atkins and Perez (1988) and is a "waterfall plot" for a five-stage centrifugal compressor which becomes unstable at a running speed around 8500 rpm. Atkins and Perez identified high-pressure oil seals as the source of destabilizing forces. This type of figure is obtained by plotting Fourier spectra at successive running speeds. Lower-running-speed spectra contain Fourier components at running speed (synchronous response) and at

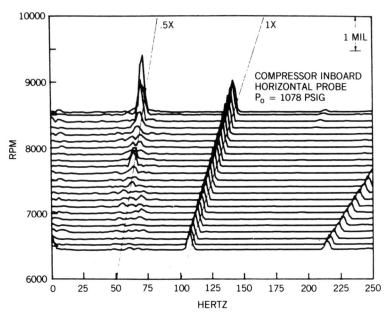

Figure 1.16 Cascade or waterfall plot of a rotordynamic instability case [Atkins and Perez (1988)].

twice running speed. As the running speed increases small subsynchronous "ripples" begin at approximately 50% of running speed. Eventually, a sharp peak develops at slightly less than 50% of running speed and grows until the subsynchronous component exceeds the synchronous component. Although not provided in Figure 1.6, spectra at lower running speeds would have shown a peak in the synchronous response around 4200 rpm coinciding with the rotor's critical speed.

A variety of physical mechanisms,[*] [Ehrich and Childs (1984)], have been proposed and experimentally verified as causes for unstable motion of flexible rotating equipment. Historically, the first recognized mechanism for rotor instability was internal rotor damping. Insight into the different nature of internal and external damping and the mechanism of rotor instability that is controlled by the relative availability of these two types of damping can be obtained from the Jeffcott model of Section 1.2. The effect of external damping on synchronous rotor motion was shown in Section 1.2 to be consistent with the familiar beneficial effect of viscous damping in linear vibrations. However, this is not the case with the force due to internal rotor

---

[*]Rotor instabilities due to hydrodynamic bearings are introduced in Chapter 3, and additional mechanisms for rotor instabilities are introduced in Chapters 4–6. The reference cited provides an overview of rotor instability mechanisms.

damping, which is defined in terms of its rotor-fixed components by

$$\mathbf{F}_{dr} = -c_r\left(\mathbf{i}\dot{R}_x + \mathbf{j}\dot{R}_y\right) = -c_r\hat{\dot{\mathbf{R}}}. \tag{1.49}$$

The vector $\hat{\dot{\mathbf{R}}}$ is the time rate of change of $\mathbf{R}$ relative to the x, y, z system; for strictly synchronous motion of the rotor it is zero. In physical terms, the bending stress of the rotor must change with time to develop an internal rotor damping force. Alternating bending stresses and strain in rotating equipment can give rise to internal damping forces due to either (a) material hysteresis effects or (b) rubbing at the interface between shrink-fitted parts. We saw in Section 1.3 that gravity loading or a similar stationary lateral load causes an alternating stress in the rotor; hence, an internal damping force similar to Eq. (1.49) can normally be expected with rotating equipment.

Figure 1.17 is adapted from Gasch and Pfützner (1975) and provides a visual realization of internal viscous damping. Observe that if the rotor precesses synchronously, without alternating bending stresses, there is no energy dissipation.

The velocity vectors $\dot{\mathbf{R}}$ and $\hat{\dot{\mathbf{R}}}$ are related by

$$\dot{\mathbf{R}} = \hat{\dot{\mathbf{R}}} + \mathbf{k}\dot{\phi} \times \mathbf{R}, \tag{1.50}$$

where $\mathbf{k}\dot{\phi}$ is the angular velocity of the x, y, z coordinate system relative to the X, Y, Z system. Hence, from the coordinate transformation of Eq. (1.17), the components of the force $\mathbf{F}_{dr}$ in the stationary X, Y, Z system are

$$F_{drX} = -c_r\left(\dot{R}_X + \omega R_Y\right),$$
$$F_{drY} = -c_r\left(\dot{R}_Y - \omega R_X\right). \tag{1.51}$$

Hence, with internal damping, the homogeneous version of Eq. (1.9) is

$$\begin{Bmatrix} \ddot{R}_X \\ \ddot{R}_Y \end{Bmatrix} + 2\zeta \begin{Bmatrix} \dot{R}_X \\ \dot{R}_Y \end{Bmatrix} + \begin{bmatrix} \lambda^2 & 2\zeta_r\lambda\omega \\ -2\zeta_r\lambda\omega & \lambda^2 \end{bmatrix}\begin{Bmatrix} R_X \\ R_Y \end{Bmatrix} = 0, \tag{1.52}$$

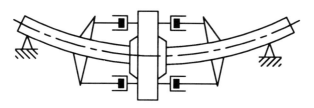

Figure 1.17   Internal-friction representation; adapted from Gasch and Pfützner (1975).

where

$$2\zeta_r\lambda = c_r/m, \qquad \zeta = \zeta_r + \zeta_e. \tag{1.53}$$

The characteristic equation for Eq. (1.52) can be stated

$$\hat{s}^4 + 4\zeta\hat{s}^3 + 2(1 + 2\zeta^2)\hat{s}^2 + 4\zeta\hat{s} + (1 + 4\zeta_r^2\hat{\omega}^2) = 0, \tag{1.54}$$

where $\hat{\omega} = \omega/\lambda$, and $\hat{s}$ is the normalized Laplace variable defined by $\hat{s} = s/\lambda$. Equation (1.54) is of the form

$$\hat{s}^4 + a_1\hat{s}^3 + a_2\hat{s}^2 + a_3\hat{s} + a_4 = 0, \tag{1.55}$$

and the Routh-Hurwitz sufficient conditions for system stability are (a) the coefficients a_i must be positive, and (b) the following two conditions must be satisfied:

$$\Delta_2 = \begin{vmatrix} a_1 & 1 \\ a_3 & a_2 \end{vmatrix} > 0, \qquad \Delta_3 = \begin{vmatrix} a_1 & 1 & 0 \\ a_3 & a_2 & a_1 \\ 0 & a_4 & a_3 \end{vmatrix} > 0. \tag{1.56}$$

All these conditions are satisfied except the second of (1.56), which yields

$$(1 + \zeta_e/\zeta_r)^2 - \hat{\omega}^2 > 0. \tag{1.57}$$

The conclusion to be reached from this result is that the model is unstable for running speeds greater than the onset speed of instability defined by

$$\hat{\omega}_s = \omega_s/\lambda = 1 + (\zeta_e/\zeta_r). \tag{1.58}$$

Smith (1933) initially obtained this result.

An alternative and instructive view of the destabilizing influence of internal damping is provided by the following complex-variable version of these equations

$$\ddot{R} + 2\zeta\lambda\dot{R} + (\lambda^2 - j2\zeta_r\lambda\omega)R = 0,$$

which yields the characteristic equation

$$s^2 + 2\zeta\lambda s + (\lambda^2 - j2\zeta_r\lambda\omega) = 0$$

and root definition

$$s = -\zeta\lambda \pm \lambda(1 - \zeta^2)^{1/2}(-1 + j\alpha)^{1/2}, \qquad \alpha = 2\zeta_r\hat{\omega}/(1 - \zeta^2). \tag{1.59}$$

By assuming that both α and ζ are small, the following two roots are obtained from Eq. (1.59):

$$s_1 = -\lambda(\zeta - \zeta_r\hat{\omega}) + j\lambda(1 - \zeta^2)^{1/2} = -\sigma_1 + j\omega_1,$$

$$s_2 = -\lambda(\zeta + \zeta_r\hat{\omega}) - j\lambda(1 - \zeta^2)^{1/2} = -\sigma_2 - j\omega_1. \qquad (1.60)$$

The real part of the s_1 root again yields the onset-speed-of-instability definition, Eq. (1.58). Specifically, for $\omega > \omega_s$, the real part of s_1 is positive, and its associated motion is divergent. For $\omega < \omega_s$ both roots define damped orbital motion at the frequency ω_1. In comparison to the assumed, positive direction of rotation for the shaft, the direction of rotation for the orbits associated with s_1 and s_2 are, respectively, positive and negative. Note that increasing the running speed increases the stability of the backward-processing orbits associated with s_2, while decreasing the stability of the forward-precessing orbits associated with s_1. This is a typical result in that forces which tend to destabilize forward-precessing modes of a rotor generally enhance the stability of backward-precessing modes.

Returning to Eq. (1.52), note that internal damping yields skew-symmetric terms in the stiffness matrix. Castigliano's theorem demonstrates that a neutrally stable elastic system must have a symmetric stiffness matrix; hence, the presence of skew-symmetric or "cross-coupled" stiffness coefficients demonstrates the presence of destabilizing forces. This type of destabilizing force is characteristic of virtually all rotor destabilizing forces.

The rotor-fixed version of Eq. (1.52) is

$$\begin{Bmatrix} \ddot{R}_x \\ \ddot{R}_y \end{Bmatrix} + \begin{bmatrix} 2\zeta\lambda & -2\omega \\ 2\omega & 2\zeta\lambda \end{bmatrix} \begin{Bmatrix} \dot{R}_x \\ \dot{R}_y \end{Bmatrix} + \begin{bmatrix} (\lambda^2 - \omega^2) & -2\zeta_e\lambda\omega \\ 2\zeta_e\lambda\omega & (\lambda^2 - \omega^2) \end{bmatrix} \begin{Bmatrix} R_x \\ R_y \end{Bmatrix} = 0,$$

which has the following characteristic equation:

$$\hat{s}^4 + 4\zeta\hat{s}^3 + 2(1 + 2\zeta^2 + \hat{\omega}^2)\hat{s}^2 + 4[\zeta + \hat{\omega}^2(\zeta_e - \zeta_r)]s$$
$$+ \left[(\hat{\omega}^2 - 1)^2 + 4\zeta_e^2\omega^2 \right] = 0. \qquad (1.61)$$

Application of the Routh-Hurwitz criteria to the coefficients of this polynomial again yields the stability criteria of Eq. (1.57).

The value of the Jeffcott model with internal and external damping is subject to a variety of interpretations. From an analytical viewpoint, it is of extraordinary value in specifying the mathematical characteristics of phenomena which lead to rotor instabilities. Specifically, the displacement cross-coupling terms $(2\zeta_r\lambda\omega R_Y, -2\zeta_r\lambda\omega R_X)$ are characteristic of virtually all physical mechanisms which have been identified as the causes of subsynchronous rotor motion. Further, the qualitative prediction for motion is consistent with most rotor instabilities encountered in practice. However, for a variety of technological and theoretical reasons, the *present* practical value of the model is minimal. From a theoretical viewpoint, the dominant nature of shrink-fit rubbing as a mechanism for internal damping (as compared to material hysteresis) has long been recognized. Hence, Coulomb damping is a more appropriate model for internal damping than the viscous model of Eq. (1.49), and various investigators have analyzed the Jeffcott model for more realistic Coulomb-type damping models, obtaining both more realistic and complicated stability criteria. From a practical viewpoint, existing design practices have minimized the potential rubbing at shrink-fit interfaces to such a degree that internal damping is not a likely explanation for rotor instabilities occurring in practice, although incidents of rotor whirling due to this phenomenon are occasionally reported for lightly damped rotors. Units employing spline couplings may also develop sufficient Coulomb damping at spline-teeth interfaces to become unstable [Williams and Trent (1970)].

1.6   INFLUENCE OF BEARINGS AND BEARING SUPPORTS ON STABILITY AND RESPONSE

Orthotropic Bearing Carrier Stiffness

In our preceding discussion of the Jeffcott model we have assumed that the bearings are rigid. Actually, the equations stated apply equally well for flexible bearings and/or bearing supports, if the flexibility is both linear and circumferentially symmetric. From this viewpoint, the stiffness coefficient k of the Jeffcott model is the equivalent spring constant resulting from the series connection of the shaft, bearing, and bearing-support stiffnesses. Further, if the bearing-support structure has a different stiffness* in the X and Y directions, the following governing equations result:

$$
m \begin{Bmatrix} \ddot{R}_X \\ \ddot{R}_Y \end{Bmatrix} + (c_e + c_r) \begin{Bmatrix} \dot{R}_X \\ \dot{R}_Y \end{Bmatrix} + \begin{bmatrix} k_X & c_r\omega \\ -c_r\omega & k_Y \end{bmatrix} \begin{Bmatrix} R_X \\ R_Y \end{Bmatrix} = ma\omega^2 \begin{Bmatrix} \cos \omega t \\ \sin \omega t \end{Bmatrix},
$$

which can be restated as

$$
\begin{Bmatrix} \ddot{R}_X \\ \ddot{R}_Y \end{Bmatrix} + 2\zeta\lambda \begin{Bmatrix} \dot{R}_X \\ \dot{R}_Y \end{Bmatrix} + \begin{bmatrix} \lambda^2(1+q) & 2\zeta_r\lambda\omega \\ -2\zeta_r\lambda\omega & \lambda^2(1-q) \end{bmatrix} \begin{Bmatrix} R_X \\ R_Y \end{Bmatrix} = a\omega^2 \begin{Bmatrix} \cos \omega t \\ \sin \omega t \end{Bmatrix},
$$

$$(1.62)$$

*Note that the shaft stiffness remains isotropic. See Section 1.7 for the effects of shaft orthotropy.

where

$$\lambda^2 = \bar{k}/m, \qquad 2\bar{k} = (k_X + k_Y), \qquad 2q = (k_X - k_Y)/\bar{k},$$
$$2\zeta_e\lambda = c_e/m, \qquad 2\zeta_r\lambda = c_r/m, \qquad \zeta = \zeta_e + \zeta_r.$$

The characteristic equation for Eq. (1.62) is

$$\hat{s}^4 + 4\zeta\hat{s}^3 + 2(1 + 2\zeta^2)\hat{s}^2 + 4\zeta\hat{s} + \left(1 + 4\zeta_r^2\hat{\omega}^2 - q^2\right) = 0.$$

A comparison of this result with the symmetric stiffness case of Eq. (1.54) shows a difference only in the last term due to the stiffness asymmetry coefficient q. Routh-Hurwitz stability analysis of this characteristic equation yields the stability criterion

$$4\zeta^2 + q^2 - 4\zeta_r^2\hat{\omega}^2 > 0,$$

and the consequent onset-speed-of-instability definition

$$\hat{\omega}_s = \omega_s/\lambda = \left\{[1 + (\zeta_e/\zeta_r)]^2 + (q/2\zeta_r)^2\right\}^{1/2}. \qquad (1.63)$$

This result was initially obtained by Smith (1933) and demonstrates that support-stiffness orthotropy can be used to increase the onset speed of instability. Rotor support-stiffness orthotropy can be specifically introduced to improve the stability of a ball-bearing-supported rotor [Williams and Trent (1970)], and can be developed in journal-bearing-supported units by increasing the nominal eccentricity ratio of the journal bearings.*

To examine the influence of support-stiffness orthotropy on response, Eq. (1.62) is restated in the following nondimensional form:

$$\begin{Bmatrix} r_X'' \\ r_Y'' \end{Bmatrix} + 2\zeta\begin{Bmatrix} r_X' \\ r_Y' \end{Bmatrix} + \begin{bmatrix} (1 + q) & 2\zeta_r\hat{\omega} \\ -2\zeta_r\hat{\omega} & (1 - q) \end{bmatrix}\begin{Bmatrix} R_X \\ R_Y \end{Bmatrix} = \hat{\omega}^2\begin{Bmatrix} \cos\hat{\omega}\tau \\ \sin\hat{\omega}\tau \end{Bmatrix}, \qquad (1.64)$$

where the prime denotes differentiation with respect to the dimensionless time variable $\tau = \lambda t$, and

$$\hat{\omega} = \omega/\lambda,$$
$$r_X = R_X/a, \qquad r_y = R_Y/a.$$

Substituting the assumed steady-state-solution format,

$$\begin{Bmatrix} r_X \\ r_Y \end{Bmatrix} = \begin{Bmatrix} a_1 \\ a_2 \end{Bmatrix}\cos\hat{\omega}\tau + \begin{Bmatrix} b_1 \\ b_2 \end{Bmatrix}\sin\hat{\omega}\tau,$$

*See Section 3.6.

into Eq. (1.64) yields a result of the form

$$
\begin{bmatrix} -\hat{\omega}^2[I] + [K] & 2\zeta\hat{\omega}[I] \\ -2\zeta\hat{\omega}[I] & \hat{\omega}^2[I] + [K] \end{bmatrix} \begin{pmatrix} a_1 \\ a_2 \\ b_1 \\ b_2 \end{pmatrix} = \begin{pmatrix} \hat{\omega}^2 \\ 0 \\ 0 \\ \hat{\omega}^2 \end{pmatrix},
\tag{1.65}
$$

where $[I]$ is the identity matrix and $[K]$ is the stiffness matrix. The solution from this equation may be stated

$$
\begin{aligned}
r &= r_X + jr_Y \\
&= \tfrac{1}{2}[(a_1 + b_2) + j(a_2 - b_1)]e^{j\omega t} + \tfrac{1}{2}[(a_1 - b_2) + j(a_2 + b_1)]e^{-j\omega t} \\
&= f e^{j\omega t} + g e^{-j\omega t}.
\end{aligned}
$$

This solution format, including forward- and backward-precession terms, was obtained earlier for the Stodola-Green model with orthotropic supports.

Figures 1.18(a) and 1.18(b) provide frequency-response plots for $|r|_{\max}$ from Eq. (1.65) for $q = 0$ and 0.25, with a range of internal and external damping. These results show that orthotropy tends to reduce the maximum amplitudes at resonance, but yields a broader speed range of high-amplitude response. Further, the results show that increasing the rotor's internal damping couples the response in the X-Z and Y-Z planes, eventually replacing the two peaks with a single-peak response curve comparable to that obtained for a symmetrically supported rotor. Gyroscopic coupling caused a comparable merging of the two peaks for the Stodola-Green model considered earlier.

Figure 1.18(c) illustrates the maximum amplitude $|r|_{\max}$, the synchronous amplitude f, and asynchronous amplitude g as a function of $\hat{\omega}$ for $\zeta_r = 0.005$, $\zeta_e = 0.025$, and $q = 0.5$. The results of this figure demonstrate that the forward and backward critical speeds coincide; i.e., peaks occur in both f and g at the normalized speeds $\hat{\omega}_1 = \sqrt{1 - q}$, $\hat{\omega}_2 = \sqrt{1 + q}$. This result should be compared to the separated forward and backward critical-speed peaks of Figure 1.13 for the Stodola-Green model with asymmetry. Note further that the asynchronous amplitude g is greater than the synchronous amplitude f between the two critical speeds. Hence, between the two critical speeds, the net rotor precession is backward. The results of this figure can be explained in terms of the following solution to Eq. (1.65) for $\zeta_r = 0$:

$$
\begin{aligned}
r =\ & \frac{\hat{\omega}^2}{D_1 D_2} \left\{ \begin{array}{l} (1 - \hat{\omega}^2)\left[4\zeta_e^2\hat{\omega}^2 + (1 - \hat{\omega}^2)^2 - q^2\right] \\ -j2\zeta_e\hat{\omega}\left[4\zeta_e^2\hat{\omega}^2 + (1 - \hat{\omega}^2)^2 + q^2\right] \end{array} \right\} e^{j\hat{\omega}\tau} \\
&+ \frac{\hat{\omega}^2 q}{D_1 D_2} \left\{ \begin{array}{l} \left[4\zeta_e^2\hat{\omega}^2 - (1 - \hat{\omega}^2)^2 + q^2\right] \\ -j4\zeta_e\hat{\omega}(1 - \hat{\omega}^2) \end{array} \right\} e^{-j\hat{\omega}\tau},
\end{aligned}
\tag{1.66}
$$

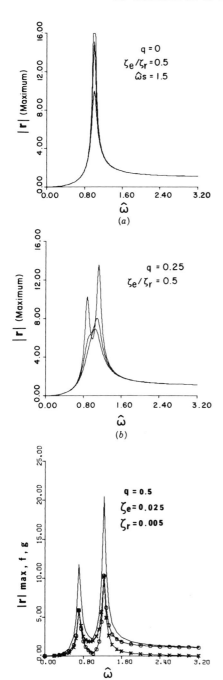

Figure 1.18 Maximum $|r|$ imbalance response from Eq. (1.66) for (a) $q = 0$; $\zeta_r = 0.03, 0.066, 0.1, \overline{\omega}_s = 1.50$ and (b) $q = 0.25$; $\zeta_r = 0.33$ ($\overline{\omega}_s = 4.43$); $\zeta_r = 0.066$ ($\overline{\omega}_s = 2.42$); $\zeta = R = 0.1$ ($\overline{\omega}_s = 1.95$). (c) Maximum $|r|$ imbalance response, synchronous amplitude f, and asynchronous amplitude g; $\zeta_r = 0.005$, $\zeta_e = 0.025$, $q = 0.5$.

where

$$D_1 = \{(1 + q) - \hat{\omega}^2\}^2 + 4\zeta_e^2\hat{\omega}^2,$$

$$D_2 = \{(1 - q) - \hat{\omega}^2\}^2 + 4\zeta_e^2\hat{\omega}^2.$$

For $\zeta_e/q \ll 1$, the response at critical speeds may be stated

$$r(\hat{\omega}_1) = \frac{(1 - q)^{1/2}}{2\zeta_e}\left\{\frac{\zeta_e}{q}(1 - q)^{1/2} - j\right\}\cos\hat{\omega}\tau,$$

$$r(\hat{\omega}_2) = \frac{(1 - q)^{1/2}}{2\zeta_e}\left\{1 - j\frac{\zeta_e}{q}(1 + q)^{1/2}\right\}\sin\hat{\omega}\tau.$$

Hence, at critical speeds the orbit degenerates to a line. At the upper critical speed, motion consists of a nearly horizontal oscillation, while at the lower speed an oscillation in a nearly vertical plane results. Returning to Eq. (1.66), the solution between criticals at $\hat{\omega} = 1$ is

$$r(1) = -j\frac{2\zeta_e}{D}e^{j\hat{\omega}\tau} + \frac{q}{D}e^{-j\hat{\omega}\tau}, \qquad D = 4\zeta_e^2 + q^2.$$

Hence, the backward component exceeds the forward component if $q > 2\zeta_e$, and backward precession results between the critical speeds.

Bearing-Support Damping

In the preceding paragraphs, we have observed the favorable consequences of external damping in both reducing synchronous rotor response, and in elevating a rotor's onset speed of instability. However, one frequently encounters serious practical problems in supplying external damping to a rotor. For example, in many applications significant damping may only be introduced at a rotor's bearings. Sections 3.4 and 3.5 demonstrate that hydrodynamic bearings provided damping directly, and squeeze-film dampers may be provided to supply damping for ball-bearing-supported rotors. As the following analysis due to Black (1976) reveals, the degree to which external damping can be provided for a rotor at bearing locations is limited.

Figure 1.19 illustrates a planar model for a flexible-bearing-support/flexible-shaft rotor model. The equations for free motion for this system are

$$m\ddot{y} + ky - ky_1 = 0,$$

$$c_s\dot{y}_1 - ky + (k + k_s)y_1 = 0. \tag{1.67}$$

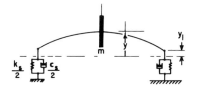

Figure 1.19 Black's (1976) flexible-bearing-support/flexible-shaft rotor model.

The undamped natural frequency for these equations is $\sqrt{k_e/m}$, where $k_e = kk_s/(k + k_s)$. By assuming that $k_s \gg k$, the natural frequency is approximately equal to $\lambda = \sqrt{k/m}$, and the following assumed solution format applies:

$$y = Y(t)e^{j\lambda t}, \qquad y_1 = Y_1(t)e^{j\lambda t}, \qquad \lambda^2 = k/m,$$

where $Y(t)$ and $Y_1(t)$ are assumed to be slowly varying parameters. Substituting the assumed solution into Eq. (1.67) yields

$$\ddot{Y} + j2\lambda\dot{Y} + \lambda^2 Y_1 = 0,$$

$$\dot{Y}_1(c_s/k) + (1 + h + j2\xi)Y_1 - Y = 0$$

$$h = k_s/k, \qquad 2\xi = c_s\lambda/k.$$

Neglecting the higher derivatives of these slowly varying variables ($\ddot{Y}$ in the first equation, $\dot{Y}_1$ in the second) and eliminating Y_1 yields

$$\dot{Y} + \frac{\lambda(j + \eta)}{2(1 + h)(1 + \eta^2)}Y = 0, \qquad \eta = 2\xi/(1 + h).$$

Hence, the (dimensionless) damped forward-precession root for Eqs. (1.67) is approximately defined by

$$\hat{s} = (s/\lambda) = -\hat{\sigma} + j\hat{\gamma},$$

where

$$\hat{\sigma} = \eta/2(1 + h)(1 + \eta^2) = \eta/d, \qquad \hat{\gamma} = 1 - 1/d. \qquad (1.68)$$

The significance of this result is that an optimum damping factor may be defined for a given stiffness ratio h. Specifically from $\partial\hat{\sigma}/\partial\eta = 0$, the optimum value for η is one, and the maximum possible damping ratio for the rotor is

$$\hat{\sigma}_{max} \simeq 1/4(1 + h).$$

The roots for Eq. (1.67) are exactly defined by the dimensionless characteristic equation

$$\hat{s}^3 + \hat{s}^2 \eta^{-1} + \hat{s} + h/\eta(1 + h) = 0. \tag{1.69}$$

Figure 1.20 compares the solution to this exact equation with the approximate solutions of Eq. (1.68) and demonstrates that the effectiveness of damping is reduced as the stiffness ratio h is increased. In physical terms, the effectiveness of bearing-support damping is decreased if the amplitude of rotor motion at the bearing supports is decreased by increasing k_s. Barret et al. (1978) have demonstrated the validity of Black's results for a

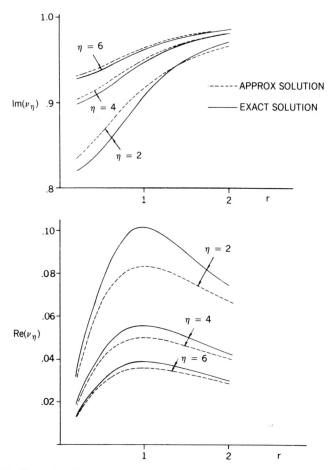

Figure 1.20 Damping factor and frequency of oscillatory root for the model of Figure 1.17; approximate solution from Eq. (1.68), exact solution for Eq. (1.69) [Black (1976)].

wide range of rotor configurations; viz., for a given rotor and support stiffness, there is an optimum value of bearing-support damping.

From a practical viewpoint, the following options are generally available for the attempted elevation of an existing rotor's onset speed of instability to a point above its operating range:

(a) The effect of the destabilizing force can be reduced, which for Eq. (1.52) implies a reduction in the internal damping factor ζ_r.

(b) The external damping can be increased. However, the above results demonstrate that the degree to which external damping can be provided at bearing locations is limited.

(c) The critical speed λ can be increased by increasing the stiffness of the bearing, the bearing support, or the shaft. Unfortunately, increasing λ frequently yields an unwanted critical-speed location at or near a steady operating speed of the rotor.

(d) Orthotropy can be introduced in the bearing-support stiffness, keeping in mind that this may yield a broader speed range of larger-amplitude synchronous response and may occasionally provide a mechanism for exciting asynchronous critical speeds. When feasible, this approach has the advantage of increasing the onset speed of instability without markedly altering critical-speed locations. The author has not found this theoretical approach for increasing the rotor stability to be very practical. Generally, physical considerations limit the amount of stiffness orthotropy which can be introduced.

1.7 PARAMETRIC INSTABILITY: SHAFT-STIFFNESS ORTHOTROPY

The preceding sections of this chapter have demonstrated that large rotor amplitudes can result from resonance phenomena associated with the operation of a rotor at or near a critical speed, and internal damping or similar mechanisms resulting in stiffness cross-coupling can lead to linear instabilities if a rotor is operated above a limiting onset speed of instability. We are concerned in this section with a second type of rotor instability associated with parametric excitation due to (a) stiffness orthotropy of the rotor, as opposed to the support-stiffness orthotropy of the preceding section, and (b) partial rubbing conditions.

Shaft-Stiffness Orthotropy

From Eq. (1.18) the following constant-running-speed equations of motion can be stated for a rotor mounted on a shaft with orthotropic stiffness characteristics:

$$
\begin{Bmatrix} \ddot{R}_x \\ \ddot{R}_y \end{Bmatrix} + \begin{bmatrix} 2\zeta\bar{\lambda} & -2\omega \\ 2\omega & 2\zeta\bar{\lambda} \end{bmatrix} \begin{Bmatrix} \dot{R}_x \\ \dot{R}_y \end{Bmatrix} + \begin{bmatrix} \left[\bar{\lambda}^2(1+q) - \omega^2 \right] & -2\zeta_e\bar{\lambda}\omega \\ 2\zeta_e\bar{\lambda}\omega & \left[\bar{\lambda}^2(1-q) - \omega^2 \right] \end{bmatrix}
$$

$$
\times \begin{Bmatrix} R_x \\ R_y \end{Bmatrix} = \omega^2 \begin{Bmatrix} a_x \\ a_y \end{Bmatrix}, \tag{1.70}
$$

where

$$
2\bar{\lambda}^2 = (k_x + k_y)/m, \qquad 2q\bar{\lambda}^2 = (k_x - k_y)/m,
$$

$$
2\zeta_e\bar{\lambda} = c_e/m, \qquad 2\zeta_r\bar{\lambda} = c_r/m, \qquad \zeta = \zeta_e + \zeta_r. \tag{1.71}
$$

The stiffness orthotropy of the rotor is defined by the shaft-stiffness coefficients k_x, k_y for the x, z and y, z planes. The governing equations apply equally well for rigid bearing supports or supports with circumferential stiffness symmetry. The possibility of parametric excitation due to shaft-stiffness orthotropy becomes more evident in the following stationary-coordinate version of these equations:

$$
\begin{Bmatrix} \ddot{R}_X \\ \ddot{R}_Y \end{Bmatrix} + 2\zeta\bar{\lambda} \begin{Bmatrix} \dot{R}_X \\ \dot{R}_Y \end{Bmatrix} + \begin{bmatrix} \bar{\lambda}^2[1 + qc(2\omega t)] & \left[2\zeta_r\bar{\lambda}\omega + q\bar{\lambda}^2 s(2\omega t) \right] \\ \left[-2\zeta_r\bar{\lambda}\omega + q\bar{\lambda}^2 s(2\omega t) \right] & \bar{\lambda}^2[1 - qc(2\omega t)] \end{bmatrix}
$$

$$
\times \begin{Bmatrix} R_X \\ R_Y \end{Bmatrix} = \omega^2 \begin{Bmatrix} a_X \\ a_Y \end{Bmatrix}. \tag{1.72}
$$

The stiffness coefficients are obviously harmonic in the stationary X-Y-Z coordinate system, indicating the possibility of a parametric instability of the Mathieu or Hill type. However, stability analysis is much more easily accomplished in the rotor-fixed version of Eq. (1.70), which has the following characteristic equation:

$$
\bar{s}^4 + 4\zeta\bar{s}^3 + 2\left(1 + 2\zeta^2 + \bar{\omega}^2\right)\bar{s}^2 + 4\left[\zeta + (\zeta_e - \zeta_r)\bar{\omega}^2\right]\bar{s}
$$

$$
+ \left[(\bar{\omega}^2 - 1)^2 + 4\zeta_e\bar{\omega}^2 - q^2\right] = 0, \tag{1.73}
$$

where $\bar{\omega} = \omega/\bar{\lambda}$, $\bar{s} = s/\bar{\lambda}$. This result differs from the corresponding symmetric-rotor result of Eq. (1.61) only in the last (zeroth-order) coefficient. Application of the Routh-Hurwitz stability criteria to the roots of Eq. (1.72) yields the following requirements for stability:

$$\zeta + (\zeta_e - \zeta_r)\overline{\omega}^2 > 0,$$

$$-(\zeta_r/\zeta)^2\overline{\omega}^4 + (1 - \zeta_r^2)\overline{\omega}^2 + \zeta^2 - (q/2)^2 > 0,$$

$$(\overline{\omega}^2 - 1)^2 + 4\zeta_e^2\overline{\omega}^2 - q^2 > 0. \tag{1.74}$$

Violation of the first two conditions of (1.74) yields instabilities due to internal damping similar to those previously obtained for symmetric rotors. The first condition may be disregarded, since the second always yields a lower onset-speed-of-instability definition. For small q, the onset speed of instability is approximately defined by

$$\overline{\omega}_s^2 \simeq (1 + \zeta_e/\zeta_r)^2 - (q/2)^2/(1 + \zeta_r^2). \tag{1.75}$$

By comparison to Eq. (1.58), orthotropy lowers the onset speed of instability due to internal damping.

The last condition from (1.74) does not involve internal damping and is always satisfied for $q = 0$; hence, an instability occasioned by violation of this condition is solely the result of parametric excitation due to q. This condition predicts instability within the following speed range:

$$1 - A < \frac{\overline{\omega}^2}{1 - 2\zeta_e^2} < 1 + A, \qquad A^2 = 1 - (1 - q^2)/(1 - 2\zeta_e^2)^2. \tag{1.76}$$

For zero external damping, the unstable speed range is a maximum defined by

$$\lambda_x < \omega < \lambda_y, \qquad \lambda_x^2 = \lambda^2(1 + q) = k_x/m, \qquad \lambda_y^2 = \lambda^2(1 - q) = k_y/m;$$

i.e., the rotor is unstable at running speeds between the undamped natural frequencies in the x-z and y-z planes. From (1.76), the possibility of a parametric instability is eliminated for ζ_e large enough to cause A to be zero; i.e., no parametric instability occurs if $2\zeta_e^2 \geq 1 - (1 - q^2)^{1/2}$.

The motion illustrated in Figure 1.21 is a solution to the following nondimensional version of Eq. (1.72):

$$\begin{Bmatrix} r_X'' \\ r_Y'' \end{Bmatrix} + 2\zeta \begin{Bmatrix} r_X' \\ r_Y' \end{Bmatrix} + \begin{bmatrix} [1 + qc(2\overline{\omega}\tau)] & [2\zeta_r\overline{\omega} + qs(2\overline{\omega}\tau)] \\ [-2\zeta_r\overline{\omega} + qs(2\overline{\omega}\tau)] & [1 - qc(2\tilde{\omega}\tau)] \end{bmatrix} \begin{Bmatrix} r_X \\ r_Y \end{Bmatrix}$$

$$= \overline{\omega}^2 \begin{Bmatrix} c(\overline{\omega}\tau) \\ s(\tilde{\omega}\tau) \end{Bmatrix}, \tag{1.77}$$

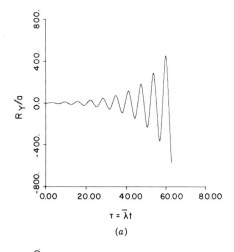

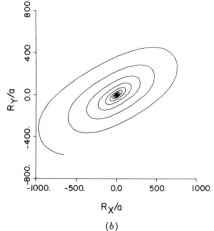

Figure 1.21 Transient solution for Eq. (1.77) for $\zeta_e = \zeta_r = 0.01$, $\overline{\omega} = 1$. (a) $r_Y = R_Y/a$ versus $\tau = \omega t$. (b) r_Y versus r_X.

where $a_x = a$, $a_y = 0$, $r_X = R_X/a$, $r_Y = R_Y/a$, and the prime denotes differentiation with respect to the dimensionless time variable $\tau = \overline{\lambda} t$. The solution presented is for $q = 0.25$, $\zeta_e = \zeta_r = 0.01$, $\overline{\omega} = 1$, and illustrates the type of unstable motion associated with running speeds which lie within the range defined by (1.76). The motion illustrated is similar to that which results from the linear instability of Section 1.5, in that both involve forward-precessional motion. However, the precessional frequency of the parametrically induced instability is approximately the running speed, while the precessional frequency of the internal-friction-induced instability of Section 1.5 is characteristically subharmonic at a rotor natural frequency. An additional difference between these two types of instabilities is the theoretical possibility of "running through" a parametric instability; i.e., the system defined by Eq. (1.70) is stable at running speeds above the range defined by (1.76), if the second condition of (1.74) is satisfied.

The theoretical potential for parametric instabilities due to a lack of rotor symmetry has long been recognized. The analysis provided above was initially presented by Smith (1933), who also derived equations of motion for a system in which both the rotor and the support-structure stiffness are orthotropic. As a rule, rotors are symmetric, and there is no reason to be concerned with a parametric instability due to stiffness orthotropy. Two-pole turbogenerators can present an exception to this rule [Bishop and Parkinson (1965)]; however, the external damping in these systems due to hydrodynamic bearings has historically been sufficient to preclude parametric instability.

Orthotropic rotors are subject to dynamic excitation from gravity in addition to the steady-state Jeffcott gravity solution of Section 1.3. To examine this phenomenon, we consider the following undamped version of Eq. (1.70), without imbalance, and with the weight of the rotor directed along the $-Y$ axis:

$$\ddot{R}_x - 2\omega \dot{R}_y + \left[\bar{\lambda}^2(1 + q) - \omega^2 \right] R_x = -g \sin \omega t,$$

$$\ddot{R}_y + 2\omega \dot{R}_y + \left[\bar{\lambda}^2(1 - q) - \omega^2 \right] R_y = -g \cos \omega t. \tag{1.78}$$

The complex-variable version of this equation is

$$\ddot{r} + j2\omega \dot{R} + (\bar{\lambda}^2 - \omega^2)r + q\bar{\lambda}^2 \bar{r} = -jge^{-j\omega t},$$

where $r = R_x + jR_y$, and $\bar{r}$ is its conjugate. We assume a perturbation solution of the form

$$r = Ae^{-j\omega t} + qBe^{j\omega t},$$

with q treated as a small parameter to obtain the approximate solution

$$r \simeq -j(g/\bar{\lambda}^2)e^{-j\omega t} + \frac{jqg}{(\bar{\lambda}^2 - 4\omega^2)}e^{j\omega t}.$$

The stationary-reference version of this solution is

$$R = re^{j\omega t} = -j(g/\bar{\lambda}^2) + \frac{jqg}{(\bar{\lambda}^2 - 4\omega^2)}e^{j2\omega t}. \tag{1.79}$$

Comparing this result to that of Eq. (1.23) demonstrates that shaft orthotropy introduces a twice-frequency solution component due to gravity, which is unbounded at a running speed that is one-half the nominal rotor natural frequency. If external damping is retained in the analysis, the predicted rotor amplitude at an $\omega = \lambda/2$ running speed is bounded. The large-amplitude motion associated with the second term in Eq. (1.79) is a forced-harmonic-response phenomenon, comparable to a critical-speed response, and should not be confused with either a linear or parametric instability.

Despite the limited practical interest in asymmetric rotors, numerous investigators have presented both theoretical and experimental results concerning rotor/bearing orthotropy. Bishop and Parkinson (1965) and Black (1969) provide a review of analyses based on extensions of the Jeffcott model. Parametric stability analyses have also been performed on versions and extensions of the Stodola-Green model of Section 1.4. Specifically, if $J_x \neq J_y$ in this model, the moments of inertia J_X, J_Y are harmonic, and provide another mechanism for parametric excitation. The introduction of dissimilar principal moments of inertia sharply expands the possible complications of analytical rotor models. For example, one can consider the stability of a rotor having dissimilar moments of inertia, carried by a shaft with stiffness orthotropy, supported by orthotropic bearings, with rotor-fixed and external damping. If desired, one can additionally consider the consequences of rotating the principal stiffness axes of the shaft relative to the principal inertia axes of the rotor. The initial analysis of parametric excitation due to dissimilar principal moments of inertia was performed by Brosens and Crandall (1961). All of the analysis results demonstrate that dissimilar moments of inertia yield parametric instabilities over restricted speed ranges in a manner which is analogous to the results of Eq. (1.74). Again, these results are of limited practical consequence, since rotors of rotating machinery are customarily designed to be symmetric.

On occasion, a rotor which is initially symmetric becomes unsymmetric due to a transverse crack. Cases have occurred where a transverse crack relieves more than half of a shaft's cross-sectional area. When a horizontal shaft with a substantial crack rotates, its weight causes the crack to alternately open and close, yielding a periodic stiffness variation analogous to an orthotropic rotor. Cracked rotors typically have a large response component at a twice-running-speed frequency, as suggested by Eq. (1.79). Grabowski (1979) developed a transient procedure for the analysis of cracked shaft, and provides a summary of references on the subject.

1.8 FRACTIONAL-FREQUENCY ROTOR MOTION DUE TO NONSYMMETRIC CLEARANCE EFFECTS

Introduction

Bently (1974) has proposed and demonstrated experimentally that large subsynchronous rotor motion can result from the following types of nonsymmetric clearance effects:

(a) The rotor's synchronous motion causes a rubbing condition with a stationary surface over a portion of the rotor's orbit. Contact with the stationary surface causes a periodic increase in the rotor's stiffness yielding what Bently calls a "normal-tight" condition. A jammed or off-centered seal generally provides the physical mechanism for this condition.

(b) A "normal-loose" rotor-stiffness condition is also possible if the rotor's radial stiffness is reduced over a fraction of its orbit. This is an abnormal circumstance in units supported by hydrodynamic bearings, which can result from excessive bearing clearances or inadequate vertical bearing restraints. However, in rotors supported by rolling-element bearings, a small radial clearance is normally provided to allow axial shaft motion. This clearance, in combination with a fixed-direction side load, provides the continuing possibility for a normal-loose condition.

Subsynchronous motion resulting from either of these conditions is an exact fraction of running speed, at predominantly one-half running speed in field experience but occasionally one-third or one-fourth running speed. The subsynchronous motion is most easily excited when the rotor running speed is a corresponding multiple of the rotor's critical speed; i.e., half-speed motion is the general result of running speeds which are approximately twice the first rotor critical speed. In contrast to rotor instabilities due to hydrodynamic bearings,[*] this motion tends to be exactly one-half running speed rather than slightly less than one-half running speed. Stated differently, motion due to nonsymmetric clearance effects is a fractional-frequency phenomenon, while the frequency of unstable motion due to hydrodynamic bearings is at the rotor critical speed, which depends less directly on running speed. Motion from either source may be large and potentially damaging.

Bently interprets his experimental findings in terms of analytical results for both linear parametric-excitation phenomena, modeled by the Mathieu equation, and nonlinear subharmonic motion modeled by Duffing's equation, but presents neither models nor formal analysis to support his very insightful conclusions. Child's analysis (1982), based on the Jeffcott model, provided a linear parametric-excitation explanation for the half-speed fractional-frequency response results for the loose-tight rub condition and a nonlinear-analysis explanation for the one-third-speed response. The linear analysis includes the effect of Coulomb friction rub; the nonlinear does not.

Physical Model

The Jeffcott model of Figure 1.1 is employed for the present analysis, and has the following constant-running-speed equations of motion:

$$\ddot{R}_X + 2\zeta_e \lambda_0 \dot{R}_X + \lambda_0^2 R_X = a\omega^2 \cos \phi,$$

$$\ddot{R}_Y + 2\zeta_e \lambda_0 \dot{R}_Y + \lambda_0^2 R_X = a\omega^2 \sin \phi - g, \qquad (1.80)$$

[*]See Section 3.5.

where, for the purposes of this section, the notation

$$\lambda_0^2 = k/m, \qquad 2\zeta_e\lambda_0 = c_e/m$$

is employed. The solution may be stated

$$R_X = A \cos(\phi + \psi),$$
$$R_Y = A \sin(\phi + \psi) - D, \tag{1.81}$$

where $D = k/W$ is the static rotor deflection due to gravity, and

$$A = a\omega^2 / \left\{ \left(\lambda_0^2 - \omega^2\right)^2 + 4\zeta_e^2\lambda^2\omega^2 \right\}^{1/2},$$
$$\psi = \tan^{-1}\left\{ -2\zeta_e\lambda_0\omega / \left(\lambda_0^2 - \omega^2\right) \right\}. \tag{1.82}$$

Normal-Loose Radial-Stiffness Model

Figure 1.22 illustrates the solution to Eq. (1.82), and may be used to explain the discontinuous radial-stiffness model used in the present analysis. The assumption is made that the rotor's *radial* shaft-bearing stiffness is reduced for $R_Y > 0$. More specifically, the nominal stiffness value holds when the solution amplitude A is less than the static deflection amplitude D; however, when $A > D$ the stiffness becomes $k(1 - \varepsilon)$. Figure 1.23 illustrates the dependence of the Y component of radial spring force on R_Y.

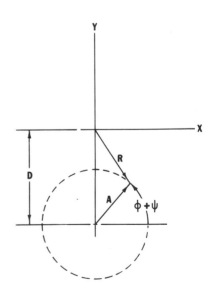

Figure 1.22   Steady-state solution for the Jeffcott model with gravity [Childs (1982)].

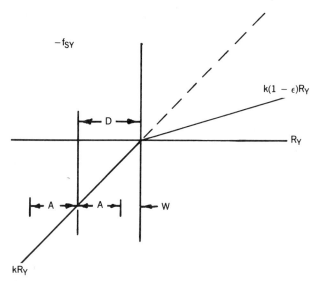

Figure 1.23   The Y component of radial spring force as a function of R_Y [Childs (1982)].

The differential equations of motion including the stiffness variation are

$$\ddot{R}_X + 2\zeta_e\lambda_0\dot{R}_X + \lambda_0^2[1 - \varepsilon U(R_Y)]R_X = a\omega^2 \cos\phi,$$

$$\ddot{R}_Y + 2\zeta_e\lambda_0\dot{R}_Y + \lambda_0^2[1 - \varepsilon U(R_Y)]R_Y = a\omega^2 \sin\phi - g, \qquad (1.83)$$

where $U(R_Y)$ is the unit step function defined by

$$U(R_Y) = 0, \qquad R_Y \leq 0,$$
$$U(R_Y) = 1, \qquad R_Y > 0.$$

Normal-Loose Parametric-Excitation Analysis

The parameter ε in Eq. (1.83) lies in the range $0 \leq \varepsilon \leq 1$, with $\varepsilon = 1$ corresponding to a complete loss of radial stiffness. In the present analysis, we assume that ε is small in comparison to unity and examine the following perturbation solution

$$R_X = D(2 + \varepsilon)\cos(\phi + \psi) + x_1,$$
$$R_Y = D(1 + \varepsilon)\sin(\phi + \psi) - D + y_1, \qquad (1.84)$$

where x_1 and y_1 are assumed to be small compared to D. Observe that the amplitude of the synchronous portion of this solution is slightly larger than the static amplitude D; hence, the synchronous orbit periodically traverses the region of reduced radial stiffness. The present analysis seeks to determine the influence of this periodic stiffness variation on the perturbation variables x_1, y_1.

Substituting Eq. (1.84) into Eq. (1.83) and discarding second-order terms in ε yields the perturbation differential equations

$$\ddot{x}_1 + 2\zeta_e\lambda_0\dot{x}_1 + \lambda_0^2 x_1[1 - \varepsilon U(R_Y)] = \varepsilon g \cos(\phi + \psi)U(R_Y),$$

$$\ddot{y}_1 + 2\zeta_e\lambda_0\dot{y}_1 + \lambda_0^2 y_1[1 - \varepsilon U(R_Y)] = -\varepsilon g[1 - \sin(\phi + \psi)]U(R_Y). \tag{1.85}$$

We assume that the system is lightly damped and operating well above its critical speed; hence $\psi \simeq -\pi$. For convenience, the initial rotation angle is chosen* to be $\phi_0 = -\pi/2$. Hence, the differential equations become

$$\ddot{x}_1 + 2\zeta_e\lambda_0\dot{x}_1 + \lambda_0^2 x_1[1 - \varepsilon U(R_Y)] = -\varepsilon g \sin \omega t U(R_Y),$$

$$\ddot{y}_1 + 2\zeta_e\lambda_0\dot{y}_1 + \lambda_0^2 y_1[1 - \varepsilon U(R_Y)] = -\varepsilon g(1 - \cos \omega t)U(R_Y). \tag{1.86}$$

As noted above, the present analysis seeks to determine the influence on the perturbation variables x_1, y_1 of a periodic stiffness variation due to synchronous motion. Further, we have assumed that y_1 is small compared to D. Hence, the following additional assumption is made:

$$U(R_Y) \cong U(R_{Y0}),$$

where R_{Y0} is the base synchronous solution,

$$R_{Y0} = D(1 + \varepsilon)\cos \omega t - D. \tag{1.87}$$

The functions R_{Y0} and $U(R_{Y0})$ are both illustrated in Figure 1.24. The function R_{Y0} is positive over the rotation angle 2β, and yields $U(R_{Y0}) = 1$ over this interval. The angle β is related to ε by

$$\cos \beta = 1/(1 + \varepsilon),$$

and for small β is defined by

$$\beta \simeq \left(\frac{2\varepsilon}{1 + \varepsilon}\right)^{1/2} = (2\varepsilon)^{1/2}(1 - \varepsilon + \varepsilon^2 \cdots). \tag{1.88}$$

---

*This is equivalent to the choice of the initial time.

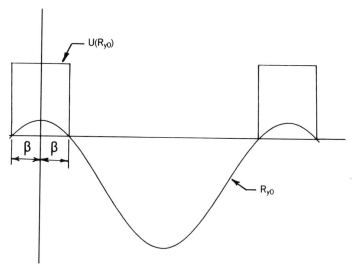

Figure 1.24   The function R_{Y0} from Eq. (1.87) and the associated function $U(R_{Y0})$ [Childs (1982)].

The Fourier-series definition for $U(R_{Y0})$ is

$$U(R_{Y0}) = \frac{\beta}{\pi}[1 + f(t)],$$

$$f(t) = 2 \sum_{j=1}^{\infty} c_j \cos j\omega t, \qquad c_j = \frac{\sin j\beta}{j\beta}. \tag{1.89}$$

The functions $(1 - \cos \omega t)U(R_{Y0})$ and $\sin \omega t U(R_{Y0})$ in Eq. (1.86) yield Fourier series on the order of $(\beta/\pi)^3$ and are assumed to represent a negligible direct excitation of perturbation equations. In any case, the homogeneous portion of Eq. (1.86) is of interest. The homogeneous equations for x_1 and y_1 in Eq. (1.86) are equivalent, and the equation for y_1 may be stated

$$\ddot{y}_1 + 2\zeta_e \lambda_0 \dot{y}_1 + \lambda_0^2\{1 - q[1 + f(t)]\}y_1 = 0, \qquad q = \frac{\varepsilon\beta}{\pi}. \tag{1.90}$$

The looseness model yields both a reduction in the apparent natural frequency and a harmonic oscillation in the coefficient of y_1. The equation may be restated as a Hill equation [Hsu (1965)] of the form

$$y_1'' + 2\zeta\left(\frac{\lambda}{\omega}\right)y_1' + \left(\frac{\lambda}{\omega}\right)^2\left[1 - 2q\sum_{j=1}^{\infty} c_j \cos j\tau\right]y_1 = 0;$$

the prime denotes differentiation with respect to the dimensionless time variable, $\tau = \omega t$, and

$$\lambda^2 = \lambda_0^2(1 - q); \qquad \zeta\lambda = \zeta_e\lambda_0.$$

The *stability* requirements for this equation are [Hsu (1965)],

$$\left[\left(\frac{\lambda}{\omega}\right)^2 - \left(\frac{j}{2}\right)^2\right]^2 + 2\left[\left(\frac{\lambda}{\omega}\right)^2 + \left(\frac{j}{2}\right)^2\right]\zeta^2\left(\frac{\lambda}{\omega}\right)^2 + \zeta^4\left(\frac{\lambda}{\omega}\right)^4 \geq q^2 c_j^2\left(\frac{\lambda}{\omega}\right)^4;$$

$$j = 1, 2, \ldots. \quad (1.91)$$

In general this inequality yields the possibility of an instability at the running speeds

$$\omega = 0, 2\lambda, \lambda, \frac{2\lambda}{3}, \ldots, \frac{2\lambda}{j}, \qquad j = 1, 2, \ldots.$$

Our interest is with the $\omega = 2\lambda$ parametric excitation frequency, since the experimental data cited and provided by Bently demonstrates a half-frequency whirl associated with normal/loose bearing characteristics. For small β, the *unstable* speed range about $\omega = 2\lambda$ is defined from the $j = 1$ condition of (1.91) by

$$\left[\left(\frac{\omega}{2\lambda}\right)^2 - 1 + \zeta^2\right]^2 \leq q^2 - 4\zeta^2.$$

For zero damping this reduces to

$$1 - q \leq \left(\frac{\omega}{2\lambda}\right)^2 \leq 1 + q,$$

while for $\omega = 2\lambda$ and small q, the peak damping required for stability is

$$\zeta(\omega = 2\lambda) \cong q/2. \quad (1.92)$$

Figure 1.25 illustrates the stability curves in terms of q and the new parameter p, defined by

$$\frac{\omega}{2\lambda} = 1 + p, \quad (1.93)$$

where p represents the fractional change in running speed about $\omega = 2\lambda$.

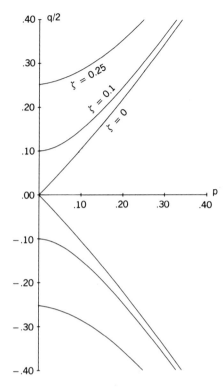

Figure 1.25 Stability curves for normal-loose ($q > 0$) and normal-tight ($q < 0$) parametric excitation [Childs (1982)].

Our normal-loose model has been predicated upon a positive value for ε (the increase/decrease in radial-stiffness parameter) and β/π (the fraction of the rotor's whirl orbit for which an increase/decrease in radial stiffness results). Hence, increases in either ε or β/π broaden the range of unstable speeds centered about $\omega = 2\lambda$.

Normal-Loose Nonlinear Analysis

The analysis of the preceding section provides a linear parametric-excitation explanation for the half-frequency whirl motion cited and experimentally produced by Bently, but fails to explain his one-third-frequency motion. The following nonlinear analysis provides an explanation for the one-third- and one-half-frequency motion as stable subharmonic motion.

Following conventional nonlinear analysis procedures, the following perturbation solution is assumed for Eq. (1.83):

$$R_X = x_0 + \varepsilon x_1, \qquad R_Y = y_0 + \varepsilon y_1, \qquad \lambda^2 = \lambda_0^2 - \varepsilon h. \qquad (1.94)$$

Substitution from this equation yields the following perturbation differential equations:

$$\varepsilon^0: \ddot{x}_0 + \lambda_0^2 x_0 = a\omega^2 \cos \phi,$$

$$\ddot{y}_0 + \lambda_0^2 y_0 = a\omega^2 \sin \phi - g, \qquad (1.95)$$

$$\varepsilon^1: \ddot{x}_1 + \lambda_0^2 x_1 = -hx_0 + \lambda_0^2 U(R_Y) x_0,$$

$$\ddot{y}_1 + \lambda_0^2 y_1 = -hy_0 + \lambda_0^2 U(R_Y) y_0, \qquad (1.96)$$

where, for our present purposes, damping has been dropped.

The half-frequency subharmonic solution will be investigated first, and we accordingly set $\omega = 2\lambda_0$, and for convenience choose $\phi_0 = -\pi/2$ to obtain the following ε^0 differential equations:

$$\ddot{x}_0 + \lambda_0^2 x_0 = 4a\lambda_0^2 \sin 2\lambda_0 t,$$

$$\ddot{y}_0 + \lambda_0^2 y_0 = -4a\lambda_0^2 \cos 2\lambda_0 t - g.$$

The particular solution for y_0 is

$$y_{0p} = A \cos 2\lambda_0 t - D; \qquad A = \frac{4a}{3}.$$

As noted earlier, if $A \leq D$, the solution is not influenced by the radial-stiffness discontinuity. If $A = D(1 + \varepsilon)$ is slightly larger than D, the synchronous motion itself provides a periodic activation of the nonlinearity, and leads to the parametric excitation situation of the preceding section. To investigate the possibility of subharmonic motion, we consider a complete solution for y_0 of the form

$$y_0 = b \cos \lambda_0 t + A \cos 2\lambda_0 t - D.$$

The additional term of this solution would decay exponentially in a linear damped system. In the present nonlinear system, it may reinforce itself and be sustained in a steady-state condition.

To consider this possibility, we assume $U(R_Y) \simeq U(y_0)$. The nonlinearity is activated by the motion providing $(b + A - D) > 0$, and is always activated if $A \simeq D$. For convenience, we assume* that $A = D$; i.e.,

$$y_0 = b \cos \lambda_0 t + D(\cos 2\lambda_0 t - 1). \qquad (1.97)$$

With this solution, any finite-amplitude transient solution amplitude $b > 0$ will periodically activate the nonlinearity at the fundamental frequency λ. The functions y_0 and $U(y_0)$ are illustrated in Figure 1.26. The assumed solution yields a periodic variation in radial stiffness at the frequency λ over

---

*This solution corresponds to the initial conditions $y_0 = b$, $\dot{y}_0 = 0$.

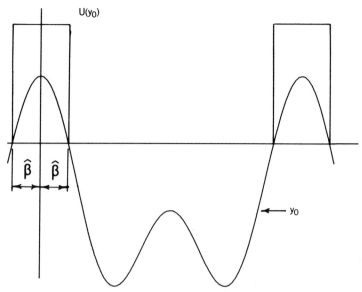

Figure 1.26 The function y_0 from Eq. (1.97) and the associated function $U(y_0)$ [Childs (1982)].

a $\hat{\beta}/\pi$ fraction of the $2\pi/\lambda$ period. The angle $\hat{\beta}$ is defined by

$$b \cos \hat{\beta} + D \cos 2\hat{\beta} = D. \tag{1.98}$$

The function $y_0 U(y_0)$ may be expanded in the Fourier series

$$y_0 U(y_0) = \frac{a_0}{2} + \sum_{i=1}^{\infty} a_i \cos i\lambda_0 t; \qquad a_i = \frac{2}{\pi} \int_0^{\hat{\beta}} y_0 \cos i\lambda_0 t d\lambda_0 t.$$

Substituting this result, together with y_0 from Eq. (1.97), into Eq. (1.96) yields

$$\ddot{y}_1 + \lambda_0^2 y_1 = \left(Dh + \frac{a_0 \lambda_0^2}{2} \right) + \cos \lambda_0 t \left(-bh + a_1 \lambda_0^2 \right)$$

$$+ \cos 2\lambda_0 t \left(-Dh + a_2 \lambda_0^2 \right) + a_3 \lambda_0^2 \cos 3\lambda_0 t + \cdots . \tag{1.99}$$

The constant term on the right indicates that the normal-loose radial-stiffness model causes the average rotor position to be above the static deflection position $y_0 = -D$. The secular term on the right is removed by requiring

$$h = a_1 \lambda_0^2 / b = \hat{Q} \lambda_0^2,$$

where

$$\hat{Q} = \frac{1}{\pi} \left[1 + \frac{\sin 2\hat{\beta}}{2} + \frac{D}{3b} (\sin 3\hat{\beta} - 2 \sin \hat{\beta}) \right].$$

For small $\hat{\beta}$ and (b/D), one obtains from this result and Eq. (1.98),

$$\hat{Q} \simeq \frac{1}{\pi} \left(1 + \frac{\hat{\beta}}{3} \right); \qquad \hat{\beta} \simeq \left(\frac{b}{2D} \right)^{1/2}. \tag{1.100}$$

Hence, h is always positive and is well behaved for reasonable values of b/D. Substituting from Eq. (1.99) into Eq. (1.94) yields the frequency relationship

$$\lambda^2 = \lambda_0^2 (1 - \varepsilon \hat{Q}). \tag{1.101}$$

This result indicates that a half-frequency subharmonic solution can be generated for a range of running speeds at or below the value $\omega = 2\lambda_0$, depending upon the ratios (b/D) and ε.

In a comparable analysis of Duffing's equation or similar nonlinear differential equations, the stability of the nonlinear solution provided by Eq. (1.97) is investigated by examining a perturbation about the assumed solution of the form $R_y = y_0 + \varepsilon u$. However, this is not possible for the differential equation (1.83) because the functions $\lambda_0^2 [1 - \varepsilon U(R_Y)] R_X$, $\lambda_0^2 [1 - \varepsilon U(R_Y)] R_Y$ have discontinuous derivatives with respect to the coordinates R_X, R_Y. The close resemblance between the solution of Eq. (1.97) and Bently's experimental results provides the strongest evidence for the stability of the subharmonic solution.

The one-third-running-speed solution is investigated by setting $\omega = 3\lambda$ in Eq. (1.95) to obtain

$$\ddot{x}_0 + \lambda_0^2 x_0 = 9 a \lambda_0^2 \sin 3\lambda_0 t,$$

$$\ddot{y}_0 + \lambda_0^2 y_0 = -9 a \lambda_0^2 \cos 3\lambda_0 t - g,$$

where once again $\phi_0 = \pi/2$. The assumed solution for y_0 is

$$y_0 = b \cos \lambda_0 t + D(\cos 3\lambda_0 t - 1), \qquad D = \frac{9a}{8}. \tag{1.102}$$

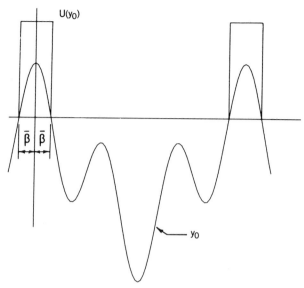

Figure 1.27 The function y_0 from Eq. (1.102) and the associated function $U(y_0)$ [Childs (1982)].

As illustrated in Figure 1.27, this function is positive for $\bar{\beta}/\pi$ of the fundamental period where $\bar{\beta}$ is defined by

$$b \cos \bar{\beta} + D \cos 3\bar{\beta} = D.$$

The remaining analysis for the one-third-frequency motion is the same as that employed for the half-frequency motion, with $\bar{\beta}$ replacing $\hat{\beta}$. However, the small $\bar{\beta}$ and (b/D) solutions of Eq. (1.100) now reduce to

$$\bar{Q} \simeq \frac{1}{\pi}\left(1 + \frac{19}{27}\bar{\beta}\right), \qquad \bar{\beta} = \frac{1}{3}\left(\frac{2b}{D}\right)^{1/2}. \qquad (1.103)$$

This result demonstrates that $\bar{\beta}$ is smaller than $\hat{\beta}$ for the same b/D ratio, and the corresponding frequency reduction is also smaller.

Normal-Tight Rubbing with Coulomb Damping

The preceding analysis for the normal-loose motion is mathematically equivalent to a normal-tight condition if Coulomb damping is neglected. Simply changing the sign of ε in Eq. (1.85) yields an increase in stiffness rather than a loss of stiffness. However, the author's analysis demonstrated that the

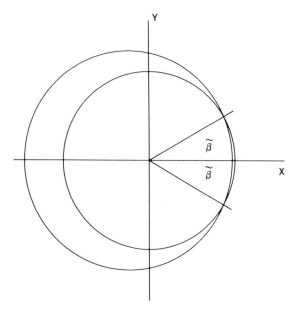

Figure 1.28   Radial clearance function for normal-tight interference [Childs (1982)].

presence of Coulomb damping during rubbing can cause markedly different results between the normal-loose and normal-tight conditions.

As illustrated in Figure 1.28, the radial clearance function between the shaft equilibrium position and its stator at a specified axial location is assumed to be of the form

$$H(\theta) = B - C \cos \theta.$$

Hence the clearance varies from a minimum of $B - C$ on the right to $B + C$ on the left, and the shaft is in contact with the stator if $R \simeq H$. For an assumed forward-precessional motion, the equations of motion are

$$\ddot{R}_X + 2\zeta_e\lambda_0 + \lambda_0^2 R_X[1 + \varepsilon U(R - H)]$$

$$= a\omega^2 \cos \phi - \mu\lambda_0^2(1 + \varepsilon)(R - H)\frac{R_Y}{R}U(R - H),$$

$$\ddot{R}_Y + 2\zeta_e\lambda_0 + \lambda_0^2 R_Y[1 + \varepsilon U(R - H)]$$

$$= a\omega^2 \sin \phi + \mu\lambda_0^2(1 + \varepsilon)(R - H)\frac{R_X}{R}U(R - H), \quad (1.104)$$

where μ is the Coulomb damping factor. These equations state that rubbing causes the radial stiffness to increase from k to $k(1 + \varepsilon)$ and also yields a *tangential* Coulomb damping force.

The normal-loose model in the preceding portion of this discussion permitted a planar analysis. However, the Coulomb damping term in Eq. (1.105) couples the motion in the $X - Z$ and $Y - Z$ plane. The author (1982) carried out a parametric-excitation analysis of the equations, the results of which will be summarized here. The parameter $\tilde{q} = \varepsilon\tilde{\beta}/\pi$ is an indication of the severity of contact for the present problem, since ε is a measure of the change in stiffness due to contact, and $\tilde{\beta}/\pi$ is a relative indication of the duration of contact.

Figure 1.29 illustrates the amount of damping required for stability as a function of p [as defined in Eq. (1.93)] for selected values of μ and $\tilde{q}$. Motion is stable above the stability boundaries, and unstable below them. The figure demonstrates that the maximum required damping at $p = 0$ is not appreciably influenced by Coulomb damping. However, damping is required over a much broader speed range to suppress an instability due to partial rubbing.

Discussion of Results

The models and accompanying analysis presented here for normal-loose and normal-tight radial-stiffness conditions in rotors basically confirm the parametric-excitation and nonlinear mechanisms proposed by Bently as explanations for fractional frequency whirl; viz.,

(a) For normal-loose conditions, the excitation frequency (running speed) must be twice or somewhat lower than the rotor critical speed.

(b) For normal-tight conditions, the excitation frequency must be twice or about anything higher than twice the critical speed.

The nonlinear results of Eq. (1.101) explain these shifts upwards/downwards of the base rotor critical speed with running speed as a result of the discontinuous radial stiffness model. This result basically agrees with Bently's view of the normal-tight conditions; viz., an increase in effective rotor stiffness accounts for the persistence of half-speed whirl motion with increasing speed. However, the results of Figure 1.29 indicate that the persistence of this motion may be equally the result of Coulomb damping.

One aspect of these results which should be considered carefully is the magnitude of damping required to suppress fractional-frequency motion resulting from nonsymmetric clearance effects. The values of $\tilde{q} = \varepsilon(\tilde{\beta}/\pi)$, 0.1 and 0.25, used in Figure 1.29 are comparatively modest since the relative-stiffness parameter ε can range up to unity for normal-loose conditions and achieve much larger values for normal-tight conditions. Moreover, rotors operating at twice their first critical speed are very likely to be lightly damped and not have a first mode damping factor on the order of 5 or 10 percent of critical. Hence, as Bently emphasizes, fractional-frequency whirl due to nonsymmetric clearance effects represents a serious source of large and potentially damaging vibration levels.

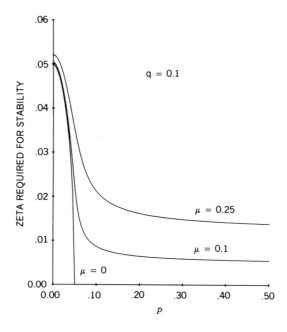

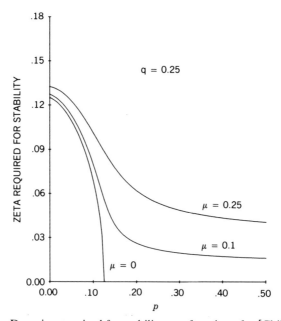

Figure 1.29   Damping required for stability as a function of p [Childs (1982)].

As noted in the introduction of this section, radial-bearing clearances in roller-element rotors present a continuing possibility of normal-loose radial-stiffness conditions, depending on the relative amplitudes of the static deflection D and synchronous amplitude A of Figure 1.22. Specifically, if the rotor imbalance is large enough to drive the rotor orbit through the "deadband" region of the radial stiffness function, half-frequency whirl motion is likely to result over substantial speed ranges, irrespective of the unit's "classical" linear stability. Since the occurrence of this motion depends both on the state of rotor balance and the tolerance stack-up defining bearing clearances, a considerable unit-to-unit variation in vibration characteristics of ball-bearing units is likely.

The partial-rubbing mechanism discussed here is distinct from rubbing over a complete precessional rotor orbit, which is discussed in Section 1.10.

1.9 SYNCHRONOUS MOTION WITH BEARING CLEARANCES

The preceding section has demonstrated that rotor motion which moves in and out of roller-element bearing clearances can yield fractional-frequency resonant phenomena predominantly at one-half of rotor running speed, but also at one-third and one-fourth of running speed. The analysis of this section shows that the bearing-clearance nonlinearity can also sharply modify the synchronous vibration characteristics of the rotor.

Bearing-Clearance Effects: Synchronous Response with the Jeffcott Model

The results of this section are generally based on Yamamoto's (1954) work. As illustrated in Figure 1.30(a), the principal modification introduced into the Jeffcott model by Yamamoto is the radial clearance e at the bearing locations. The bearing axis is drawn vertically to emphasize the absence of a gravity side load. The principal assumption involved in our analysis is that the displacement vector **R** is the sum of (a) the elastic deflection vector of the shaft and (b) the clearance vector **e**, and that these vectors are colinear, as illustrated in Figure 1.30(b). Assuming further that $R > e$, the constant-running-speed equations of motion can be stated:

$$\ddot{R} - R\dot{\theta}^2 + 2\zeta\lambda\dot{R} + \lambda^2(R - e) = a\omega^2\cos(\theta - \omega t),$$

$$R\ddot{\theta} + 2\dot{R}\dot{\theta} + 2\zeta\lambda(R\dot{\theta}) = -a\omega^2\sin(\theta - \omega t). \quad (1.105)$$

The synchronous steady-state solution to these equations may be stated:

$$R = R_0, \qquad \theta = \omega t + \psi.$$

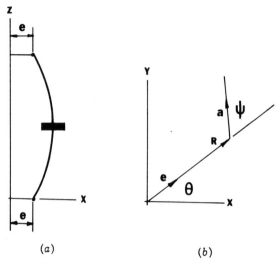

Figure 1.30   The Jeffcott model with bearing clearances.

The nondimensional radius is defined by the quadratic equation

$$dr^2 - 2(1 - \hat{\omega}^2)\hat{e}r + (\hat{e}^2 - \hat{\omega}^4) = 0, \qquad (1.106)$$

where

$$r = R_0/a, \qquad \hat{e} = e/a, \qquad \hat{\omega} = \omega\lambda,$$
$$d = (1 - \hat{\omega}^2)^2 + 4\zeta^2\hat{\omega}^2.$$

The phase angle ψ is defined by

$$\sin \psi = -2\zeta\hat{\omega}r/\hat{\omega}^2,$$
$$\cos \psi = [r(1 - \hat{\omega}^2) - \hat{e}]/\hat{\omega}^2. \qquad (1.107)$$

Yamamoto used Cartesian coordinates (and considerably more algebra) to obtain the definition for r provided by Eq. (1.106). The solution to (1.106) is

$$r = \left\{ (1 - \hat{\omega}^2)\hat{e} + \left[\hat{\omega}^4 d + \hat{e}^2[(1 - \hat{\omega}^2) - d] \right]^{1/2} \right\} \Big/ d. \qquad (1.108)$$

This solution is not defined if the quantity within the square-root term is negative; i.e., a necessary condition for the solution to exist is

$$\hat{\omega}^4 d + \hat{e}^2[(1 - \omega^2) - d] > 0. \qquad (1.109)$$

Further, if this condition is met, r must be greater than $\hat{e}$ to assure that the assumed differential equations (1.105) are correct.

If no solution exists to Eq. (1.105), the motion takes place within the clearance circle, and the appropriate equations of motion are

$$\ddot{R} - R\dot{\theta}^2 + 2\zeta\lambda\dot{R} = a\omega^2 \cos(\theta - \omega t),$$

$$R\ddot{\theta} + 2\dot{R}\dot{\theta} + 2\zeta\lambda(R\dot{\theta}) = -a\omega^2 \sin(\theta - \omega t), \qquad (1.110)$$

with the solution

$$r = \left(1 + 4\zeta^2/\hat{\omega}^2\right)^{-1}$$

$$\cos\psi = -r, \qquad \sin\psi = 2\zeta\hat{\omega}r/\hat{\omega}^2. \qquad (1.111)$$

Figure 1.31 illustrates r versus $\hat{\omega}$ solutions from Eqs. (1.108) and (1.111) for $\zeta = 0.025, 0.06$ and a range of $\hat{e}$. The value of r defined by Eq. (1.108) is plotted when it exists (satisfies inequality (1.109)], and is greater than $\hat{e}$. When either of these conditions is violated, the solution corresponding to Eq. (1.111) is plotted instead. The solution corresponding to Eq. (1.108) breaks down at low speeds due to a violation of condition (1.109), and breaks down at higher speeds by failing to satisfy $r > \hat{e}$. The results of these two figures demonstrate that increasing the clearance drops the apparent critical-speed location, while reducing the peak amplitudes. Further, bearing clearances cause a more rapid drop in amplitude with increasing speed after the apparent critical speed has been traversed. Increasing the damping reduces the speed range over which solution Eq. (1.108) is defined. Figure 1.32 illustrates the phase definition corresponding to the amplitude curves of Figure 1.31.

Note that for most of the speed range in Figure 1.31, either of the two solutions corresponding to $r > \hat{e}$ and $r < \hat{e}$ is possible. Hence, the rotor could be precessing at low-vibration amplitudes within the clearance circle, and then, due to an external disturbance, "jump" to the high-vibration, $r > \hat{e}$, solution. The possibility of jump phenomena is examined below in discussing bilinear bearing characteristics.

The practical consequences of the results illustrated in Figure 1.31 are (a) bearing clearances may sharply reduce the apparent locations of critical speeds, and (b) with bearing clearances, vibration amplitudes become erratically dependent on imbalance magnitudes. The results of Figure 1.31(b) support both of these points, since for $\hat{e} = 0$ and $\hat{e} = 3.0$, peak amplitudes are located at $\hat{\omega} = 1.0$ and $\hat{\omega} = 0.7$, respectively. Further, for a given clearance, the vibration amplitudes may *increase* at some speeds as the imbalance magnitude is reduced. The normalized speed $\hat{\omega} = 0.6$ provides an illustration of this type of behavior.

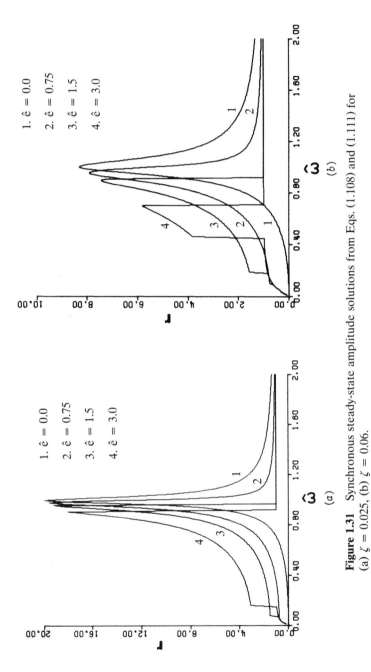

Figure 1.31 Synchronous steady-state amplitude solutions from Eqs. (1.108) and (1.111) for (a) $\zeta = 0.025$, (b) $\zeta = 0.06$.

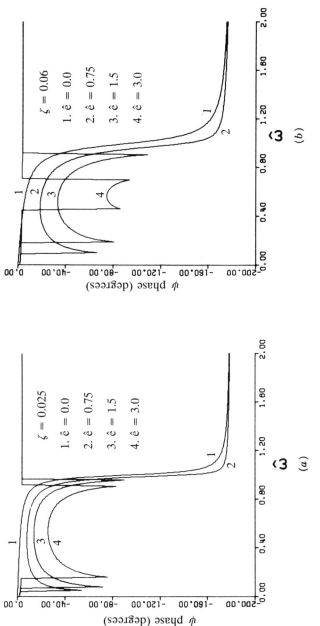

Figure 1.32 Synchronous steady-state phase solutions corresponding to the amplitude solutions of Figure 1.29.

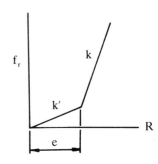

Figure 1.33   Bilinear bearing-stiffness support.

Figure 1.33 illustrates a bilinear stiffness support arrangement for the bearings, which yields the following bearing reaction-deflection relationship:

$$f_r = k'R, \qquad R \le e,$$
$$f_r = k(R - e'), \qquad R > e, \tag{1.112}$$

where

$$e' = e[1 - (k'/k)]. \tag{1.113}$$

In words, the rotor has a soft-support stiffness for deflections smaller than e, with synchronous response defined by the conventional Jeffcott solution of Eqs. (1.12) and (1.13). For $R > e$, the solution is defined by Eqs. (1.107) and (1.108) with e replaced by e' of Eq. (1.113).

This arrangement leads to the solution possibilities illustrated by Figure 1.34, for which $k'/k = 0.04$, $\hat{e} = 1.5$, and $\xi = 0.025$. For $r < \hat{e}$, the softly supported rotor has a critical speed at $\hat{\omega} = 0.20$. Hence, as $\hat{\omega}$ increases, r also increases until it reaches $\hat{e}$ (point B in Figure 1.34), and the $r > \hat{e}$ stiffly-supported-rotor solution is encountered. For increasing $\hat{\omega}$ above this point, the solution for r follows the curve of Eq. (1.108) through the peak at D and subsequent "crash" to point E. If the speed is further increased the $r < \hat{e}$ softly-supported-rotor solution curve is followed. More interestingly, if the speed is reduced, the softly-supported-rotor solution is also followed until point C is encountered. For a further reduction in speed, the solution follows the $r > \hat{e}$ solution to point A with a subsequent smaller "crash" to the softly supported solution curve. Either the $r > \hat{e}$ or $r < \hat{e}$ solution is possible between the points (A, B), or the points (C, D). Within these speed ranges, external disturbances may induce jumps from the low-amplitude $r < \hat{e}$ curve to the high-amplitude $r > \hat{e}$ curve.

At noted in the beginning of this discussion, the solutions of Figures 1.31–134 are for a vertical rotor without fixed-direction side load such as the rotor weight. If side loads are admitted, the situation of Section 1.8 holds with possible rotor motion in and out of the bearing clearance circle. However, for a sufficiently large side load, the clearance effect is completely

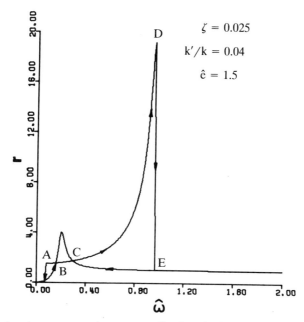

Figure 1.34   Synchronous steady-state amplitude solutions for the bilinear stiffness characteristics of Figure 1.31; $\zeta = 0.025$; $k' = -0.04$; $\hat{e} = 1.5$.

eliminated, since the motion is always outside the clearance circle. At present, nonlinear transient simulations are required for the analysis of real rotors supported with bearings in clearances, and acted upon by side loads. Section 1.11 introduces the transient simulation approach for a critical-speed-transition example.

Recent work by Choi and Noah (1987) and Kim and Noah (1991) considerably generalize Yamamoto's model and analysis by including a fixed-direction side load. Their analysis procedure accounts for strong nonlinearities versus the small-parameter perturbation approach of the preceding section.

1.10   ROTOR-HOUSING RESPONSE ACROSS
AN ANNULAR CLEARANCE

Figure 1.35 illustrates a simple shaft-disk system with an annular clearance C_r between the shaft and a stator element. As discussed earlier, the disk imbalance excites synchronous motion, and as the speed increases, the possibility arises for contact between the shaft and the stator with a consequent coupled motion of the shaft and stator across the annular clearance. The current problem differs from the nonsymmetric clearance effects of Section 1.8; specifically, if contact is established, it is maintained throughout an orbit. Two quite different types of motion can result from the proposed

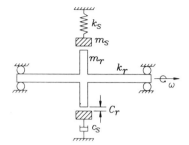

Figure 1.35 Simple rotor-stator-interaction model.

continuous contact between the shaft and the stator. First, synchronous, forwardly precessing motion, which is driven by the rotor imbalance, can occur with various possible rotor-stator "interaction" zones. Second, "dry-friction" whip or whirl can develop due to Coulomb friction forces at the point of contact. Motion driven by dry friction involves reverse precessional motion. This section considers both types of motion and is based primarily on Black (1968).

Figure 1.36 illustrates general positions for the rotor, stator, and radial clearance vectors $\mathbf{r}_s$, $\mathbf{r}_r$, and $\mathbf{C}_r$ for the idealized model of Figure 1.35. These vectors are related by

$$\mathbf{r}_s + \mathbf{C}_r = \mathbf{r}_r. \tag{1.114}$$

The normal contact force N between the rotor and stator is colinear with $\mathbf{C}_r$; the friction force F_f is normal to N and $\mathbf{C}_r$. The equations of motion can be stated:

$$m_r \ddot{x}_r + c_r \dot{x}_r + k_r x_r = ma\omega^2 \cos \omega t - N \cos \gamma + F_f \sin \gamma,$$

$$m_r \ddot{y}_r + c_r \dot{y}_r + k_r y_r = ma\omega^2 \sin \omega t - N \sin \gamma - F_f \cos \gamma,$$

$$m_s \ddot{x}_s + c_s \dot{x}_s + k_s x_s = N \cos \gamma - F_f \sin \gamma,$$

$$m_s \ddot{y}_s + c_s \dot{y}_s + k_s y_s = N \sin \gamma + F_f \cos \gamma.$$

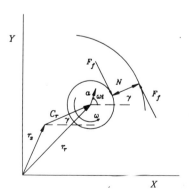

Figure 1.36 Rotor-stator-interaction motion.

By introducing the complex notation

$$\boldsymbol{r}_r = x_r + jy_r, \qquad \boldsymbol{r}_s = x_s + jy_s,$$

the following differential equations result:

$$m_r\ddot{\boldsymbol{r}}_r + c_r\dot{\boldsymbol{r}}_r + k_r\boldsymbol{r}_r = ma\omega^2 e^{j\omega t} - (N + jF_f)e^{j\gamma},$$

$$m_s\ddot{\boldsymbol{r}}_s + c_s\dot{\boldsymbol{r}}_s + k_s\boldsymbol{r}_s = (N + jF_f)e^{j\gamma}. \qquad (1.115)$$

The complex governing equations are completed by the following complex version of Eq. (1.114):

$$\boldsymbol{C}_r = C_{rX} + jC_{rY} = \boldsymbol{r}_r - \boldsymbol{r}_s = C_r e^{j\gamma}. \qquad (1.116)$$

A transient solution for these equations (during contact) could best be accomplished by eliminating the contact force to yield

$$(m_r + m_s)\ddot{\boldsymbol{r}}_s + (c_r + c_s)\dot{\boldsymbol{r}}_s + (k_r + k_s)\boldsymbol{r}_s$$
$$= ma\omega^2 e^{j\omega t} + \left(m_r\ddot{\boldsymbol{C}}_r + c_r\dot{\boldsymbol{C}}_r + k_r\boldsymbol{C}_r\right),$$

$$(m_r + m_s)\ddot{\boldsymbol{r}}_r + (c_r + c_s)\dot{\boldsymbol{r}}_r + (k_r + k_s)\boldsymbol{r}_r$$
$$= ma\omega^2 e^{j\omega t} - \left(m_r\ddot{\boldsymbol{C}}_r + c_r\dot{\boldsymbol{C}}_r + k_r\boldsymbol{C}_r\right),$$

$$\ddot{\boldsymbol{C}}_r - \ddot{\boldsymbol{r}}_r + \ddot{\boldsymbol{r}}_s = 0,$$

with the constraints

$$\boldsymbol{C}_r = \boldsymbol{r}_r - \boldsymbol{r}_s, \qquad \dot{\boldsymbol{C}}_r = \dot{\boldsymbol{r}}_r - \dot{\boldsymbol{r}}_s.$$

At present, however, we want the steady-state synchronous solution to Eqs. (1.115) and (1.116) due to imbalance. Note before beginning the solution for the model of Figure 1.35 that the rotor, stator, and combined (zero clearance) natural frequencies are

$$\omega_r = \sqrt{k_r/m_r},$$

$$\omega_s = \sqrt{k_s/m_s},$$

$$\omega_c = \sqrt{\frac{k_r + k_s}{m_r + m_s}}.$$

Synchronous Response

We assume solutions of the form

$$\boldsymbol{r}_r = \boldsymbol{r}_{r0}e^{j\omega t}, \qquad \boldsymbol{r}_s = \boldsymbol{r}_{s0}e^{j\omega t}. \qquad (1.117)$$

for the rotor and stator vectors. Hence, the vectors C_r, N, and F_f also must precess at the frequency ω, and the resulting solution format is

$$r_{r0} = |\rho|e^{j\xi} - \alpha_{11}(\omega)(N + jF_f),$$

$$r_{s0} = \beta_{11}(\omega)(N + jF_f),$$

$$C_r = r_{r0} - r_{s0}, \qquad (1.118)$$

where $\alpha_{11}(j\omega)$ and $\beta_{11}(j\omega)$ are "receptances" for the rotor and stator, respectively, and $\rho = |\rho|e^{j\xi}$ is the response of the rotor due to imbalance without contact. Note that ξ is the phase of ρ relative to C_r, not relative to the imbalance vector. The solution for r_{r0} simply states that the rotor response is the response due to imbalance (without contact) minus the response due to the contact force. The stator response is entirely due to the contact force. Note that the phase of C_r has been arbitrarily set to zero and forms the phase reference.

The rotor and stator receptances are easily stated:

$$\alpha_{11}(\omega) = 1/\left[(k_r - m_r\omega^2) + jc_r\omega\right],$$

$$\beta_{11}(\omega) = 1/\left[(k_s - m_s\omega^2) + jc_s\omega\right]. \qquad (1.119)$$

Receptance definitions are not always so obvious, as can be illustrated by the example of Figure 1.37. Assuming axisymmetric properties for the rotor and stator means that the receptances can be defined by a planar model. The

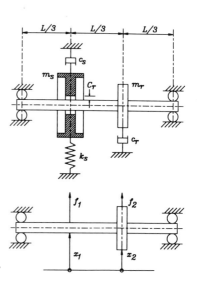

Figure 1.37 Rotor-stator-interaction example 2.

rotor governing equations are

$$k_{11}x_1 + k_{12}x_2 = f_1,$$

$$m_r\ddot{x}_2 + c_r\dot{x}_2 + k_{21}x_1 + k_{22}x_2 = 0.$$

Assuming harmonic motion at the frequency ω and eliminating $x_2(\omega)$ yields

$$\frac{x_1(\omega)}{f_1(\omega)} = \alpha_{11}(\omega) = \frac{\left[(k_{22} - m_r\omega^2) + jc_r\omega\right]}{k_{11}\left[(k_{22} - m_r\omega^2) + jc_r\omega\right] - k_{12}k_{21}}.$$

The receptance $\alpha_{11}(\omega)$ relates the rotor's harmonic response of the point of contact due to the contact force. It is a dynamic flexibility coefficient and can be related to the static flexibility coefficients

$$\begin{bmatrix} a_{11} & a_{12} \\ a_{21} & a_{22} \end{bmatrix} = \begin{bmatrix} k_{11} & k_{12} \\ k_{21} & k_{22} \end{bmatrix}^{-1} = \frac{1}{D}\begin{bmatrix} k_{22} & -k_{21} \\ -k_{12} & k_{11} \end{bmatrix},$$

where

$$D = k_{11}k_{22} - k_{12}k_{21}.$$

Specifically,

$$\alpha_{11}(\omega) = a_{11}\frac{\left[1 + (jc_r\omega - m_r\omega^2)/a_{11}D\right]}{a_{22}(jc_r\omega - m_r\omega^2) + 1}. \tag{1.120}$$

At zero frequency, $\alpha_{11}(\omega)$ simply reduces to the flexibility coefficient a_{11}. The stator receptance can be stated directly as

$$\frac{x_s(\omega)}{f_1(\omega)} = \beta_{11}(\omega) = \frac{1}{(k_s - m_s\omega^2) + jc_s\omega}. \tag{1.121}$$

Although Eqs. (1.118) were derived for the model of Figure 1.35, they apply equally for the model of Figure 1.37, or for any other circumferentially axisymmetric rotor/housing problem.

Synchronous Response Due to Imbalance

To obtain a solution, the governing equations (1.118) can be combined to obtain

$$C_r + |\alpha_{11} + \beta_{11}|Pe^{j(\psi + \nu_m)} = |\rho|e^{j\xi}, \tag{1.122}$$

where

$$\alpha_{11} + \beta_{11} = |\alpha_{11} + \beta_{11}|e^{j\psi}, \tag{1.123}$$

$$P = N + jF_f = N(1 + j\mu_m) = Pe^{j\nu_m}. \tag{1.124}$$

Because synchronous motion is assumed, slipping occurs continuously at the point of contact; hence, $F_f = \mu_m N$, where μ_m is the Coulomb friction factor, and

$$\tan \nu_m = \mu_m. \tag{1.125}$$

The two unknowns in Eq. (1.122) are P, the contact force magnitude, and ξ, the phase angle of ρ relative to C_r. P is of immediate interest and satisfies the quadratic equation

$$\left(\frac{P}{C_r}\right)^2 + 2\left(\frac{P}{C_r}\right)\frac{\cos(\psi + \nu_m)}{|\alpha_{11} + \beta_{11}|} + \frac{\left[1 - (|\rho|/C_r)^2\right]}{|\alpha_{11} + \beta_{11}|^2} = 0, \tag{1.126}$$

with the following two solutions:

$$\frac{P}{C_r} = \frac{-\cos(\psi + \nu_m) + \left\{(|\rho|/C_r)^2 - \sin^2(\psi + \nu_m)\right\}^{1/2}}{|\alpha_{11} + \beta_{11}|}, \tag{1.127a}$$

$$\frac{P}{C_r} = \frac{-\cos(\psi + \nu_m) - \left\{(|\rho|/C_r)^2 - \sin^2(\psi + \nu_m)\right\}^{1/2}}{|\alpha_{11} + \beta_{11}|}. \tag{1.127b}$$

Black demonstrated that the second of these solutions is unstable.

For a running-speed region in which $|\rho| > C_r$, engagement between the rotor and stator must occur. As we shall see, engagement is also possible for $|\rho| < C_r$, and Figure 1.38 illustrates the solution to the vector Eq. (1.122) for this situation. For P to be positive, $\cos(\psi + \nu_m)$ must be negative in Eq. (1.127a). P must also be real; hence, when $|\rho| < C_r$ the following two conditions must be met for a real and positive P:

$$\frac{\pi}{2} < \psi + \nu_m < \frac{3\pi}{2}, \qquad |\rho| \geq C_r \sin(\psi + \nu_m). \tag{1.128}$$

From the lower diagram of Figure 1.38, tangency of the line BD with the circle would yield $|\rho| = C_r \sin(\psi + \nu)$; hence, the second condition of (1.128) requires that BD intersect the circle. Having obtained P from Eq. (1.127a), ξ is easily obtained from Eq. (1.122) and r_{r0} and r_{s0} from Eq. (1.118).

Although not immediately obvious, when $|\rho| < C_r$ the receptances α_{11}, β_{11}, and their sum $\alpha_{11} + \beta_{11}$ actually define the running-speed regions for which engagement between the rotor and stator is possible. To clarify the

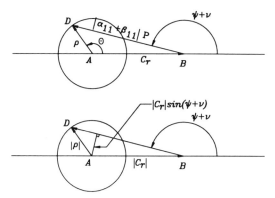

Figure 1.38   Rotor-stator-interaction solution for $|\rho| < C_r$ from Eq. (1.127).

explanation, we will consider motion in the absence of Coulomb friction and external damping. Without Coulomb friction, $\nu_m = 0$, and, without damping of the rotor or stator, $\alpha_{11} + \beta_{11}$ is real and either positive or negative. Hence, the requirements for engagement between the rotor and stator provided by (1.128) reduce to

$$\alpha_{11} + \beta_{11} < 0 \Rightarrow \psi = \pi. \tag{1.129}$$

To appreciate this requirement, consider Eq. (1.122) for the example of Figure 1.35. Substitution for α_{11} and β_{11} yields

$$
\begin{aligned}
C_r &= \left(\frac{m_r a \omega^2}{k_r - m_r \omega^2} \right) - \left(\frac{1}{k_r - m_r \omega^2} + \frac{1}{k_s - m_s \omega^2} \right) P \\
&= \left(\frac{m_r a \omega^2}{k_r - m_r \omega^2} \right) - \frac{[(k_r + k_s) - (m_r + m_s)\omega^2]}{(k_r - m_r \omega^2)(k_s - m_s \omega^2)} P.
\end{aligned}
$$

The solution for P is then

$$P = \frac{m_r a \omega^2 (k_s - m_s \omega^2) - C_r (k_r - m_r \omega^2)(k_s - m_s \omega^2)}{(k_r + k_s) - (m_s + m_r)\omega^2}. \tag{1.130}$$

Observe from this result that

$$\alpha_{11}(\omega) + \beta_{11}(\omega) = 0 \tag{1.131}$$

defines the natural frequency of the combined system with zero clearance; viz., $\omega = \omega_c = \sqrt{(k_s + k_r)/(m_r + m_s)}$. Further, $\omega = \omega_r$ yields

$$\alpha_{11} = \infty, \tag{1.132}$$

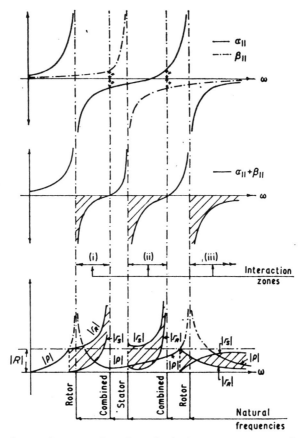

Figure 1.39 Interaction zones for a hypothetical rotor-stator system [Black (1968)].

while $\omega = \omega_s$ gives

$$\beta_{11} = \infty. \tag{1.133}$$

Figure 1.39 shows a hypothetical situation in which the stator's natural frequency lies between the rotor's two natural frequencies. Observe that as the running speed increases, $|\rho|$ grows and then engages the stator. As the running speed traverses the rotor's first critical speed, $|\rho|$ (and α_{11}) go from positive ∞ to negative ∞.* As the speed increases in region (i), α_{11} is negative and decreasing, β_{11} is positive and increasing, and $\alpha_{11} + \beta_{11}$ is negative and decreasing. As ω continues to increase, $|\rho|$ eventually becomes less than C_r; however, engagement is maintained until $\alpha_{11} + \beta_{11} = 0$ at $\omega = \omega_c$.

---

*Observe from the example of Eq. (1.130) that unbounded $|\rho|$ ($\omega = \omega_r$), does not yield an unbounded P. Also, $\beta_{11} = \infty$ ($\omega = \omega_s$), does not yield unbounded P. Only a running speed at the combined natural frequency yields an unbounded P.

When the running speed traverses the stator natural frequency, β_{11} goes from $+\infty$ to $-\infty$. Between the stator-only natural frequency and the second combined rotor-stator natural frequency, $(\alpha_{11} + \beta_{11})$ is negative and engagement is possible. This is region (ii) in Figure 1.39. Observe, however, that $|\rho| < C_r$ in this region; hence, imbalance could not trigger coupled rotor-housing motion, and an external disturbance would be required to start engagement.

As the speed traverses the second, rotor-only, critical speed, α_{11} again goes from $+\infty$ to $-\infty$, and engagement is again possible for $|\rho| < C_r$ in region (iii). As with region (i), engagement would be caused by imbalance in this region.

Figure 1.40 shows the probable response, due to imbalance, for runup and rundown of the hypothetical rotor-stator response. Observe that motion in region (ii) is not excited by imbalance on either runup or rundown. Further, portions of region (i) which are excited during runup are not excited during rundown.

Returning to the model of Figure 1.37, the receptances of Eqs. (1.120) and (1.121) are

$$\alpha_{11} = a_{11} \frac{[1 - 0.234\bar{\omega}^2]}{1 - \bar{\omega}^2}, \qquad \beta_{11} = a_{11}\left[\frac{0.1}{1 - 0.1\bar{\omega}^2}\right],$$

$$\bar{\omega} = \omega/\omega_r, \qquad\qquad \omega_r = 1/a_{11}m_r, \qquad (1.134)$$

where $a_{11} = a_{22}$, $a_{12} = 7a_{11}/8$, and b_{11} has been set to $a_{11}/10$. The natural

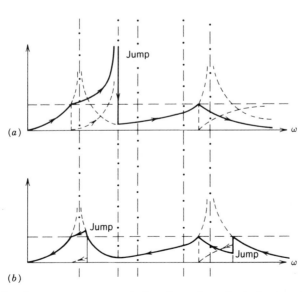

Figure 1.40 Synchronous response solution for a hypothetical rotor-stator system [Black (1968)].

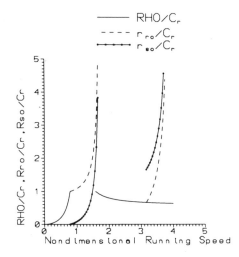

Figure 1.41 Interaction zones for example 2 of Figure 1.37 [Black (1968)].

frequencies are

$$\text{rotor alone,} \quad \bar{\omega}_r = 1,$$

$$\text{stator alone,} \quad \bar{\omega}_s = 3.16,$$

$$\text{combined,} \quad \bar{\omega}_c = 1.74, 3.94.$$

From a sketch similar to that of Figure 1.39, if $|\rho| < C_r$ interaction can only occur for the speed ranges

$$1 < \bar{\omega} < 1.74, \qquad 3.16 < \bar{\omega} < 3.94.$$

Figure 1.41 illustrates the response for the imbalance solution:

$$\frac{|\rho|}{C_r} = \frac{0.6\bar{\omega}^2}{1 - \bar{\omega}^2}. \tag{1.135}$$

Black also presents solutions including damping and Coulomb friction for this case; however, the frequency ranges for which engagement is possible are only slightly reduced.

Dry-Friction Whip and Whirl

Figure 1.42 illustrates a reverse-precession situation which can arise if the contact friction force is large enough to prevent slipping between the rotor and stator. A no-slip condition at the point of contact requires the rotor to precess at a frequency Ω which is opposite in direction to the shaft rotation ω.

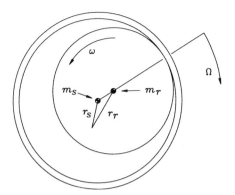

Figure 1.42 Kinematics for reverse whirl due to Coulomb friction without slipping.

The two frequencies are related through the kinematic requirement

$$\Omega = \frac{r\omega}{C_r}, \tag{1.136}$$

where r is the radius of the contacting shaft. The investigation of motion arising from dry-friction contact is the subject of this discussion.

Figure 1.43 illustrates a test apparatus and some test results from Lingener (1990) for a case when $r/C_r = 2.66$. In the absence of any external distur-

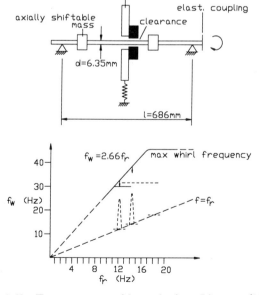

Figure 1.43   Test apparatus with results from Lingener (1990).

bance, the rotor-stator response is well predicted by our earlier analysis for *synchronous* interaction motion. Specifically, as speed is increased and approaches the rotor's first critical speed, contact is established between the rotor and the stator, and coupled motion results. Coupled synchronous motion between the rotor and stator continues with increasing speed until the running speed exceeds the coupled rotor-stator natural frequency. However, the upper curve of the lower plot of Figure 1.43 shows a quite different result when the rotor and stator are forced into contact at speeds substantially below the rotor's natural frequency. Specifically, above a certain limiting lower running speed, contact between the rotor and stator (imparted by an impact) yields supersynchronous, reverse, precession at Ω as defined by Eq. (1.136). Increasing the running speed yields an increase in Ω in accordance with Eq. (1.136) until Ω reaches the natural frequency of the coupled rotor-stator system. For further increases of the running speed, Ω persists at this limiting frequency. Precessional motion at the "tracking" frequency of Eq. (1.136) is generally referred to as dry-friction *whirling*.* Precessional motion at the limiting fixed frequency equal to the coupled, rotor-stator natural frequency is called dry-friction *whipping*.*

Dry-Friction Whirling

Black's basic analysis continues to apply to the present situation; however, the assumed harmonic motion is now at the precessional frequency Ω (instead of the running speed ω). To facilitate analysis, the effect of imbalance is neglected. Further, since the contact point is presumed not to slip, the contact force is now stated:

$$P = \left(N + jF_f \right) = N(1 + j\mu) = Pe^{j\nu},\qquad(1.137a)$$

where

$$\tan \nu = \mu.\qquad(1.137b)$$

In the following developments, ν and μ will be considered variables in addressing the question, "How much friction is needed to maintain dry-friction whirl?"

Returning to Eqs. (1.118), without imbalance, yields

$$C_r + |\alpha_{11}(\Omega) + \beta_{11}(\Omega)|Pe^{j(\psi + \nu)} = 0.\qquad(1.138)$$

---

*These designations of whipping and whirling are commonly used in the literature for describing motion driven by dry friction and are used in Section 3.6 for "oil whirl". In the balance of this text, whirl refers to pressional motion at a rotor's natural frequency.

To satisfy this equation,

$$\nu = \pi - \psi. \tag{1.139}$$

Rolling without slipping is possible as long as $\nu < \nu_m = \tan^{-1} \mu_m$ where μ_m is, in the present sense, the maximum available Coulomb friction factor. Slipping initiates when $\psi - \pi = \nu_m$.

The requirement that the friction force oppose slipping (or pending slipping) motion restricts ν as follows:

$$0 < \nu < \frac{\pi}{2}.$$

This is equivalent to requiring that μ satisfy $0 < \mu < \infty$. Hence, from Eq. (1.139), dry-friction whirl is possible when

$$0 < \pi - \psi < \frac{\pi}{2},$$

and the conditions under which dry-friction whip can exist depend entirely on the phase of $\alpha_{11}(\Omega) + \beta_{11}(\Omega)$ and the available Coulomb friction.

We apply these results to the example of Figure 1.35 with receptances given by Eq. (1.119). To obtain ψ, note that $\alpha_{11} + \beta_{11}$ can be written

$$\alpha_{11} + \beta_{11} = \frac{1}{A} + \frac{1}{B} = \frac{|B|^2\overline{A} + |A|^2\overline{B}}{|B|^2|A|^2}$$

$$= \frac{|B|^2(k_r - m_r\Omega^2) + |A|^2(k_s - m_s\Omega^2) - j\Omega(c_r|B|^2 + c_s|A|^2)}{|B|^2|A|^2},$$

where

$$|A|^2 = (k_r - m_r\Omega^2)^2 + c_r^2\Omega^2,$$

$$|B|^2 = (k_s - m_s\Omega^2)^2 + c_s^2\Omega^2.$$

Hence,

$$\tan\psi = \frac{-\Omega(c_r|B|^2 + c_s|A|^2)}{|B|^2(k_r - m_r\Omega^2) + |A|^2(k_s - m_s\Omega^2)} = \frac{-a}{b} \tag{1.140}$$

or

$$\pi - \psi = \tan^{-1}\left(\frac{a}{-b}\right). \tag{1.141}$$

From Eqs. (1.138) and (1.139), *the condition for interaction due to dry friction can be stated*:

$$\mu = \tan \nu = \tan(\pi - \psi) = \frac{a(\Omega)}{-b(\Omega)}. \tag{1.142}$$

Figure 1.42 illustrates the solution for required μ using the following data from Crandall (1990):

$$m_r = 0.418 \text{ Kg}, \qquad \omega_r = 77.2 \text{ rd/sec},$$
$$m_s = 0.858 \text{ Kg}, \qquad \omega_s = 210.9 \text{ rd/sec},$$
$$\omega_c = 178.5 \text{ rd/sec},$$
$$\zeta_r = c_r/2m_r\omega_r = 0.02,$$
$$\zeta_s = c_s/2m_s\omega_s = 0.01. \tag{1.143}$$

Coupled rotor-housing motion is only possible for positive μ in Figure 1.44. Limiting values of Ω which require infinite values for μ are obtained from Eq. (1.142) by setting $b(\Omega) = 0$. Neglecting damping, this yields

$$b(\Omega) = 0 = \left(k_s - m_s\Omega^2\right)\left(k_r - m_r\Omega^2\right)\left[(k_s + k_r) - (m_s + m_r)\Omega^2\right],$$

with the roots

$$\overline{\Omega} = \Omega/\omega_r = 1, \overline{\omega}_c, \overline{\omega}_s.$$

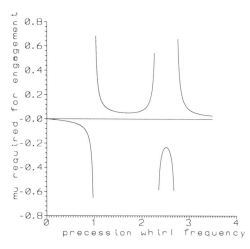

Figure 1.44 Required μ versus Ω for the model of Figure 1.33 and the data of Eq. (1.143).

From the data of Eq. (1.143), limiting values occur at

$$\overline{\Omega} = 1, 2.31, 2.73.$$

Returning to Figure 1.44, note that μ is positive for $1 < \overline{\Omega} < 2.31$ and $\overline{\Omega} > 2.73$. Lingener's tests demonstrated dry-friction whirl in the range $1 < \overline{\Omega} < 2.31$ but not in the range $\overline{\Omega} > 2.73$. When dry-friction whirl is possible, Eq. (1.136) can be used to find corresponding running-speed ranges; i.e., $\omega = C_r \Omega / r$.

Dry-Friction Whipping

The plot of Figure 1.44 and the above discussion has the viewpoint of using Eq. (1.142) to calculate μ values which are sufficient to prevent slipping for specified whirl frequencies Ω. However, the equation applies equally well for the calculation of limiting whirl frequencies for specified values of μ. Specifically, setting $\mu = \mu_m$ (the available Coulomb damping coefficient) and solving for Ω yields the whirl frequencies Ω_1, Ω_2 which will exist *with slipping*. Ω_1 denotes the lowest frequency at which backwards whirl can be initiated. The upper frequency Ω_2 is the shaft whipping frequency of Lingener's experiment, which persists as the running speed is increased beyond $\omega_2 = C_r \Omega_2 / r$.

The solution for rotor and stator response proceeds as follows:

(a) Calculate ψ which is the phase of $\alpha_{11} + \beta_{11}$.
(b) Calculate ν from Eq. (1.142).
(c) Calculate P from Eq. (1.138).
(d) Calculate r_{r0} and r_{s0} from Eq. (1.118) with $|\rho| = 0$.

Figure 1.45 is taken from Crandall and shows a solution for r_{s0}/C_r and Ω/ω_r versus ω/ω_r for $\mu = 0.3$ and the data (1.143). The circles on the figure denote Lingener's data.

A question which arises in viewing dry-friction whirl experiments is, "Given that rubbing contacts are common in rotating machinery, why doesn't dry-friction whirl or whip occur more often?" One quick answer to this question might be that most rubbing in turbomachinery has a lubricant present so the available Coulomb friction values are inadequate. However, several experimenters, including Lingener, have found that the whirl induced by rubbing is insensitive to the addition of lubricants to the point of contact. A more likely answer is provided by the C_r/r ratio used in Lingener's experiment. For most seals in turbomachinery where rubbing is likely, the C_r/r ratio is on the order of 0.01 or smaller, versus 0.375 for Lingener's

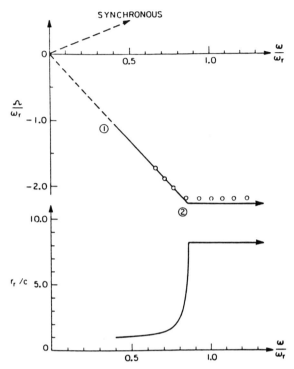

Figure 1.45   Test results for Lingener (1990), versus theory, Crandall (1990).

example. Hence for Lingener, the limiting whirl frequencies

$$\Omega_1 = \omega_r, \qquad \Omega_2 = 2.31\omega_r$$

yield the limiting running speeds

$$\omega_1 = 0.375\omega_r, \qquad \omega_2 = 0.864\omega_r.$$

However, for $C_r/r = 0.01$, the limiting running speeds would be

$$\omega_1 = 0.01\omega_r, \qquad \omega_2 = 0.0231\omega_r.$$

In practical terms, rubbing is not likely to occur at running speeds that are a tiny fraction of the rotor's first critical speed; in any case, the energy available at these low speeds is too small to cause destructive motion.

Summary, Conclusions, and Extensions

Full 360° rubbing of the type described here occurs commonly in turboma-chinery, particularly in multistage pumps. The motion is almost always synchronous, with dry-friction-induced reverse precessional motion occurring rarely. Synchronous forwardly precessional interaction motion of a rotor and stator involves continuous slipping at the contact surface which will predictably yield wear.

Dry-friction *whirl* and *whip* are only likely to occur for contact of a small-diameter shaft at a (very) large clearance. In addition, triggering of contact normally requires an outside disturbance. Although unlikely to occur, dry-friction whirl and whip can be intensely destructive.

Although not considered here, a third phenomenon arising from rubbing is "spiral vibration" as analyzed by Kellenberger (1980). Rubbing in this case causes a thermal bow in the rotor. In Kellenberger's words,

> This thermal bowing moves very gradually round the shaft circumference and can steadily increase in magnitude. Mechanical unbalance and thermal bow can thus lead to a slowly rotating (relative to the shaft) whirl vector whose tip describes a spiral (spiral vibration).

1.11   TRANSIENT RESPONSE: CRITICAL-SPEED TRANSITIONS

The analysis of the preceding sections in calculating critical speeds, synchronous response, and stability is generally linear and tends to obscure the basic nonlinearity of the governing Eqs. (1.1). However, there are a number of rotordynamic problems (albeit a distinct minority) for which linear analysis is not appropriate, and direct integration of nonlinear governing equations is required. As an example of this type of problem, we examine the transient response of the Jeffcott model of Eq. (1.9) during an acceleration through its critical speed. A nondimensional version of these equations is

$$r_X'' = (1 + bs^2\phi)(r_X + 2\zeta_e r_X') + bs\phi c\phi(r_Y + 2\zeta_e r_Y') + c\phi(\phi')^2 + c\overline{M}_z s\phi,$$

$$r_Y'' = bs\phi c\phi(r_X + 2\zeta_e r_X') - (1 + bs^2\phi)(r_Y + 2\zeta_e r_Y') + s\phi(\phi')^2 + c\overline{M}_z c\phi,$$

$$\phi'' = -bs\phi(r_X + 2\zeta_e r_X') + bc\phi(r_Y + 2\zeta_e r_Y') + c\overline{M}_z, \tag{1.144}$$

where $c\phi = \cos\phi$, $s\phi = \sin\phi$, $r_X = R_X/a$, $r_Y = R_Y/a$, $b = c(a/\bar{r})^2$, $c = 1/\{1 - (a/\bar{r})^2\}$, and $\overline{M}_z = M_z/J_z\lambda^2$. In the above $a = a_x$, $a_y = 0$, $\bar{r}$ is the rotor radius of gyration (i.e., $J_z = m\bar{r}^2$), and the prime denotes differentiation with respect to the nondimensional time variable $\tau = \lambda t$; i.e., $\phi' = \dot{\phi}/\lambda$, etc.

The three variable parameters in these equations are the external damping factor ζ_e, the ratio of the imbalance vector magnitude and the radius of gyration $(a/\bar{r})$, and the nondimensional spin-axis torque $\overline{M}_z$. The results of Figure 1.46 illustrate the solution to these equations for the parameter set, $\zeta_e = (a/\bar{r}) = \overline{M}_z = 0.01$, for zero initial conditions, and were obtained by

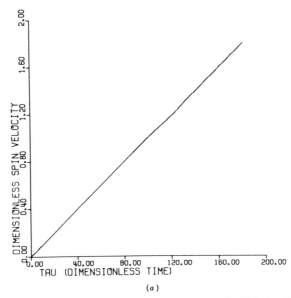

(a)

Figure 1.46(a) Transient critical-speed transition for Eq. (1.144) for $\zeta_e = (A/\bar{r}) = \overline{M}_z = 0.01$; ϕ' versus τ.

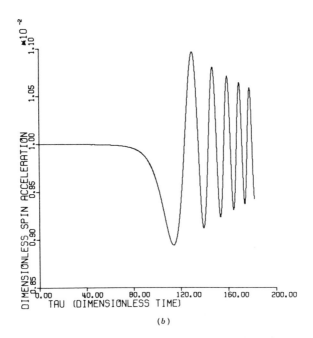

(b)

Figure 1.46(b) Transient critical-speed transition for Eq. (1.144) for $\zeta_e = (A/\bar{r}) = \overline{M}_z = 0.01$; ϕ'' versus τ.

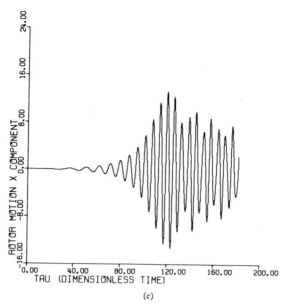

(c)

Figure 1.46(c)   Transient critical-speed transition for Eq. (1.144) for $\zeta_e = (A/\bar{r}) = \overline{M}_z = 0.01$; r_X versus τ.

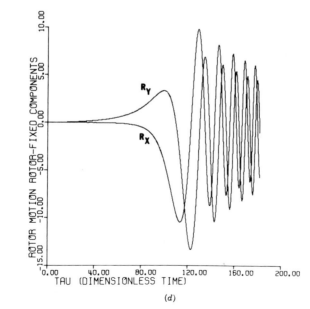

(d)

Figure 1.46(d)   Transient critical-speed transition for Eq. (1.144) for $\zeta_e = (A/\bar{r}) = \overline{M}_z = 0.01$; r_x and r_y versus τ.

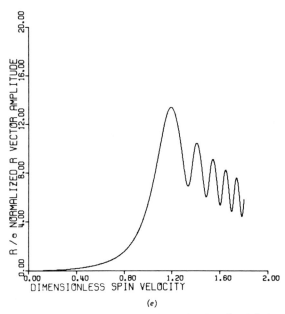

(e)

Figure 1.46(e)   Transient critical-speed transition for Eq. (1.144) for $\zeta_e = (A/\bar{r}) = \overline{M}_z = 0.01$; R/a versus ϕ'.

numerical integration. Figure 1.46(a) illustrates the expected nominal increase in ϕ' due to a constant applied torque; however, Figure 1.46(b) clearly shows the variation in ϕ'' due to coupling with the transverse motion. Figure 1.46(c) illustrates the normalized displacement components r_x, r_y, which were obtained from the stationary components r_X, r_Y via Eq. (1.17). A comparison of Figures 1.46(b) and 1.46(d) demonstrates a clear similarity in the dynamic behavior of ϕ'' and the components r_x, r_y. Figure 1.46(e) illustrates the variation of R/a as a function of $\phi' = \phi/\lambda$. Note that the steady-state response results of Figure 1.3 predict a peak response at $\phi' = 1$; however, the transient solution yields a maximum at $\phi' \simeq 1.2$. If a rotor is decelerated through a critical speed, the maximum amplitudes will occur below the critical speed.

Note that the peak nondimensional amplitude $|R|/a$ experienced during acceleration through the critical speed was about 14, as compared to a predicted steady operating speed response on the critical speed of $1/2$ $\xi_e = 50$. This result is *misleading* for commercial turbomachinery which will normally not be accelerated through the critical speed rapidly enough to yield any significant reductions in amplitude. Figure 1.46(b) shows the nominal dimensionless acceleration, $\phi'' \cong .01$. For a rotor with a 3000-rpm critical speed ($\lambda = 314$ rd/sec), the corresponding physical acceleration is $\ddot{\phi} = \lambda^2\phi'' \cong 987$ rd/sec^2 = 9,425 rpm/sec.

Approximate analytic solutions for Eq. (1.144) have been developed by various analysts; however, the excessive dimensionality of transient models for actual rotors precludes analytic solution approaches and dictates instead the development of efficient simulation models for solution via digital computation.

1.12 SUMMARY AND EXTENSIONS

This section is provided to review the lessons which have (hopefully) been learned in the preceding sections, and to indicate the connection of this material to the balance of the book.

Summary

The general lessons to be learned from the preceding sections concerning the dynamic characteristics of rotors are as follows:

(a) Linear-Frequency-Response Characteristics The customary and desired motion of a rotor is synchronous; i.e., the precessional motion of the rotor has the same direction as its rotation and the same frequency. This response is the result of excitation due to rotor imbalance or product-of-inertia disturbances. Rotors have natural frequencies associated with forward (synchronous) and backward (asynchronous) precessional motion, and these natural frequencies generally depend on running speed. Rotor running speeds which coincide with forward and backward precessional natural frequencies are referred to, respectively, as forward and backward critical speeds. Large synchronous amplitudes can be anticipated for running speeds near a rotor's forward critical speeds due to imbalance; however, backward critical speeds are (generally) not excited by imbalance. Shaft bow or distortion also provide synchronous excitation for rotors, and the combination of imbalance and shaft bow can yield synchronous response characteristics which are quite different from the response due to imbalance alone.

(b) Rotor Instabilities Internal rotor damping is representative of various external rotor forces which yield nonsymmetric stiffness cross-coupling in a rotor's equations of motion and lead to linear rotor instabilities. Specifically, above a given running speed ω_s, called the onset speed of instability, the rotor's motion diverges exponentially. The precessional motion associated with this instability is in the direction of rotor rotation, but the precessional frequency is subharmonic and approximately equal to the rotor's natural frequency. The onset speed of instability can be elevated by (a) increasing the rotor's stiffness or external damping, (b) decreasing the destabilizing forces, or (c) introducing asymmetry in the bearing support system.

Rotors are also theoretically subject to parametric instabilities due to stiffness or inertia asymmetry of the rotating element. The precessional motion associated with these parametric instabilities is forward; i.e., it is in the same direction as the rotor's rotation and is at approximately rotor running speed. However, this type of unstable motion is only possible over restricted speed bands, and a rotor's motion is stable between or above these bands. Since rotors are customarily designed to be symmetric, parametric instabilities due to stiffness or inertial asymmetry are of limited practical interest. However, cracked shafts in which undesired and unanticipated asymmetry is developed provide a notable exception to this generality.

(c) Nonsymmetric Clearance Effects Large and potentially damaging vibrations which are an exact fraction (generally one-half, occasionally one-third or one-fourth) of rotor running speed can result from either excessive bearing clearances or rotor rubbing contact over a portion of a rotor's orbit. Bearing clearances are normally present in rotors supported by rolling-element bearings to permit axial movement. A normal-loose effect can result in rotors supported by hydrodynamic bearings if the vertical bearing restraint is loose. Partial rubbing generally results when a seal is installed in an off-centered position. The half-frequency motion occurs in the vicinity of a twice-natural-frequency running speed.

(d) Symmetric Rotor-Stator Interaction across a Clearance Coupled rotor-housing motion is possible due to contact across a clearance. When driven by imbalance, the resulting precessional motion is forward and synchronous. When driven by dry-friction forces, the precessional motion is backward and supersynchronous. Its precessional frequency is, at first, a fixed factor times running speed (dry-friction whirl); however, above a limiting running speed the precessional frequency "locks in" to the coupled rotor-stator natural frequency (dry-friction whip). Synchronous coupled motion is only possible over specified running-speed ranges and is accompanied by continuous slipping at the contact surfaces with obvious consequences for wear. Synchronous coupled motion between the rotor and stator can yield quite different response curves on runup and rundown in speed. Dry-friction whirl and whip are only likely to cause problems when the rotor is flexible and contact arises between a small-diameter rotor and a stator across a large clearance. Although unlikely to occur, dry-friction whip and whirl can be extremely destructive.

(e) Nonlinear Response to Imbalance The dynamic characteristics of rotor response are generally explainable in terms of linear models and analysis. The two examples presented in this chapter which involve nonlinearities are (a) bearing-clearance effects, and (b) acceleration (or deceleration) of a rotor through a critical speed. Bearing clearances are shown to yield (a) reduction in apparent critical-speed location, (b) a sharp drop in rotor amplitudes with

increasing speed following traversal of the critical speed, and (c) discontinuous jumps in rotor vibration amplitudes. The transient simulations presented in the preceding section for an acceleration (or deceleration) through a critical speed demonstrate that a rotor's spin acceleration causes the peak amplitude to appear at running speeds which are higher than the critical speed, with a converse result for rotor deceleration.

Extensions

The balance of this book is devoted to the development of models for flexible rotors and appropriate techniques for their analysis. Models will be developed for rotors which are representative of typical rotating machinery, i.e., compressors, turbines, power-transmission shafting, etc., and are more complicated, in the following respects, than the simplified models of the present chapter:

(a) The dimensionality of structural dynamic models is significantly greater.
(b) The rotor's dynamic behavior is influenced by fluid-structure interaction forces.

Chapter 2, which follows, defines the basic structural dynamic models to be employed. Chapter 3–6 define models for the fluid forces which act on the rotor, and Chapter 7 explains how system models are developed by adding fluid-structure interaction forces to the basic structural dynamic model.

Although a system rotor model which results from the development of Chapters 2–7 is more complicated and larger than the simplified models of the present chapter, the same basic questions are to be resolved. Specifically, in the design of high-speed rotating machinery, the following questions must be addressed:

(a) For a given running speed, what are a rotor's natural frequencies?
(b) What are the rotor's critical speeds?
(c) What are the anticipated steady-state response levels over the rotor's operating range?
(d) Will the rotor be stable over its operating range?

Chapter 8 provides an example rotor problem which illustrates answers to these questions.

REFERENCES: CHAPTER 1

Atkins, K., and Perez, R. (1988), "Influence of Gas Seals on Rotor Stability of a High Speed Hydrogen Recycle Compressor, in *Proceedings of the 17th Turbomachinery Symposium*, Texas A&M University, pp. 9–18.

Barrett, L., Gunter, E., and Allaire, P. (1978), "Optimum Bearing Support Damping for Unbalance Response and Stability of Rotating Machinery," *Journal of Engineering for Power*, 89–94.

Bently, D. (1974), "Forced Subrotative Speed Dynamic Action of Rotating Machinery," ASME Paper No. 74-PET-16, Petroleum Mechanical Engineering Conference, Dallas, TX.

Biezeno, C., and Grammel, R. (1959), *Engineering Dynamics*, Vol. III. of *Steam Turbines*, D. Van Nostrand Co., Inc., New York.

Bishop, R., and Parkinson, A. (1965), "Second Order Vibrations of Flexible Shafts," *Philosophical Transactions of the Royal Society of London*, **259**, 1–31.

Black, H. (1976), "The Stabilizing Capacity of Bearings for Flexible Rotors with Hysteresis," *Journal of Engineering for Industry*, 87–91.

Black, H. (1989), "Parametrically Excited Lateral Vibrations of an Asymmetrically Slender Shaft in Asymmetrically Flexible Bearings," *Journal of Mechanical Engineering Science*, **11**(1), 57–67.

Black, H. (1968), "Interaction of a Whirling Rotor with a Vibrating Stator Across a Clearance Annulus," *Journal of Mechanical Engineering Science*, **10**(1), pp. 1–12.

Brosens, P., and Crandall, S. (1961), "Whirling of Unsymmetrical Rotors," *Journal of Applied Mechanics*, 355–362.

Childs, D. (1982), "Fractional Frequency Rotor Motion Due to Non-Symmetric Clearance Effects," *Journal of Engineering for Power*, 533–541.

Choi, Y.-S., and Noah, S. (1987), "Nonlinear Steady-State Response of a Rotor-Support System," *Journal of Vibration, Stress, and Reliability in Design*, **109**, 255–261.

Crandall, S. (1990), "From Whirl to Whip in Rotordynamics," in *Transactions, ITFoMM Third International Conference on Rotordynamics*, Lyon, France, pp. 19–26.

Crandall, S. (1983), "The Physical Nature of Rotor Instability Mechanisms," in M. Adams (Ed.), *Rotor Dynamical Instability*, ASME, New York.

Den Hartog, J. (1952), *Mechanical Vibrations* (4th ed.), McGraw-Hill, New York.

Dimentberg, F. (1961), *Flexural Vibrations of Rotating Shafts*, Butterworths, London.

Ehrich, F. (1976), "Self Excited Vibrations," in C. M. Harris and C. E. Credel (Eds.), *Shock and Vibration Handbook* (2nd ed.), McGraw-Hill, New York, Chap. 5.

Ehrich, F., and Childs, D. (1984), "Identification and Avoidance of Instabilities in High Performance Turbomachinery," *Mechanical Engineering*, 66–79.

Föppl, A. (1895), Das Problem der Laval'schen Turbinewelle, *Civilingenieur*, **41**, 332–342.

Gasch, R., and Pfützner, H. (1975), *Rotordynamik*, Springer Verlag, Berlin.

Grabowski, B. (1980), "The Vibrational Behavior of a Turbine Rotor Containing a Transverse Crack," *Journal of Mechanical Design*, **102**, 140–146.

Green, R. (1948), "Gyroscopic Effects of the Critical Speeds of Flexible Rotors," *Journal of Applied Mechanics*, **15**, 369–376.

Gunter, E. (1966), "Dynamic Stability of Rotor Bearing Systems," NASA Paper No. SP113.

Hsu, C. (1965), "Further Results on Parametric Excitation of a Dynamic System," *Journal of Applied Mechanics*, 373–377.

Jeffcott, N. (1919), "Lateral Vibration of Laded Shafts in the Neighbourhood of a Whirling Speed—The Effect of Want of Balance," *Philosophical Magazine*, **37**, 304–314.

Kellenberger, W. (1980), "Spiral Vibrations Due to the Seal Rings on Turbogenerators Thermally Induced Interaction Between Rotor and Stator," *Journal of Mechanical Design*, Vol. 102, 177.

Kim, B., and Noah, S. (1991), *Nonlinear Dynamics*, **2**, pp. 215–234.

Lingener, A. (1990), "Experimental Investigation of Reverse Whirl of a Flexible Rotor," in *Transactions, ITFoMM Third International Conference on Rotordynamics*, Lyon, France, pp. 13–18.

Lund, J. (1974), "Stability and Damped Critical Speeds of a Flexible Rotor in Fluid Film Bearings," *Journal of Engineering for Industry*, 509–517.

Smith, D. (1933), "The Motion of a Rotor Carried by a Flexible Shaft in Flexible Bearings," *Proc. of the Royal Society of London*, 92–119.

Stodola, A. (1927), *Steam and Gas Turbines*, McGraw-Hill, New York.

Timoshenko, S., and Young, D. (1955), *Vibration Problems in Engineering*, Van Nostrand, Princeton, NJ.

Williams, R., Jr., and Trent, R. (1970), "The Effects of Nonlinear Rotor Stability," SAE Paper No. 700320, National Air Transportation Meeting, New York.

Yamamoto, T. (1954), "On the Critical Speeds of a Shaft," *Memoirs of the Faculty of Engineering, Nagoya Univ.*, **6**(2).

2

STRUCTURAL-DYNAMIC MODELS AND EIGENANALYSIS FOR UNDAMPED FLEXIBLE ROTORS

2.1 INTRODUCTION

We are concerned in this chapter with the nature of structural-dynamic models for flexible-rotor/case systems. The modeling requirements for rotating-equipment cases are generally identical to conventional structural-dynamic applications. However, the need to account for the rotational motion of rotors yields distinctive rotor structural-dynamic models, and the development and subsequent eigenanalysis of these models is the subject of our present discussion.

The selection of generally appropriate and adequate structural-dynamic models for flexible rotors is best arrived at by a review of bending and shear deflections of beams [Timoshenko (1955)], transverse and rotary inertia, and "gyroscopic" effects.

Lumped-parameter structural-dynamic models have traditionally been employed to account for the distributed elastic and inertial properties of rotors. Specifically, rotors are modeled as a collection of n rigid bodies connected by massless elastic beam elements, as illustrated in Figure 2.1. Appropriate equations of motion for lumped-parameter models of this system are developed in the next section. Historically, both transfer-matrix formulations due to Prohl (1945) and Myklestad (1944) and general mass-stiffness matrix formulations [Biezeno and Grammel (1959)] have been employed to develop flexible-rotor models from the component rigid-body equations of Section 2.2, and distinct eigenanalysis procedures have been developed based on these two approaches. As used in this chapter, the term eigenanalysis refers to the calculation of undamped natural frequencies, critical speeds, and mode shapes. Section 2.3 examines the general-stiffness-matrix approach for

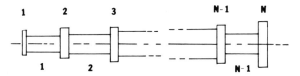

Figure 2.1   An *n*-body lumped-parameter representation for a flexible rotor.

modeling and eigenanalysis of flexible rotors, while Section 2.4 considers the same subjects from the Myklestad-Prohl transfer-matrix approach.

As noted, *n*-body lumped-parameter models have historically been employed to model the distributed elastic and inertial properties of flexible rotors. Ruhl and Booker (1972) introduced finite-element models for flexible rotors, and have employed general-stiffness versions of these models to calculate rotor natural frequencies and mode shapes. Their analysis accounts only for bending deflections and transverse inertia. Nelson and McVaugh (1976) subsequently provided more general finite-element models which account for rotary inertia, gyroscopic coupling, and axial loads, and Nelson (1980) has developed a finite-element model which also accounts for shear deflections and the effect of axial torque. Section 2.5 provides an introduction to finite-element models for (slender) flexible rotors. Finite-element models are developed based on slender beam theory, and general matrix procedures are demonstrated for developing governing system models from these component element models. Section 2.6 introduces finite-element models, which are also based on slender beam theory, but additionally accounts for rotary inertia and gyroscopic coupling. The element equations of motion derived in this section provide an alternative basis for system simulation and response models, as compared to the lumped-parameter models of Sections 2.2–2.4.

2.2   LUMPED-PARAMETER VIBRATION MODELS

As noted in the preceding section, the distributed elastic and stiffness properties of flexible rotors have traditionally been modeled as a collection of *n* rigid bodies connected by massless elastic beam elements. Figure 2.1 illustrates this modeling approach. We are concerned in this section with the development of appropriate governing vibration equations for this type of system. We begin below by deriving equations of motion for a representative component rigid body, taking advantage of the "small" deflections enforced on the body by bearings and rotor-stiffness constraints. Relative elastic axial and torsional deflections between the component rigid bodies are assumed to be negligible.

The equations to be derived are for an axisymmetric rotor, and will be stated only in an inertial (nonspinning) reference system. Comparable governing equations may readily be derived in a rotor-fixed (spinning) reference system; however, these equations are more complicated, and not as attractive for computational work. A rotor-fixed reference is preferable in modeling asymmetric rotors. However, as noted in Section 1.7, the only commonly recurring orthotropic rotor example is provided by two-pole turbogenerator units. Persons desiring information concerning simulation models for orthotropic rotors may consult Childs (1976b).

Stationary-Reference Vibration Equations

Figure 2.2 illustrates the required kinematics. The xi, yi, zi axes are fixed in rigid body i, with zi the nominal axis of symmetry. The vector R_i locates the origin of the xi, yi, zi system in the (inertial) X, Y, Z system, and the vector a_i locates the mass center of rigid body i in the xi, yi, zi system. The Z axis defines the nominal spin-axis of the rotor, and we are interested in the transverse components R_{iX}, R_{iY} of the displacement vector R_i. The linear momentum equation for rigid body i yields

$$m_i \ddot{R}_{iX} = f_{iX} + \bar{f}_{iX} + m_i a_{iX} \dot{\phi}^2 + m_i a_{iY} \ddot{\phi},$$

$$m_i \ddot{R}_{iY} = f_{iY} + \bar{f}_{iY} + m_i a_{iY} \dot{\phi}^2 - m_i a_{iX} \ddot{\phi}, \qquad (2.1)$$

where

$$a_{iX} = a_{ix} c\phi - a_{iy} s\phi, \qquad a_{iY} = a_{ix} s\phi + a_{iy} c\phi, \qquad a_{iz} = 0.$$

In the above $c\phi = \cos \phi$, $s\phi = \sin \phi$, m_i is the mass of body i, and

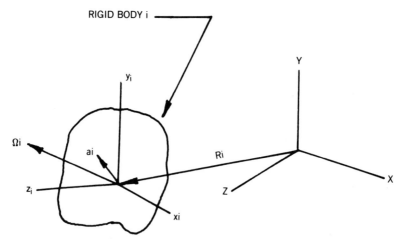

Figure 2.2 Kinematic variables for a component rigid body.

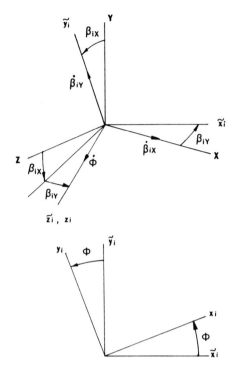

Figure 2.3 Euler angles for a component rigid body.

$(f_{iX}, f_{iY}), (\bar{f}_{iX}, \bar{f}_{iY})$ are the components, respectively, of the external and the elastic reaction forces acting on rigid body i.

The angular velocity of rigid body i relative to the X, Y, Z system is defined by the vector $\mathbf{\Omega}_i$, and its orientation is defined by the Euler angle set $\beta_{iX}, \beta_{iY}, \phi$ illustrated in Figure 2.3. The components of a vector V in the X, Y, Z and xi, yi, zi systems are denoted, respectively, by $(V)_I$ and $(V)_i$. These components are related by

$$(V)_i = [A_i](V)_I, \qquad (V)_I = [A_i]^T(V)_i, \qquad (2.2)$$

where the T denotes the transpose operation, and

$$[A_i] = \begin{bmatrix} c\phi & s\phi & \beta_{iX}s\phi - \beta_{iY}c\phi \\ -s\phi & c\phi & \beta_{iX}c\phi + \beta_{iY}s\phi \\ \beta_{iY} & -\beta_{iX} & 1 \end{bmatrix}. \qquad (2.3)$$

Bearing constraints which restrict the rotor's transverse rotation justify dropping second-order terms in β_{iX}, β_{iY} in this definition for $[A_i]$.

The governing, rotational equations of motion for rigid body i are obtained by stating a moment-of-momentum equation for the body about the origin of the xi, yi, zi system to obtain

$$[J_i](\dot{\Omega})_i + [(\Omega)_i][J_i](\Omega)_i + m_i[(a)_i](\ddot{R})_i = (M)_i, \qquad (2.4)$$

where $[J_i]$ is the inertia matrix for the rigid body in the xi, yi, zi system, and **M** is the resultant moment vector about the origin of the xi, yi, zi system. The notation $[(V)_i]$ in Eq. (2.4) implies

$$[(V)_i] = \begin{bmatrix} 0 & -V_{zi} & V_{yi} \\ V_{zi} & 0 & -V_{xi} \\ -V_{yi} & V_{xi} & 0 \end{bmatrix},$$

and performs the matrix equivalent of the vector-cross-product operation; i.e., $(A \times B)_i = [(A)_i](B)_i$.

In rigid-body dynamics, one normally employs the body-fixed expression of Eq. (2.4). However, in the present situation, the elastic-reaction-moment components are defined in the reference X, Y, Z system, and the following restatement of Eq. (2.4) in this system will be employed:

$$[J_I](\dot{\Omega})_I + [(\Omega)_I][J_I](\Omega)_I + m_i[(a)_I](\ddot{R})_I = (M)_I,$$
$$[J_I] = [A_i]^T[J_i][A_i]. \qquad (2.5)$$

The appropriate definitions for the components of Ω and $\dot{\Omega}$ are

$$\Omega_X = \dot{\beta}_{iX} + \phi\dot{\beta}_{iY}, \qquad \Omega_Y = \dot{\beta}_{iY} - \phi\dot{\beta}_{iX}, \qquad \Omega_Z = \dot{\phi}, \qquad (2.6)$$

and

$$\dot{\Omega}_X = \ddot{\beta}_{iX} + \ddot{\phi}\beta_{iY} + \dot{\phi}\dot{\beta}_{iY},$$
$$\dot{\Omega}_Y = \ddot{\beta}_{iY} - \ddot{\phi}\beta_{iX} + \dot{\phi}\dot{\beta}_{iX},$$
$$\dot{\Omega}_Z = \ddot{\phi}. \qquad (2.7)$$

Substituting from these kinematic equations into Eq. (2.5), and retaining only those nonlinear terms involving ϕ yields the following defining equations for β_{iX}, β_{iY}:

$$J_i\ddot{\beta}_{iX} = M_{iX} + \overline{M}_{iX} - \dot{\phi}\overline{J}_i\dot{\beta}_{iY} - \dot{\phi}^2 J_{iYZ} + \ddot{\phi}J_{iXZ},$$
$$J_i\ddot{\beta}_{iY} = M_{iy} + \overline{M}_{iY} + \dot{\phi}\overline{J}_i\dot{\beta}_{iX} + \dot{\phi}^2 J_{iXZ} + \ddot{\phi}J_{iYZ}, \qquad (2.8)$$

where J_i and $\overline{J}_i$ are the diametral and polar moments of inertia; (M_{iX}, M_{iY})

and $(\overline{M}_{iX}, \overline{M}_{iY})$ are the components, respectively, of the external applied moment and the elastic reaction moment acting on the body. The stationary-reference definition of the products of inertia in Eq. (2.8) are related to their rotor-fixed counterparts by

$$J_{iXZ} = J_{ixz}c\phi - J_{iyz}s\phi,$$
$$J_{iYZ} = J_{ixz}s\phi + J_{iyz}c\phi. \tag{2.9}$$

Note that the transverse acceleration term $m_i[(a)_I](\ddot{R})_I$ of Eq. (2.5) does not contribute to the governing Eq. (2.8), because of the elimination of second-order terms.

The governing vibration equations (2.1) and (2.8) can be arranged into vector differential equations, defining motion in the X-Z and Y-Z planes. For motion in the X-Z plane, the vectors $(R_X)^T = (R_{1X}, R_{2X}, \ldots, R_{nX})$, $(\beta_Y)^T = (\beta_{1Y}, \beta_{2Y}, \ldots, \beta_{nY})$ are used, while for the Y-Z plane, one has $(R_Y)^T = (R_{1Y}, R_{2Y}, \ldots, R_{nY})$, $(\beta_X)^T = (\beta_{1X}, \beta_{2X}, \ldots, \beta_{nX})$. Comparable vectors of elastic reactions can be assembled, and are related to these deflection vectors by the following stiffness-matrix definitions:

$$\left\{\begin{matrix}(\bar{f}_X)\\(\overline{M}_Y)\end{matrix}\right\} = -[K_X]\left\{\begin{matrix}(R_X)\\(\beta_Y)\end{matrix}\right\}, \quad \left\{\begin{matrix}(\bar{f}_Y)\\(\overline{M}_X)\end{matrix}\right\} = -[K_Y]\left\{\begin{matrix}(R_Y)\\(\beta_X)\end{matrix}\right\}. \tag{2.10}$$

These matrices define the *linear* axisymmetric structural properties of both the rotor and the bearings, and their derivation is the subject of Section 2.3. From Eqs. (2.1), (2.8), and (2.11), the vibration equations for the X-Z and Y-Z planes have the form

$$\begin{bmatrix}[m] & 0\\0 & [J]\end{bmatrix}\left\{\begin{matrix}(\ddot{R}_X)\\(\ddot{\beta}_Y)\end{matrix}\right\} + [K_X]\left\{\begin{matrix}(R_X)\\(\beta_Y)\end{matrix}\right\} = \left\{\begin{matrix}(f_X)\\(M_Y)\end{matrix}\right\} + \cdots,$$

$$\begin{bmatrix}[m] & 0\\0 & [J]\end{bmatrix}\left\{\begin{matrix}(\ddot{R}_Y)\\(\ddot{\beta}_X)\end{matrix}\right\} + [K_Y]\left\{\begin{matrix}(R_Y)\\(\beta_X)\end{matrix}\right\} = \left\{\begin{matrix}(f_Y)\\(M_X)\end{matrix}\right\} + \cdots, \tag{2.11}$$

where $[m]$ and $[J]$ are diagonal matrices whose entries are m_i and J_i, respectively. The additional terms indicated on the right side of these equations are dynamic terms involving $\dot{\phi}$ and $\ddot{\phi}$.

The equation of motion for ϕ is obtained by stating the zi component of Eq. (2.5) for rigid body i, and summing this component over all n rigid bodies to obtain

$$\bar{J}\ddot{\phi} = M_z + \sum\left(\ddot{\beta}_{iX}J_{iXZ} + \ddot{\beta}_{iY}J_{iYZ}\right) - \sum M_i\left(\ddot{R}_{iY}a_{iX} - \ddot{R}_{iX}a_{iY}\right), \tag{2.12}$$

where

$$\bar{J} = \sum \left(\bar{J}_i + m_i |a_i|^2 \right), \qquad M_z = \sum M_{iz}.$$

The desired flexible-rotor lumped-parameter dynamics model is (conceptually) provided by Eqs. (2.11) and (2.12). The structural-dynamics portion of the model is provided in Eq. (2.11), while Eq. (2.12) defines the rotational motion about the bearing's axis. We will be largely concerned in the balance of this chapter with the structural-dynamics portion of the dynamics model, specifically, nonspinning versions of Eq. (2.11). Since the rotors to be considered have axisymmetric stiffness and inertial properties, we can additionally restrict our attention to motion in the X-Z plane defined by

$$\begin{bmatrix} [m] & 0 \\ 0 & [J] \end{bmatrix} \begin{Bmatrix} (\ddot{R}_X) \\ (\ddot{\beta}_Y) \end{Bmatrix} + [K] \begin{Bmatrix} (R_X) \\ (\beta_Y) \end{Bmatrix} = \begin{Bmatrix} (f_X) \\ (M_Y) \end{Bmatrix}. \qquad (2.13)$$

The additional dynamic terms, which are indicated on the right-hand side of Eq. (2.11), have been discarded in this nonspinning version.

Historically, the "structural" aspect of structural-dynamics analysis has dealt with the development of the stiffness matrix $[K]$, and the "dynamics analysis" has been concerned with eigen- and frequency-response analysis of the linear vibration problem. More recently, as a result of finite-element developments, structural-dynamics analysis has also involved the development of generalized inertia matrices for rotor models. A rotor's inertia and stiffness-matrix definitions represent the basic skeleton of any rotordynamics analysis, and both their development and subsequent eigenanalysis are reviewed in the balance of this chapter. In these discussions, we will restrict our attention to the relatively small and fundamental area of structural dynamics with immediate applicability to rotordynamics.

2.3 MATRIX STIFFNESS METHODS FOR ROTORDYNAMICS ANALYSIS

In the preceding section, we were generally explicit in defining the governing equations of motion to be used in a lumped-parameter representation of a flexible rotor. However, the stiffness-matrix definitions provided were only conceptual in nature, and we propose in this section to explain in detail the derivation of stiffness matrices that are appropriate for the n-body lumped-parameter rotor model of Figure 2.1. We will be primarily concerned with static conditions and will refer to the rigid body lying between two members as a joint.* The stiffness method provides a systematic approach for proceed-

---

*This is the customary structural-analysis designation; however, in the next section dealing with transfer-matrix analysis, this same entity is referred to as a station.

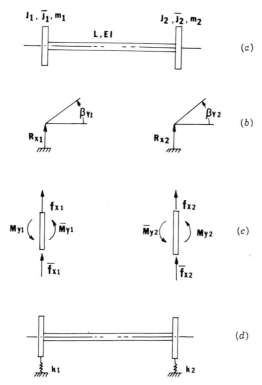

Figure 2.4 A two-body lumped-parameter rotor model. (a) Free-free model with inertia, geometric, and stiffness properties; (b) degrees of freedom; (c) applied and reaction forces; and (d) elastically supported rotor.

ing from a stiffness matrix for the two-body model of Figure 2.4 to the n-body model of Figure 2.1, based on the static equations of equilibrium at joints. This procedure will be demonstrated by first deriving the stiffness matrix for the two-body model of Figure 2.4, and then proceeding to the three-body model of Figure 2.6.

Two-Body Lumped-Parameter Model

The model illustrated in Figure 2.4(a) consists of two disks connected by a massless elastic beam element, and the four degrees of freedom for this "free-free" system are shown in Figure 2.4(b). We assume here that all external forces are applied to the rigid bodies (or joints) of the model, and Figure 2.4(c) illustrates the resultant external and reaction forces and moments. Restricting our attention to static conditions, equilibrium of the rigid bodies (or joints) yields the following simple equations:

$$f_{x1} + \bar{f}_{x1} = 0, \qquad f_{x2} + \bar{f}_{x2} = 0,$$
$$M_{y1} + \overline{M}_{y1} = 0, \qquad M_{y2} + \overline{M}_{y2} = 0,$$

which state that the external and internal forces and moments add up to zero. The reaction forces and moments are related to the deflections of Figure 2.4(b) by the following stiffness-matrix definition:

$$
\begin{Bmatrix} \bar{f}_{X1} \\ \overline{M}_{Y1} \\ \bar{f}_{X2} \\ \overline{M}_{Y2} \end{Bmatrix} = - \begin{bmatrix} K_{11} & K_{12} & K_{13} & K_{14} \\ K_{21} & K_{22} & K_{23} & K_{24} \\ K_{31} & K_{32} & K_{33} & K_{34} \\ K_{41} & K_{42} & K_{43} & K_{44} \end{bmatrix} \begin{Bmatrix} R_{X1} \\ \beta_{Y1} \\ R_{X2} \\ \beta_{Y2} \end{Bmatrix}.
\tag{2.14}
$$

The physical interpretation of the K_{ij} element is that $-K_{ij}$ is the ith force (or moment) corresponding to a unit j deflection with all other deflections held to zero. For example, the first column of this matrix defines the (negative) reaction forces and moments acting on the two rigid bodies due to $(R_{X1} = 1, \ R_{X2} = \beta_{Y1} = \beta_{Y2} = 0)$. The individual K_{ij} coefficients in Eq. (2.14) may be calculated directly from their definitions, as illustrated in Figure 2.5. Specifically, internal reactions corresponding to a unit deflection of one coordinate can be obtained by a linear superposition of elementary

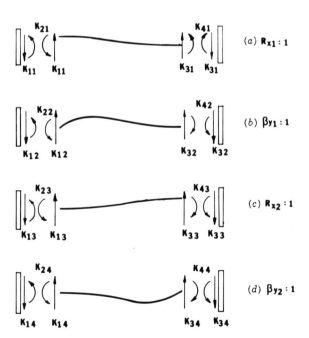

Figure 2.5 Stiffness-matrix coefficient definitions for the two-body model of Figure 2.4.

beam solutions, yielding the resultant stiffness-matrix definition stated below:

$$[K] = \frac{2EI}{L^3} \begin{bmatrix} 6 & 3L & -6 & 3L \\ 3L & 2L^2 & -3L & L^2 \\ -6 & -3L & 6 & -3L \\ 3L & L^2 & -3L & 2L^2 \end{bmatrix}. \qquad (2.15)$$

If we wish to restrain the free-free system of Figure 2.4(a) by adding the linear (displacement) springs of Figure 2.4(d), the reaction force definitions for bodies 1 and 2 simply become

$$\bar{f}_{x1} = -\{K_{11}R_{X1} + K_{12}\beta_{Y1} + K_{13}R_{X2} + K_{14}\beta_{Y2}\} - k_1 R_{X1},$$

$$\bar{f}_{x2} = -\{K_{21}R_{X1} + K_{22}\beta_{Y1} + K_{23}R_{X2} + K_{24}\beta_{Y2}\} - k_2 R_{X2}; \qquad (2.16)$$

i.e., the spring constraints k_1, k_2 yield the additional reaction forces $-k_1 R_{X1}, -k_2 R_{X2}$. The stiffness matrix definition for the elastically constrained system of Figure 2.4(d) will be denoted by $[\bar{K}]$. From Eqs. (2.15) and (2.16) it is related to the elements of the stiffness matrix for the unrestrained system $[K]$ by

$$[\bar{K}] = \begin{bmatrix} K_{11} + k_1 & K_{12} & K_{13} & K_{14} \\ K_{21} & K_{22} & K_{23} & K_{24} \\ K_{31} & K_{32} & K_{33} + k_2 & K_{34} \\ K_{41} & K_{42} & K_{43} & K_{44} \end{bmatrix}. \qquad (2.17)$$

The addition of rotationally restraining springs would yield a comparable direct addition to the K_{22} and K_{44} entries of the free-free matrix $[K]$. The ease by which spring constraints may be added to the basic rotor-stiffness definition is one of the principal attractive features of the stiffness method.

Continuing our consideration of the structural definition for the restrained model of Figure 2.4(d), suppose that Eq. (2.17) is employed to define the internal reaction forces and moments, and combined with Eq. (2.14) to yield

$$[\bar{K}] \begin{Bmatrix} R_{X1} \\ \beta_{Y1} \\ R_{X2} \\ \beta_{Y2} \end{Bmatrix} = \begin{Bmatrix} f_{X1} \\ M_{Y1} \\ f_{X2} \\ M_{Y2} \end{Bmatrix}.$$

We assume that the applied forces and moments on the right-hand side are known, and the displacements and rotations on the left are to be determined.

Premultiplying by the inverse of $[\bar{K}]$ yields the desired solution

$$
\begin{Bmatrix} R_{X1} \\ \beta_{Y1} \\ R_{X2} \\ \beta_{Y2} \end{Bmatrix} = [\bar{K}]^{-1} \begin{Bmatrix} f_{X1} \\ M_{Y1} \\ f_{X2} \\ M_{Y2} \end{Bmatrix} = [F] \begin{Bmatrix} f_{X1} \\ M_{Y1} \\ f_{X2} \\ M_{Y2} \end{Bmatrix}.
$$

The matrix $[F] = [\bar{K}]^{-1}$ is the structural flexibility matrix. The F_{ij} component of this matrix may be interpreted as the deflection resulting at the ith coordinate from a unit positive force or moment acting at the jth coordinate. Flexibility matrices are not defined for unrestrained rotor systems like that of Figure 2.4(a), since an applied load on this system yields an infinite displacement. In other words, the stiffness matrix $[K]$ of Eq. (2.16) is singular for the unrestrained system of Figure 2.4(a).

In review, the material of this subsection introduces and defines the stiffness matrix for the two-body system illustrated in Figure 2.4(a). Further, the analysis presented demonstrates that linear-spring constraints such as those illustrated in Figure 2.4(d) can be accounted for by directly adding the spring constants to the appropriate entries of the unrestrained rotor-stiffness matrix.

Three-Body Lumped-Parameter Model

Our present objective is a demonstration of the stiffness approach for proceeding from a component stiffness matrix for component beam members such as that of Eq. (2.14) to a stiffness matrix for the n-body system of Figure 2.1. The procedure will be demonstrated by developing the stiffness matrix for the three-body unrestrained system of Figure 2.6. The coordinates of the system are illustrated in Figure 2.6(b), and the applied and elastic-reaction forces and moments are illustrated in Figure 2.6(c). For simplicity, we denote the coordinates, applied forces and moments, and reaction forces and moments by

$$
(X)^T = (X_1, X_2, \dots, X_6) = (R_{X1}, \beta_{Y1}, R_{X2}, \beta_{Y2}, R_{X3}, \beta_{Y3}),
$$

$$
(F)^T = (F_1, F_2, \dots, F_6) = (f_{X1}, M_{Y1}, f_{X2}, M_{Y2}, f_{X3}, M_{Y3}),
$$

$$
(\bar{F})^T = (\bar{F}_1, \bar{F}_2, \dots, \bar{F}_6) = (\bar{f}_{X1}, \bar{M}_{Y1}, \bar{f}_{X2}, \bar{M}_{Y2}, \bar{f}_{X3}, \bar{M}_{Y3}). \quad (2.18)
$$

Further, in the balance of this discussion, the term *force* will denote both forces and moments, and *coordinate* will include both displacements and rotations.

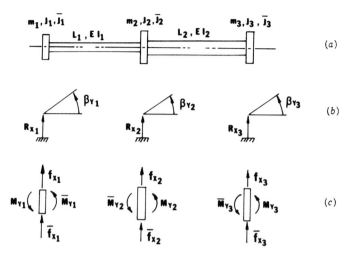

Figure 2.6 A three-body lumped-parameter rotor model. (a) Free-free model with inertia, geometric, and stiffness properties; (c) applied and reaction forces.

The stiffness matrix which we propose to develop satisfies the equations

$$(\bar{F}) = -[K](X), \qquad (F) = [K](X), \tag{2.19}$$

while the component stiffness matrices for the two beam elements are denoted by $[K^1], [K^2]$, and are stated

$$[K^1] = \begin{bmatrix} K_{11}^1 & K_{12}^1 & K_{13}^1 & K_{14}^1 \\ K_{21}^1 & K_{22}^1 & K_{23}^1 & K_{24}^1 \\ K_{31}^1 & K_{32}^1 & K_{33}^1 & K_{34}^1 \\ K_{41}^1 & K_{42}^1 & K_{43}^1 & K_{44}^1 \end{bmatrix},$$

$$[K^2] = \begin{bmatrix} K_{33}^2 & K_{34}^2 & K_{35}^2 & K_{36}^2 \\ K_{43}^2 & K_{44}^2 & K_{45}^2 & K_{46}^2 \\ K_{53}^2 & K_{54}^2 & K_{55}^2 & K_{56}^2 \\ K_{63}^2 & K_{64}^2 & K_{65}^2 & K_{66}^2 \end{bmatrix}. \tag{2.20}$$

The superscript i in K_{jk}^i identifies either beam 1 or 2. The subscripts j and k denote the reaction force $\bar{F}_j$ resulting from a unit deflection of coordinate X_k. The elements of these matrices can be immediately calculated from Eq. (2.15) using the structural properties provided in Figure 2.6(a).

Figure 2.7 illustrates the reactions developed by successive deflections of the X_i variables. From this figure, the equilibrium equations $(F_i + \bar{F}_i = 0$;

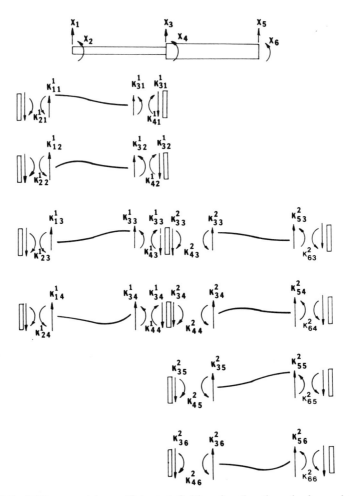

Figure 2.7 Stiffness-matrix coefficient definition for the three-body model of Figure 2.6.

$i = 1, 2, \ldots, 6$) yield

$$f_{X1} - \left\{ K_{11}^1 R_{X1} + K_{12}^1 \beta_{Y1} + K_{13}^1 R_{X2} + K_{14}^1 \beta_{Y2} \right\} = 0,$$

$$M_{Y1} - \left\{ K_{21}^1 R_{X1} + K_{22}^1 \beta_{Y1} + K_{23}^1 R_{X2} + K_{24}^1 \beta_{Y2} \right\} = 0,$$

$$f_{X2} - \left\{ K_{31}^1 R_{X1} + K_{32}^1 \beta_{Y1} + K_{33}^1 R_{X2} + K_{34}^1 \beta_{Y2} \right.$$
$$\left. + K_{33}^2 R_{X2} + K_{34}^2 \beta_{Y2} + K_{35}^2 R_{X3} + K_{36}^2 \beta_{Y3} \right\} = 0,$$

$$M_{Y2} - \left\{ K_{41}^1 R_{X1} + K_{42}^1 \beta_{Y1} + K_{43}^1 R_{X2} + K_{44}^1 \beta_{Y2} \right.$$
$$\left. + K_{43}^2 R_{X2} + K_{44}^2 \beta_{Y2} + K_{35}^2 R_{X3} + K_{36}^2 \beta_{Y3} \right\} = 0,$$

$$f_{X3} - \left\{ K_{53}^2 R_{X2} + K_{54}^2 \beta_{Y2} + K_{55}^2 R_{X3} + K_{56}^2 \beta_{Y3} \right\} = 0,$$

$$M_{Y3} - \left\{ K_{63}^2 R_{Y2} + K_{64}^2 \beta_{Y2} + K_{65}^2 R_{X3} + K_{66}^2 \beta_{Y3} \right\} = 0. \quad (2.21)$$

These equations may be restated as

$$
\begin{bmatrix}
K_{11}^1 & K_{12}^1 & K_{13}^1 & K_{14}^1 & 0 & 0 \\
K_{21}^1 & K_{22}^1 & K_{23}^1 & K_{24}^1 & 0 & 0 \\
K_{31}^1 & K_{32}^1 & (K_{33}^1 + K_{33}^2) & (K_{34}^1 + K_{34}^2) & K_{35}^2 & K_{36}^2 \\
K_{41}^1 & K_{42}^1 & (K_{43}^1 + K_{43}^2) & (K_{44}^1 + K_{44}^2) & K_{45}^2 & K_{46}^2 \\
0 & 0 & K_{53}^2 & K_{54}^2 & K_{55}^2 & K_{56}^2 \\
0 & 0 & K_{63}^2 & K_{64}^2 & K_{65}^2 & K_{66}^2
\end{bmatrix}
\begin{Bmatrix}
R_{X1} \\
\beta_{Y1} \\
R_{X2} \\
\beta_{Y2} \\
R_{X3} \\
\beta_{Y3}
\end{Bmatrix}
=
\begin{Bmatrix}
f_{X1} \\
M_{Y1} \\
f_{X2} \\
M_{Y2} \\
f_{X3} \\
M_{Y3}
\end{Bmatrix},
$$

$$(2.22)$$

and the matrix on the left is the desired "system" stiffness matrix defined in terms of the elements of the component stiffness matrices given in Eq. (2.20). The definition provided by Eqs. (2.21) and (2.22) results from a requirement of static equilibrium at each joint. However, the K_{jk} element of the resultant stiffness matrix can be obtained "mechanically" by simply summing all elements of the component stiffness matrices in Eq. (2.20) with the same subscripts; i.e.,

$$
K_{jk} = \sum_{i=1}^{2} K_{jk}^i,
\tag{2.23}
$$

and it is precisely this property which may be used to assemble stiffness matrices for n-body lumped-parameter models of flexible rotors. Specifically, all component rigid-body displacement and rotational degrees of freedom are numbered, with the same number applying to the corresponding applied and elastic-reaction forces and moments. Component stiffness matrices are then calculated based on Eq. (2.16) for each beam element, with the subscripts of these component stiffness matrices corresponding to the numbering system used in identifying coordinates and forces. The K_{jk} element of the desired system stiffness matrix is then obtained by performing the summation

$$
K_{jk} = \sum_{i=1}^{n} K_{jk}^i
\tag{2.24}
$$

over the n degrees of freedom.

The purpose of this discussion has not been to prepare the reader to formulate stiffness matrices for rotordynamics models, but rather to explain the procedures customarily employed to obtain such matrices. A variety of computer programs have been developed, and are generally available for calculation of stiffness matrices which are appropriate for rotordynamics analysis.

Eigenanalysis for Matrix Vibration Equations

In the next section, we will be discussing transfer-matrix formulations for rotordynamics analysis, including the calculation of eigenvalues and eigenvectors. Hence, before leaving this discussion of stiffness-matrix formulations, we will briefly review the customary eigenanalysis procedures employed for this type of formulation. The differential equation model of Eq. (2.13) yields a homogeneous vibration equation of the form

$$[L](\ddot{X}) + [K](X) = 0. \tag{2.25}$$

In Eq. (2.13), the inertia matrix $[L]$ is diagonal; however, as we shall see in Section 2.5, if a finite-element approach is employed, this matrix is symmetric but not diagonal. The customary analysis procedure used in calculating undamped eigenvalues and eigenvectors for Eq. (2.25) is the Jacobi method [Meirovitch (1975)], which employs a sequence of rotational transformations to diagonalize a *symmetric* matrix. This method converges simultaneously to all eigenvalues and eigenvectors.

By rewriting Eq. (2.25) as

$$(\ddot{X}) + [L]^{-1}[K](X) = 0,$$

the system eigenvalues are eigenvalues of the matrix $[L]^{-1}[K]$. However, the Jacobi procedure applies only to symmetric matrices, and a more circuitous route must be taken to preserve symmetry. First, applying the Jacobi procedure to the inertia matrix $[L]$ yields the matrix of eigenvectors $[A_L]$ which satisfies

$$[A_L]^T[L][A_L] = [\lambda_L]; \qquad [A_L]^T[A_L] = [I], \tag{2.26}$$

where $[\lambda_L]$ is the diagonal matrix of eigenvalues for the inertia matrix $[L]$. We may use Eq. (2.26) to define the symmetric matrices

$$[L]^{1/2} = [A_L][\lambda_L^{1/2}][A_L]^T,$$
$$[L]^{-1/2} = [A_L][\lambda_L^{-1/2}][A_L]^T, \tag{2.27}$$

where $[\lambda_L^{1/2}]$ and $[\lambda_L^{-1/2}]$ are diagonal matrices whose elements are, respectively, the square roots and inverse square roots of the elements of $[\lambda_L]$. From Eq. (2.26), these matrices satisfy

$$[L]^{1/2}[L]^{1/2} = [L], \qquad [L][L]^{-1/2} = [L]^{1/2}, \qquad [L]^{-1/2}[L]^{1/2} = [I].$$

Hence, the transformation

$$(X) = [L]^{1/2}(Y) \tag{2.28}$$

may be applied to Eq. (2.25), yielding

$$(\ddot{Y}) + [K_L](Y) = 0, \qquad [K_L] = [L]^{-1/2}[K][L]^{-1/2}. \qquad (2.29)$$

Applying the Jacobi transformation procedure to $[K_L]$ yields the transformation

$$(Y) = [A_K](q),$$

which satisfies

$$[A_K]^T[K_L][A_K] = [\Lambda]; \qquad [A_K]^T[A_K] = [I],$$

and yields the uncoupled matrix differential equation

$$(\ddot{q}) + [\Lambda](q) = 0. \qquad (2.30)$$

The matrix $[\Lambda]$ is the diagonal matrix of eigenvalues for the differential Eq. (2.25). The corresponding system matrix of eigenvectors $[A]$ defines the transformation

$$(X) = [A](q); \qquad [A] = [L]^{-1/2}[A_K], \qquad (2.31)$$

which satisfies

$$[A]^T[L][A] = [I], \qquad [A]^T[K][A] = [\Lambda]. \qquad (2.32)$$

Following this analysis procedure for the inertia and stiffness matrices of Eq. (2.11) will yield the following coordinate transformations:

$$(R_X) = [A_e](q_X), \qquad (\beta_Y) = [A_\beta](q_Y),$$
$$(R_Y) = [A_e](q_Y), \qquad (\beta_X) = -[A_\beta](q_Y), \qquad (2.33)$$

where $[A_e]$ and $[A_\beta]$ are subeigenvector matrices, which are assembled to define the full eigenvector matrices

$$[A_{XZ}] = \begin{bmatrix} [A_e] \\ [A_\beta] \end{bmatrix}, \qquad [A_{YZ}] = \begin{bmatrix} [A_e] \\ -[A_\beta] \end{bmatrix}. \qquad (2.34)$$

From Eqs. (2.11) and (2.32) these matrices are normalized to satisfy

$$[A_e]^T[m][A_e] + [A_\beta]^T[J][A_\beta] = [I],$$
$$[A_{XZ}]^T[K_{XZ}][A_{XZ}] + [A_{YZ}]^T[K_{YZ}][A_{YZ}] = [\Lambda]. \qquad (2.35)$$

Substitution from Eq. (2.33) into Eq. (2.11) yields the modal differential equation set

$$(\ddot{q}_X) + [\Lambda](q_X) = [A_e]^T(f_X) + [A_\beta]^T(M_Y) + \cdots,$$

$$(\ddot{q}_Y) + [\Lambda](q_Y) = [A_e]^T(f_Y) - [A_\beta]^T(M_X) + \cdots. \qquad (2.36)$$

These uncoupled equations have frequently been used for rotordynamic analysis.

The results of Section 1.4 demonstrated the influence of "gyroscopic" coupling terms in defining both forward and backward critical speeds. Matrix procedures to be employed in calculating these critical speeds will now be examined. Suppose, initially, that we wished to calculate a natural frequency and its associated eigenvector for a *homogeneous* version of Eq. (2.11). We could assume a synchronous vector solution of the form

$$(R_X) = (A_e)\cos \omega t, \qquad (\beta_Y) = (A_\beta)\cos \omega t,$$

$$(R_Y) = (A_e)\sin \omega t, \qquad (\beta_X) = -(A_\beta)\cos \omega t, \qquad (2.37)$$

and substitute into the first of Eq. (2.11) to obtain

$$\left[-\omega^2 \begin{bmatrix} [m] & 0 \\ 0 & [J] \end{bmatrix} + [K_{XZ}] \right] \left\{ \begin{matrix} (A_e) \\ (A_\beta) \end{matrix} \right\} = 0.$$

The eigenvalues are defined by the determinant requirement

$$\left| -\omega^2 \begin{bmatrix} [m] & 0 \\ 0 & [J] \end{bmatrix} + [K_{X2}] \right| = 0.$$

Now, to consider the effect of gyroscopic coupling on these results, we retain the gyroscopic coupling terms in the homogeneous version of Eq. (2.11), yielding

$$\begin{bmatrix} [m] & 0 \\ 0 & [J] \end{bmatrix} \left\{ \begin{matrix} (\ddot{R}_X) \\ (\ddot{\beta}_Y) \end{matrix} \right\} - \dot{\phi} \begin{bmatrix} 0 & 0 \\ 0 & [\bar{J}] \end{bmatrix} \left\{ \begin{matrix} (\dot{R}_Y) \\ (\dot{\beta}_X) \end{matrix} \right\} + [K_X] \left\{ \begin{matrix} (R_X) \\ (\beta_Y) \end{matrix} \right\} = 0,$$

$$\begin{bmatrix} [m] & 0 \\ 0 & [J] \end{bmatrix} \left\{ \begin{matrix} (\ddot{R}_Y) \\ (\ddot{\beta}_X) \end{matrix} \right\} + \dot{\phi} \begin{bmatrix} 0 & 0 \\ 0 & [\bar{J}] \end{bmatrix} \left\{ \begin{matrix} (\dot{R}_X) \\ (\dot{\beta}_Y) \end{matrix} \right\} + [K_Y] \left\{ \begin{matrix} (R_Y) \\ (\beta_X) \end{matrix} \right\} = 0.$$

$$(2.38)$$

Substituting from Eq. (2.37) and noting the forward-critical-speed frequency requirement, $\dot{\phi} = \omega$, yields

$$\left[-\omega^2 \begin{bmatrix} [m] & 0 \\ 0 & [J_i - \bar{J}_i] \end{bmatrix} + [K_X] \right] \left\{ \begin{matrix} (A_e) \\ (A_\beta) \end{matrix} \right\} = 0. \qquad (2.39)$$

Comparing this result with Eq. (2.37) demonstrates that forward critical speeds may be calculated by simply replacing the diagonal entry of the original problem J_i by $J_i - \bar{J}_i$.

To define the backward critical speeds we replace the synchronous solution by the following assumed asynchronous solution:

$$(R_X) = (A_e)\cos \omega t, \qquad (\beta_Y) = (A_\beta)\cos \omega t,$$
$$(R_Y) = -(A_e)\sin \omega t, \qquad (\beta_X) = (A_\beta)\sin \omega t.$$

Substituting this result into the first of Eq. (2.38), and noting the backward-whirl frequency requirement, $\dot{\phi} = -\omega$, yields

$$\left[-\omega^2 \begin{bmatrix} [m] & 0 \\ 0 & [J_i + \bar{J}_i] \end{bmatrix} + [K_X] \right] \left\{ \begin{matrix} (A_e) \\ (A_\beta) \end{matrix} \right\} = 0. \qquad (2.40)$$

Again, a comparison of Eqs. (2.37) and (2.40) demonstrates that backward critical speeds may be calculated by simply replacing the diagonal entry of the original problem J_i by $J_i + \bar{J}_i$.

One practical problem which arises in calculating forward critical speeds with the Jacobi method is that the diagonal inertia matrix $[L]$ of Eq. (2.25) now has negative entries. Hence, the matrices $[L]^{1/2}$ and $[L]^{-1/2}$ are imaginary and are no longer satisfactory for an initial diagonalizing transformation. However, one may proceed by initially calculating the eigenvalues and eigenvectors of the stiffness matrix $[K]$, which satisfy

$$[A_K]^T[K][A_K] = [\lambda_K], \qquad [A_K]^T[A_K] = [I].$$

The matrix $[\lambda_K]$ is diagonal, and its entries are the eigenvalues of $[K]$. From these results, the matrix

$$[K]^{-1/2} = [A_K][\lambda_K^{-1/2}][A_K]^T$$

is defined, which is directly comparable to the second of Eq. (2.27). We define the transformation

$$(X) = [K]^{1/2}(Y) \qquad (2.41)$$

and substitute into Eq. (2.25) to obtain

$$[L_K](\ddot{Y}) + (Y) = 0, \quad [L_K] = [K]^{-1/2}[L][K]^{-1/2}. \quad (2.42)$$

The transformation

$$(Y) = [A_L](Q), \quad [A_L]^T[A_L] = [I], \quad [A_L]^T[L_K][A_L] = [\Gamma] \quad (2.43)$$

is defined in terms of the matrix of eigenvectors $[A_L]$ and diagonal matrix of eigenvalues $[\Gamma]$ for the matrix $[L_K]$. Substituting from this result into Eq. (2.43) yields

$$[\Gamma](\ddot{Q}) + (Q) = 0.$$

By comparison to Eq. (2.30), $[\Gamma] = [\Lambda]^{-1}$; i.e., the elements of $[\Gamma]$ are the inverse eigenvalues of the problem, and the desired forward critical speeds corresponding to Eq. (2.39) are defined. To obtain the associated eigenvectors, we introduce the additional transformation

$$(Q) = [\Gamma]^{-1/2}(q) = [\Lambda]^{1/2}(q) \quad (2.44)$$

to obtain the final modal differential equation format

$$(\ddot{q}) + [\Lambda](q) = 0.$$

The net transformation from the physical variables (X) to the modal coordinates (q) is obtained from Eqs. (2.41), (2.42), and (2.44) as

$$(X) = [K]^{-1/2}[A_L][\Lambda]^{1/2}(q) = [A](q).$$

The matrix $[A]$ defines the eigenvectors corresponding to the eigenvalues of $[\Lambda]$.

2.4 MYKLESTAD-PROHL TRANSFER-MATRIX APPROACH FOR FLEXIBLE-ROTOR MODELS

As noted in this chapter's introduction, the Myklestad-Prohl transfer-matrix* format has historically been employed for analysis of lumped-parameter models of flexible rotors, viz., eigenanalysis, synchronous response, and stability analysis. In the present chapter, we will derive the transfer-matrix

---

*Pestel and Leckie (1973) provide a detailed and comprehensive survey of transfer-matrix methods.

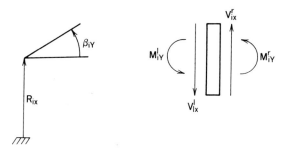

Figure 2.8 Transfer-matrix state variables for station i.

formulation for flexible axisymmetric rotors, and demonstrate its utility for the calculation of undamped eigenvalues and eigenvectors.

Myklestad-Prohl Method

Following conventional transfer-matrix notation, we denote the ith rigid body as the ith station, and the beam lying between ith and $(i + 1)$th stations as the ith field. This numbering system is illustrated in Figure 2.1. Figure 2.8 illustrates the variables of interest at station i, viz., moment, displacement, slope, and shear. The superscripts r and l in this figure denote the right- and left-hand sides of the station. We propose to derive relationships which relate the variables on the left-hand side of this station to those on the right. First, continuity requires that

$$R_{iX}^r = R_{iX}^l = R_{iX},$$
$$\beta_{iY}^r = \beta_{iY}^l = \beta_{iY}. \tag{2.45}$$

Neglecting external forces, including apparent forces due to imbalance and asymmetry, the equations of motion are

$$m_i \ddot{R}_{iX} = V_{iX}^r - V_{iX}^l,$$
$$J_i \ddot{\beta}_{iY} = M_{iY}^r - M_{iY}^l. \tag{2.46}$$

As illustrated in Figure 2.8, (V_{iX}^r, V_{iX}^l) and (M_{iY}^r, M_{iY}^l) denote, respectively, the internal shear and moments acting on the left- and right-hand sides of rigid body i. Assuming a periodic solution for the variables of the form

$$R_{iX} = R_i \cos \omega t, \qquad \beta_{iY} = \beta_i \cos \omega t,$$
$$V_{iX}^r = V_i^r \cos \omega t, \qquad V_{iX}^l = V_i^l \cos \omega t,$$
$$M_{iY}^r = M_i^r \cos \omega t, \qquad M_{iY}^l = M_i^l \cos \omega t \tag{2.47}$$

yields the following results:

$$V_i^r = -m_i \omega^2 R_i^l + V_i^l,$$
$$M_i^r = -J_i \omega^2 \beta_i^l + M_i^l. \tag{2.48}$$

Equations (2.45) and (2.48) may be combined to yield the following *station transfer-matrix* relating variables on the left- and right-hand sides of station i:

$$
\begin{Bmatrix} R_i^r \\ \beta_i^r \\ M_i^r \\ V_i^r \end{Bmatrix} =
\begin{bmatrix}
1 & 0 & 0 & 0 \\
0 & 1 & 0 & 0 \\
0 & -J_i\omega^2 & 1 & 0 \\
-m_i\omega^2 & 0 & 0 & 1
\end{bmatrix}
\begin{Bmatrix} R_i^l \\ \beta_i^l \\ M_i^l \\ V_i^l \end{Bmatrix}. \tag{2.49}
$$

We identify the state vector

$$(Q)_i^T = (R_i, \beta_i, M_i, V_i), \tag{2.50}$$

and write Eq. (2.49) more compactly as

$$(Q)_i^r = [T_{si}](Q)_i^l. \tag{2.51}$$

Comparable equations will now be developed relating the variables on the left- and right-hand sides of the ith field (beam). Considering Figure 2.9, we obtain from static-equilibrium requirements

$$M_{i+1}^l = M_i^r - V_i^r L_i, \qquad V_{i+1}^l = V_i^r. \tag{2.52}$$

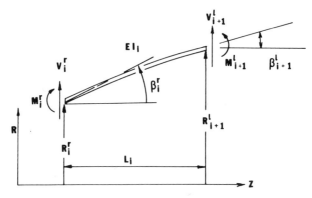

Figure 2.9   Transfer-matrix state variables for field i.

Also from this figure, we combine the initial displacement and rotation on the left with the elastic deflections caused by the loading on the right to obtain

$$R_{i+1}^l = R_i^r + L_i\beta_i^r + \frac{L_i^2}{2EI_i}M_{i+1}^l + \frac{L_i^3}{3EI_i}V_{i+1}^l,$$

$$\beta_{i+1}^l = \beta_i^r + \frac{L_i}{EI}M_{i+1}^l + \frac{L_i^2}{2EI_i}V_{i+1}^l. \tag{2.53}$$

Substitution from Eq. (2.52) into this result yields

$$R_{i+1}^l = R_i^r + L_i\beta_i^r + \frac{L_i}{2EI_i}M_i^r + \left(\frac{L_i^3}{3EI_i} - \frac{L_i^3}{2EI_i}\right)V_i^r,$$

$$\beta_{i+1}^l = \beta_i^r + \frac{L_i}{EI}M_i^r + \left(\frac{L_i^2}{2EI_i} - \frac{L_i^2}{EI_i}\right)V_i^r. \tag{2.54}$$

From Eqs. (2.52) and (2.54), the desired field transfer matrix may be stated

$$\begin{Bmatrix} R_{i+1}^l \\ \beta_{i+1}^l \\ M_{i+1}^l \\ V_{i+1}^l \end{Bmatrix} = \begin{bmatrix} 1 & L_i & b_i L_i/2 & -b_i L_i^2/6 \\ 0 & 1 & b_i & -b_i L_i/2 \\ 0 & 0 & 1 & -L_i \\ 0 & 0 & 0 & 1 \end{bmatrix} \begin{Bmatrix} R_i^r \\ \beta_i^r \\ M_i^r \\ V_i^r \end{Bmatrix}, \tag{2.55}$$

where

$$b_i = L_i/EI_i. \tag{2.56}$$

Equation (2.55) may be stated more compactly as

$$(Q)_{i+1}^l = [T_{fi}](Q)_i^r, \tag{2.57}$$

and combined with Eq. (2.51) to yield the overall transfer matrix

$$(Q)_{i+1}^l = [T_{fi}][T_{si}](Q)_i^l = [T_i](Q)_i^l, \tag{2.58}$$

relating the variables on the left-hand sides of station i and station $i + 1$. By successive matrix multiplications, one may use the transfer-matrix definition of Eq. (2.58) to "march" from the left-hand side of station one to the left-hand side of station n, as indicated below:

$$(Q)_n^l = [T_{n-1}][T_{n-2}] \cdots [T_1](Q)_1^l.$$

The left- and right-hand-side variables for station n are related by the nth station transfer matrix

$$(Q)_n^r = [T_{sn}](Q)_n^l.$$

Hence, the system transfer matrix from the left-hand side of station one to the right-hand side of station n is

$$(Q)_n^r = [T_{sn}][T_{n-1}][T_{n-2}] \cdots [T_1](Q)_1^l$$
$$= [T](Q)_1^l. \tag{2.59}$$

The expanded version of Eq. (2.59) is

$$\begin{Bmatrix} R_n^r \\ \beta_n^r \\ M_n^r \\ V_n^r \end{Bmatrix} = \begin{bmatrix} T_{11} & T_{12} & T_{13} & T_{14} \\ T_{21} & T_{22} & T_{23} & T_{24} \\ T_{31} & T_{32} & T_{33} & T_{34} \\ T_{41} & T_{42} & T_{43} & T_{44} \end{bmatrix} \begin{Bmatrix} R_1^l \\ \beta_1^l \\ M_1^l \\ V_1^l \end{Bmatrix}. \tag{2.60}$$

The matrix $[T]$ is a function of the frequency ω, and we wish to determine values of ω that correspond to rotor natural frequencies. Suppose, for example, that the natural frequencies are required for the cantilever-beam model of Figure 2.10. The appropriate boundary conditions are

$$R_1^l = \beta_1^l = 0, \qquad M_n^r = V_n^r = 0, \tag{2.61}$$

which yield from Eq. (2.60) the result

$$0 = T_{33}M_1^l + T_{34}V_1^l,$$
$$0 = T_{43}M_1^l + T_{44}V_1^l. \tag{2.62}$$

The requirement for a nontrivial solution to these equations is

$$\begin{vmatrix} T_{33} & T_{34} \\ T_{43} & T_{44} \end{vmatrix} = T_{33}T_{44} - T_{34}T_{43} = D(\omega) = 0. \tag{2.63}$$

A value of ω which causes Eq. (2.63) to be satisfied is a natural frequency of the cantilever-beam model, and the Myklestad-Prohl method is based on

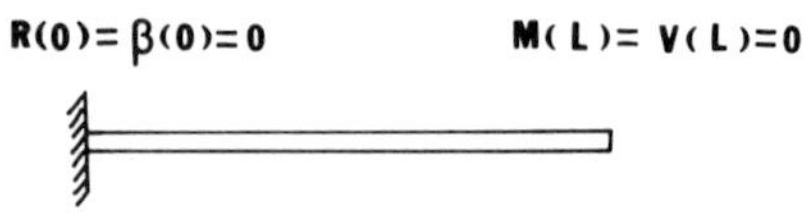

$$R(0) = \beta(0) = 0 \qquad\qquad M(L) = V(L) = 0$$

Figure 2.10 Cantilever-beam boundary conditions.

systematically varying ω and calculating $D(\omega)$ to find values which yield $D(\omega) = 0$. Reviewing, for a selected value of $\omega = \bar{\omega}$, one first calculates the system transfer matrix $[T]$ of Eq. (2.59), and then calculates $D(\bar{\omega})$ from Eq. (2.63). This procedure is repeated until a value of ω is found which satisfies Eq. (2.63). After calculation of a natural frequency ω_i, one obtains the eigenvector by solving either of Eqs. (2.62) as follows:

$$V_1^l = 1, \qquad M_1^l = -T_{34}V_1^l/T_{33} = -T_{34}/T_{33}$$

to obtain the initial state vector

$$(Q)_1^l = (0, 0, -T_{34}/T_{33}, 1). \tag{2.64}$$

The eigenvector elements at the remaining stations are calculated by successive application of Eq. (2.58). The eigenvector elements can be normalized arbitrarily but are customarily normalized with respect to either the largest displacement element or the inertia matrices; i.e.,

$$\sum_{i=1}^{n} m_i R_i^2 + \sum_{i=1}^{n} J_i \beta_i^2 = 1, \tag{2.65}$$

which is equivalent to the first of Eqs. (2.35).

The introduction of an elastically supported station is readily accounted for. Suppose the jth station has both the displacement and rotation spring supports illustrated in Figure 2.11; then the station equations (2.46) become

$$m_j \ddot{R}_{jX} + k_j R_{jX} = V_j^r - V_j^l,$$

$$J_j \ddot{\beta}_{jY} + K_j \beta_{jY} = M_j^r - M_j^l,$$

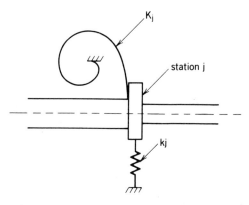

Figure 2.11 Elastically supported station.

and the station transfer matrix of Eq. (2.49) becomes

$$
\begin{Bmatrix} R_j^r \\ \beta_j^r \\ M_j^r \\ V_j^r \end{Bmatrix} = \begin{bmatrix} 1 & 0 & 0 & 0 \\ 0 & 1 & 0 & 0 \\ 0 & K_j - J_j\omega^2 & 1 & 0 \\ k_j - m_j\omega^2 & 0 & 0 & 1 \end{bmatrix} \begin{Bmatrix} R_i^l \\ \beta_i^l \\ M_i^l \\ V_i^l \end{Bmatrix}.
\tag{2.66}
$$

Accounting for the "gyroscopic" effects of Section 1.4, in calculating forward and backward critical speeds, is also managed in a straightforward manner. From Eqs. (2.8) and (2.46), the appropriate governing equations are

$$
J_i\ddot{\beta}_{iX} + \dot{\phi}\bar{J}_i\dot{\beta}_{iY} = M_{iX}^r - M_{iX}^l,
$$
$$
J_i\ddot{\beta}_{iY} - \dot{\phi}\bar{J}_i\dot{\beta}_{iX} = M_{iY}^r - M_{iY}^l.
\tag{2.67}
$$

For a synchronously precessing forward critical speed, $\omega = \dot{\phi}$, we assume the solution

$$
\begin{aligned}
R_{iX} &= R_i \cos\omega t, & R_{iY} &= R_i \sin\omega t, \\
\beta_{iY} &= \beta_i \cos\omega t, & \beta_{iX} &= -\beta_i \sin\omega t, \\
M_{iY}^r &= M_i^r \cos\omega t, & M_{iY}^l &= M_i^l \cos\omega t.
\end{aligned}
\tag{2.68}
$$

The solution for R_{iX}, R_{iY} is included to emphasize the forward precession, and should be compared to the matrix Eq. (2.37). Substitution from Eq. (2.68) into the last of Eq. (2.67) yields

$$
-\omega^2(J_i - \bar{J}_i)\beta_i = M_i^r - M_i^l.
\tag{2.69}
$$

Hence, when calculating forward critical speeds, the station transfer-matrix equation (2.49) is replaced by

$$
\begin{Bmatrix} R_i^r \\ \beta_i^r \\ M_i^r \\ V_i^r \end{Bmatrix} = \begin{bmatrix} 1 & 0 & 0 & 0 \\ 0 & 1 & 0 & 0 \\ 0 & -(J_i - \bar{J}_i)\omega^2 & 1 & 0 \\ -m_i\omega^2 & 0 & 0 & 1 \end{bmatrix} \begin{Bmatrix} R_i^l \\ \beta_i^l \\ M_i^l \\ V_i^l \end{Bmatrix}.
\tag{2.70}
$$

For an asynchronously precessing backward critical speed, $\omega = -\dot{\phi}$, we have the assumed solution format

$$
\begin{aligned}
R_{iX} &= R_i \cos\omega t, & R_{iY} &= -R_i \sin\omega t, \\
\beta_{iY} &= \beta_i \cos\omega t, & \beta_{iY} &= \beta_i \sin\omega t, \\
M_{iY}^r &= M_i^r \cos\omega t, & M_{iY}^l &= M_i^l \cos\omega t
\end{aligned}
\tag{2.71}
$$

for comparison to Eq. (2.68). Substitution from Eq. (2.71) into the last of Eq. (2.67) yields

$$-\omega^2\left(J_i + \bar{J}_i\right)\beta_i = M_i^r - M_i^l,$$

and the corresponding station transfer matrix is

$$
\begin{Bmatrix} R_i^r \\ \beta_i^r \\ M_i^r \\ V_i^r \end{Bmatrix} =
\begin{bmatrix}
1 & 0 & 0 & 0 \\
0 & 1 & 0 & 0 \\
0 & -\left(J_i + \bar{J}_i\right)\omega^2 & 1 & 0 \\
-m_i\omega^2 & 0 & 0 & 1
\end{bmatrix}
\begin{Bmatrix} R_i^l \\ \beta_i^l \\ M_i^l \\ V_i^l \end{Bmatrix}.
\tag{2.72}
$$

The procedures of Eqs. (2.70) and (2.72) for calculation of forward and backward critical speeds are identical with those followed in the general matrix formulations of Eqs. (2.39) and (2.40).

The boundary conditions of the cantilever beam in Figure 2.10 yielded the determinant definition of Eq. (2.63) from the system transfer matrix of Eq. (2.60), and there are obviously a large number of alternative boundary conditions which could be prescribed. From a computational viewpoint, a particularly attractive set are the free-free boundary conditions

$$M_1^l = M_n^r = V_1^l = V_n^r = 0,\tag{2.73}$$

which yield, from Eq. (2.60), the requirement

$$0 = T_{31}R_1^l + T_{32}\beta_1^l,$$
$$0 = T_{41}R_1^l + T_{42}\beta_1^l.$$

The frequency equation is now

$$D(\omega) = T_{31}T_{42} - T_{32}T_{41}.\tag{2.74}$$

Most rotor configurations can be modeled as having free-free boundary conditions by simply adding dummy end sections, which are massless and comparatively rigid. Hence, a computerized mechanization of the Myklestad-Prohl method can generally be simplified by using only the frequency equation (2.74).

2.5 FINITE-ELEMENT MODELS FOR (SLENDER) FLEXIBLE ROTORS

The preceding two sections have used lumped-parameter models to approximate the continuous partial-differential equation solution. The finite-element method of the present section can be viewed as a Rayleigh-Ritz approach for approximating the partial-differential-equation solution and provides both a frequency and transient model formulation.

Euler-Beam Finite-Element Model

The finite-element approach was initially developed for the vibration analysis of beams and was first applied to rotor analysis by Ruhl and Booker (1972), with subsequent contributions by Nelson and McVaugh (1976) and Nelson (1980). There is a large and rapidly expanding body of literature on the finite-element approach; however, the present developments were based on the above cited references and the text by Meirovitch (1975).

Figure 2.12 illustrates a uniform beam segment with distributed inertial and stiffness properties. This figure may be profitably compared to the massless field of Figure 2.9. The end displacements and rotations are identified similarly; however, in the present situation, the beam's mass is uniformly distributed throughout its length. The governing partial-differential equation of motion is

$$m\frac{\partial^2 r}{\partial t^2} + EI\frac{\partial^4 r}{\partial s^4} = f(s,t), \qquad (2.75)$$

where m is the mass per unit length of the beam and $f(s,t)$ is the distributed nonconservative force. We propose to develop an approximate separation-of-variable solution of the form

$$r(s,t) \simeq \sum_{i=1}^{4} \psi_i(s)u_i(t), \qquad (2.76)$$

where the $\psi_i(s)$ are referred to as *shape* functions and are required to satisfy the static homogeneous version of Eq. (2.75). The time functions $u_i(t)$ are

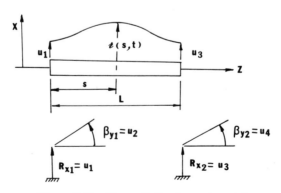

Figure 2.12   Planar finite-element model.

identified as

$$u_1 = R_{X1}, \qquad u_2 = \beta_{Y1}, \qquad u_3 = R_{X2}, \qquad u_4 = \beta_{Y2}. \qquad (2.77)$$

From Eqs. (2.75) and (2.76), the functions $\psi_i(s)$ are required to satisfy the boundary conditions

$$
\begin{array}{llll}
\psi_1(0) = 1, & \psi_2(0) = 0, & \psi_3(0) = 0, & \psi_4(0) = 0, \\
\psi_1'(0) = 0, & \psi_2'(0) = 1, & \psi_3'(0) = 0, & \psi_4'(0) = 0, \\
\psi_1(L) = 0, & \psi_2(L) = 0, & \psi_3(L) = 1, & \psi_4(L) = 0, \\
\psi_1'(L) = 0, & \psi_2'(L) = 0, & \psi_3'(L) = 0, & \psi_4'(L) = 1, \quad (2.78)
\end{array}
$$

where the prime denotes differentiation with respect to s. The static homogeneous version of Eq. (2.75) is

$$\frac{d^4 r}{ds^4} = 0,$$

and has the solution

$$r = c_4 + c_3 s + \frac{c_2 s^2}{2} + \frac{c_1 s^3}{3}.$$

Hence, from Eq. (2.78) the shape functions are defined as

$$
\begin{aligned}
\psi_1(s) &= 1 - 3s^2/L^2 + 2s^3/L^3, \\
\psi_2(s) &= s - 2s^2/L + s^3/L^2, \\
\psi_3(s) &= 3s^2/L^2 - 2s^3/L^3, \\
\psi_4(s) &= -s^2/L + s^3/L^2, \qquad\qquad\qquad (2.79)
\end{aligned}
$$

and the approximate solution of Eq. (2.76) can be stated:

$$
\begin{aligned}
r(s,t) = u_1(t)\{1 - 3s^2/L^2 + 2s^3/L^3\} + u_2(t)\{s - 2s^2/L + s^3/L^2\} \\
+ u_3(t)\{3s^2/L - 2s^3/L^3\} + u_4(t)\{-s^2/L + s^3/L^2\}. \quad (2.80)
\end{aligned}
$$

This approximate solution has the four generalized coordinates $u_i(t)$, and we propose to derive Lagrange's equations of motion for these variables. First, the kinetic energy is defined by

$$T = \frac{1}{2} \int_0^L m \left(\frac{\partial r}{\partial t} \right)^2 ds. \qquad (2.81)$$

Substitution from Eq. (2.80) into this definition yields the quadratic form

$$T = \tfrac{1}{2}(\dot{u})^T[m](\dot{u}),$$

where

$$[m] = \frac{mL}{420}
\begin{bmatrix}
156 & 22L & 54 & -13L \\
22L & 4L^2 & 13L & -3L^2 \\
54 & 13L & 156 & -22L \\
-13L & -3L^2 & -22L & 4L^2
\end{bmatrix}. \tag{2.82}$$

Similarly, substitution from Eq. (2.80) into the potential energy function

$$V = \frac{1}{2}\int_0^l EI(r'')^2 \, ds \tag{2.83}$$

yields the quadratic form

$$V = \tfrac{1}{2}(u)^T[K](u),$$

where

$$[K] = \frac{2EI}{L^3}
\begin{bmatrix}
6 & 3L & -6 & 3L \\
3L & 2L^2 & -3L & L^2 \\
-6 & -3L & 6 & -3L \\
3L & L^2 & -3L & 2L^2
\end{bmatrix}. \tag{2.84}$$

Hence, the Lagrangian is defined by

$$L = T - V = \tfrac{1}{2}(\dot{u})^T[m](\dot{u}) - \tfrac{1}{2}(u)^T[K](u),$$

and the equations of motion are

$$[m](\ddot{u}) + [K](u) = (F), \tag{2.85}$$

where the components of the force vector (F) can be identified from the following definition for virtual work:

$$\delta W = \int_0^L f(s,t)\, \delta r(s,t)\, ds + \sum_{i=1}^4 f_i\, \delta u_i(t). \tag{2.86}$$

Substitution from Eq. (2.80) into this definition yields

$$\delta W = \sum_{i=1}^4 \delta u_i(t)\left\{ \int_0^L f(s,t)\psi_i(s)\, ds + \sum_{i=1}^4 f_i \right\} = \sum_{i=1}^4 F_i\, \delta u_i, \tag{2.87}$$

where the f_i are reaction forces and moments resulting from adjacent elements, and the force distribution $f(s, t)$ results in the following nonconservative forces:

$$\bar{f}_i(t) = \int_0^L f(s, t) \psi_i(s) \, ds. \tag{2.88}$$

Hence, Eqs. (2.85) can be stated

$$[m](\ddot{u}) + [K](u) = (f) + (\bar{f}). \tag{2.89}$$

The stiffness-matrix definition of Eq. (2.84) for this finite-element model coincides with the lumped-parameter stiffness matrix of Eq. (2.15), which is to be expected, since both are based on a static beam solution. However, the "consistent-mass" matrix of Eq. (2.82) differs markedly from the lumped-parameter, diagonal matrices of Eqs. (2.11) and (2.13).

System Rotor Models Based on Finite Elements

The application of finite elements to rotor dynamics has to date been largely based on the general stiffness formulations of Section 2.3, and the demonstration of this approach to the development of a system rotor model from component finite-element models is the subject of this subsection.

Since the component stiffness matrices of the finite-element and lumped-parameter approaches coincide, the system stiffness matrices of the two methods also coincide, and the procedures of Section 2.3 apply directly. However, the consistent matrix definition of Eq. (2.82) differs substantially from traditional lumped-parameter diagonal matrices, and the statement of a system mass matrix is no longer obvious. The procedure for developing a system mass matrix will be demonstrated for the system illustrated in Figure 2.13(a). This rotor differs from the lumped-parameter model of Figure 2.6 in the following respects:

(a) The mass distribution of the elements is defined by the consistent-mass matrix definition of Eq. (2.82).
(b) There are no lumped inertia elements at beam junctions.

The first step in obtaining a system stiffness matrix is "numbering" the system variables as illustrated in Figure 2.13(c). From Eq. (2.98), the element differential equations may be stated:

$$[m^1](\ddot{u}^1) + [K^1](u^1) = (f^1),$$
$$[m^2](\ddot{u}^2) + [K^2](u^2) = (f^2), \tag{2.90}$$

where the distributed nonconservative forces of Eq. (2.88) are dropped. The

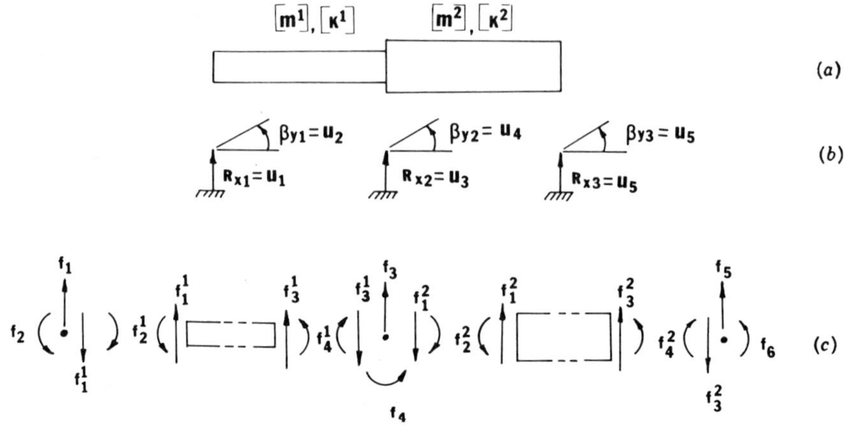

Figure 2.13 A two-element rotor model. (a) Component element definitions, (b) degrees of freedom, and (c) equilibrium requirements.

force vectors $(f^1), (f^2)$ of these equations are illustrated in Figure 2.13(c). From Figure 2.13(b), the coordinate vectors are

$$\begin{Bmatrix} u_1^1 \\ u_2^1 \\ u_3^1 \\ u_4^1 \end{Bmatrix} = \begin{Bmatrix} R_{X1} \\ \beta_{Y1} \\ R_{X2} \\ \beta_{Y2} \end{Bmatrix} = \begin{Bmatrix} u_1 \\ u_2 \\ u_3 \\ u_4 \end{Bmatrix}, \qquad \begin{Bmatrix} u_1^2 \\ u_2^2 \\ u_3^2 \\ u_4^2 \end{Bmatrix} = \begin{Bmatrix} R_{X2} \\ \beta_{Y2} \\ R_{X3} \\ \beta_{Y3} \end{Bmatrix} = \begin{Bmatrix} u_3 \\ u_4 \\ u_5 \\ u_6 \end{Bmatrix}.$$

The elements of the component mass matrices of Eq. (2.90) can be calculated from Eq. (2.82), and stated as follows:

$$[m^1] = \begin{bmatrix} m_{11}^1 & m_{12}^1 & m_{13}^1 & m_{14}^1 \\ m_{21}^1 & m_{22}^1 & m_{23}^1 & m_{24}^1 \\ m_{31}^1 & m_{32}^1 & m_{33}^1 & m_{34}^1 \\ m_{41}^1 & m_{42}^1 & m_{43}^1 & m_{44}^1 \end{bmatrix},$$

$$[m^2] = \begin{bmatrix} m_{33}^2 & m_{34}^2 & m_{35}^2 & m_{36}^2 \\ m_{43}^2 & m_{44}^2 & m_{45}^2 & m_{46}^2 \\ m_{53}^2 & m_{54}^2 & m_{55}^2 & m_{56}^2 \\ m_{63}^2 & m_{64}^2 & m_{65}^2 & m_{66}^2 \end{bmatrix}. \qquad (2.91)$$

From Figure 2.13(b), the joint equilibrium equations are

$$f_1 = f_1^1, \qquad f_2 = f_2^1, \qquad f_3 = f_3^1 + f_1^2,$$
$$f_4 = f_4^1 + f_2^2, \qquad f_5 = f_3^2, \qquad f_6 = f_4^2. \qquad (2.92)$$

From Eq. (2.90), these equations yield the system model

$$[M](\ddot{u}) + [K](u) = (f),$$

where $[K]$ is as previously defined by Eq. (2.22), and the system matrix is

$$[M] = \begin{bmatrix} m_{11}^1 & m_{12}^2 & m_{13}^1 & m_{14}^1 & 0 & 0 \\ m_{21}^1 & m_{22}^1 & m_{23}^1 & m_{24}^1 & 0 & 0 \\ m_{31}^1 & m_{32}^1 & (m_{33}^1 + m_{33}^2) & (m_{34}^1 + m_{34}^2) & m_{35}^2 & m_{36}^2 \\ m_{41}^1 & m_{42}^1 & (m_{43}^1 + m_{43}^2) & (m_{44}^1 + m_{44}^2) & m_{45}^2 & m_{46}^2 \\ 0 & 0 & m_{53}^2 & m_{54}^2 & m_{55}^2 & m_{56}^2 \\ 0 & 0 & m_{63}^2 & m_{64}^2 & m_{65}^2 & m_{66}^2 \end{bmatrix}. \quad (2.93)$$

The subscripts and superscripts here coincide with those developed for the system stiffness matrix. The conclusion to be drawn from this result is that, in finite-element analyses, the system mass matrix is developed by precisely the same procedures as those used in developing the system stiffness matrix. The analogy between mass and stiffness elements can be further demonstrated by considering the introduction of a rigid body between the two beam elements, as illustrated in Figure 2.14. The forces acting on this rigid body are also illustrated in Figure 2.14, and the equations of motion for the rigid body are

$$m_3 \ddot{u}_3 = f_3 - f_3^1 - f_1^2,$$
$$J_3 \ddot{u}_4 = f_4 - f_4^1 - f_2^2.$$

These equations replace the third and fourth entries of Eqs. (2.92) and modify the system mass matrix definition of Eq. (2.93) as follows:

$$M_{33} = m_3 + m_{33}^1 + m_{33}^2,$$
$$M_{44} = J_3 + m_{44}^1 + m_{44}^2,$$

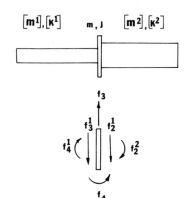

Figure 2.14 A rotor modeled with two finite elements and an intervening rigid body.

with all other elements unchanged. This result is entirely analogous to the introduction of joint springs in the development of system stiffness matrices.

2.6 FINITE-ELEMENT MODELS FOR SLENDER BEAMS INCLUDING ROTARY INERTIA AND GYROSCOPIC EFFECTS

The finite-element model, developed in the preceding section for a slender-beam segment, does not account for kinetic energy due to "rigid-body" motion of differential mass elements within the beam segment; hence, *distributed* rotary inertia and gyroscopic effects are neglected. These rigid-body effects can, however, be accounted for by modeling a rotor as a collection of uniform beams (modeled with finite elements) and rigid bodies (including rotary inertia and gyroscopic effects). This approach would be consistent with our earlier lumped-parameter development. In this section, a finite-element model is developed, based on the work of Nelson and McVaugh (1976), which accounts for distributed rotary inertia and gyroscopic coupling within the beam segment.

The same procedure will be followed in developing this "new" finite-element model as was followed in the preceding section. Specifically, kinetic and potential energy functions are to be defined in terms of the element shape functions, and Lagrangian equations of motion are then stated. Statement of the kinetic energy function is the principal additional complication, since motion must be defined in the orthogonal X-Z and Y-Z planes. The additional required kinematics are provided in Figure 2.15. The u_i coordinates at the ends of the element are defined by

$$
\begin{aligned}
u_1 &= R_{X1}, & u_5 &= R_{X2}, \\
u_2 &= R_{Y1}, & u_6 &= R_{Y2}, \\
u_3 &= \beta_{X1}, & u_7 &= \beta_{X2}, \\
u_4 &= \beta_{Y1}, & u_8 &= \beta_{Y2}.
\end{aligned}
\tag{2.94}
$$

These definitions should be compared with the planar definition provided by Eq. (2.77). The displacement functions in the X-Z and Y-Z planes are $r(s, t)$ and $p(s, t)$, respectively, and are defined in terms of the shape functions of Eq. (2.79) and the components of the element vectors of Eq. (2.94) by

$$
\begin{aligned}
r(s, t) &= u_1\psi_1 + u_4\psi_2 + u_5\psi_3 + u_8\psi_4, \\
p(s, t) &= u_2\psi_1 - u_3\psi_2 + u_6\psi_3 - u_7\psi_4.
\end{aligned}
\tag{2.95}
$$

The first of these functions repeats Eqs. (2.76) and (2.80) of our earlier development. The orientation of the differential beam element is defined by the (differential) angles α and β of Figure 2.15, which are defined in terms of

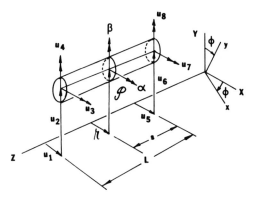

Figure 2.15   Two-plane finite-element model kinematics and degrees of freedom.

$r(s, t), p(s, t)$ by

$$\beta(s, t) = \frac{\partial r(s, t)}{\partial s} = u_1\psi'_1 + u_4\psi'_2 + u_5\psi'_3 + u_8\psi'_4,$$

$$\alpha(s, t) = \frac{-\partial p(s, t)}{\partial s} = -u_2\psi'_1 + u_3\psi'_2 - u_6\psi'_3 + u_7\psi'_4. \qquad (2.96)$$

The negative sign in Eq. (2.96) is necessary because we assume a right-hand rule for positive rotation about the X axis. This conversion also necessitates the negative signs which appear in the definition for $p(s, t)$ in Eq. (2.95). Equations (2.95) and (2.96) can be expressed in matrix format as

$$\left\{ \begin{array}{c} r(s, t) \\ p(s, t) \end{array} \right\} = [\Psi(s)]\{u(t)\}, \qquad \left\{ \begin{array}{c} \alpha(s, t) \\ \beta(s, t) \end{array} \right\} = [\Gamma(s)]\{u(t)\}, \quad (2.97)$$

where

$$[\Psi] = \begin{bmatrix} \psi_1 & 0 & 0 & \psi_2 & \psi_3 & 0 & 0 & \psi_4 \\ 0 & \psi_1 & -\psi_2 & 0 & 0 & \psi_3 & -\psi_4 & 0 \end{bmatrix},$$

$$[\Gamma] = \begin{bmatrix} \Gamma_\alpha \\ \Gamma_\beta \end{bmatrix} = \begin{bmatrix} 0 & -\psi'_1 & \psi'_2 & 0 & 0 & -\psi'_3 & \psi'_4 & 0 \\ \psi'_1 & 0 & 0 & \psi'_2 & \psi'_3 & 0 & 0 & \psi'_4 \end{bmatrix}. \quad (2.98)$$

The kinetic energy for the element is to be obtained by stating the kinetic energy of a differential element of length ds, and then integrating over the length of the finite element. The kinetic energy per unit length may be stated

$$dT = \frac{m}{2}(\dot{r}^2 + \dot{p}^2) + \frac{j}{2}(\dot{\alpha}^2 + \dot{\beta}^2) + \frac{\bar{j}}{2}(\dot{\phi}^2 - 2\dot{\phi}\alpha\dot{\beta}), \qquad (2.99)$$

where m, j, and $\bar{j}$ are the mass, diametral, and polar moments of inertia per unit length. Substitution from Eq. (2.97) into Eq. (2.99) yields

$$dT = \frac{m}{2}(\dot{u})^T[\Psi]^T[\Psi](\dot{u}) + \frac{j}{2}(\dot{u})^T[\Gamma]^T[\Gamma](\dot{u}) - \dot{\phi}\bar{j}(u)^T\Gamma_\alpha^T\Gamma_\beta(\dot{u}) + \frac{\bar{j}}{2}\dot{\phi}^2.$$
$$(2.100)$$

The first term accounts for rectilinear translation, the second for rotary inertia, the third for gyroscopic coupling, and the last for spin-axis rotation. Integration over the length of the finite element yields the kinetic-energy definition

$$T = \tfrac{1}{2}(\dot{u})^T\{[M_T] + [M_R]\}(\dot{u}) - \dot{\phi}(u)^T[N](\dot{u}) + \dot{\phi}^2\frac{\bar{j}}{2}, \qquad (2.101)$$

where

$$[M_T] = \int_0^L m[\Psi]^T[\Psi]\, ds, \qquad [M_R] = \int_0^L j[\Gamma]^T[\Gamma]\, ds,$$

$$[N] = \int_0^L \bar{j}\Gamma_\alpha^T\Gamma_\beta\, ds.$$

The potential energy function from the preceding section must be modified to account for bending energy in the Y-Z plane. The potential energy per unit length is

$$dV = EI\{(r'')^2 + (p'')^2\}/2 = \frac{EI}{2}(u)^T[\Psi'']^T[\Psi''](u),$$

and the element potential energy function is

$$V = \tfrac{1}{2}(u)^T[K_e](u), \qquad [K_e] = \int_0^L EI[\Psi'']^T[\Psi'']\, ds. \qquad (2.102)$$

Defining the Lagrangian $L = T - V$ from equations (2.101) and (2.102) yields the constant-spin-velocity ($\dot{\phi} = \omega$) equation of motion

$$\{[M_T] + [M_R]\}(\ddot{u}) - \omega[G](\dot{u}) + [K_e](u) = (F), \qquad (2.103)$$

where the gyroscopic coupling matrix $[G]$ is skew symmetric and defined by

$$[G] = [N] - [N]^T. \qquad (2.104)$$

The desired element equation, accounting for both rotary inertia and gyroscopic coupling, is provided by Eq. (2.104). The element inertia, gyroscopic, and stiffness matrices of Eq. (2.103) are stated in Appendix B. The finite-element models derived here and in the preceding section are based on the Euler beam model, which does not account for shear deflections. The development of comparable models based on the Timoshenko beam model (including shear deflections) requires the introduction of new shape functions based on static solutions for the Timoshenko beam. Element mass, gyroscopic, and stiffness matrices have been provided by Nelson (1980).

The force vector on the right-hand side of Eq. (2.103) contains terms due to both the distributed nonconservative force field acting on the element, and reaction forces from adjacent elements. Nelson and McVaugh (1976) also account for distributed imbalance acting along the length of the element by the following procedure. The imbalance distribution in the stationary X-Z and Y-Z planes may be stated

$$a_X(s) = a_x(s)c\phi - a_y(s)s\phi,$$
$$a_Y(s) = a_x(s)s\phi + a_y(s)c\phi, \qquad (2.105)$$

in terms of the rotor-fixed imbalance distributions $a_x(s), a_y(s)$ and the angle $\phi = \omega t$. The imbalance loading in the X-Z and Y-Z planes acting on a differential element ds in length is

$$dQ_X = -\omega^2 a_X(s)m\,ds,$$
$$dQ_Y = -\omega^2 a_Y(s)m\,ds. \qquad (2.106)$$

The virtual work due to this loading is

$$\delta W = -\omega^2 \int_0^L m\{a_X(s)\,\delta r(s) + a_Y(s)\,\delta p(s)\}\,ds$$
$$= -\omega^2(\delta u)^T\{(ma_c)\cos\omega t + (ma_s)\sin\omega t\},$$

where

$$(ma_c) = \int_0^L m[\Psi(s)]^T \left\{ \begin{matrix} a_x(s) \\ a_y(s) \end{matrix} \right\} ds,$$

$$(ma_s) = \int_0^L m[\Psi(s)]^T \left\{ \begin{matrix} -a_y(s) \\ a_x(s) \end{matrix} \right\} ds. \qquad (2.107)$$

Hence, the element-joint force vector due to imbalance distributions is

$$(F_a) = -\omega^2\{(ma_c)\cos\omega t + (ma_s)\sin\omega t\}. \qquad (2.108)$$

The vectors $(ma_c), (ma_s)$ of this expression and Eq. (2.107) are stated in

Appendix B for the following linear imbalance distributions:

$$a_x(s) = a_{x0}\left(1 - \frac{s}{L}\right) + a_{xL}\left(\frac{s}{L}\right),$$

$$a_y(s) = a_{y0}\left(1 - \frac{s}{L}\right) + a_{yL}\left(\frac{s}{L}\right). \tag{2.109}$$

Assembling system models based on the element equations provided in this section follows precisely the same procedure as that followed in the preceding section. Specifically, system stiffness, inertia, and gyroscopic coupling matrices are assembled by the (same) general-stiffness-matrix procedure. Lumped disks at joints are accommodated by precisely the same procedures as those used previously for point masses.

The point was made earlier that the present and preceding sections are provided to introduce the fundamentals of finite-element analysis as applied to rotordynamics. Readers who are interested in developing finite-element-analysis programs are advised to additionally read the references by Nelson and McVaugh (1976) and Nelson (1980).

2.7   NUMERICAL EXAMPLES

The contents of the preceding sections demonstrate that a flexible rotor may be modeled as (a) a collection of rigid bodies connected by massless beam elements, or (b) a collection of finite elements and rigid bodies. The resultant model may be analyzed by either transfer-matrix or general-stiffness-matrix methods. Our concern in the present section is an examination of models and eigenanalysis results for two flexible-rotor examples. The emphasis in this section is on those features of the results which tend to be of recurrent interest, and are largely independent of either the modeling or eigenanalysis approach.

Example Rotor 1

Figure 2.16 illustrates the first rotor example, which consists of a uniform rotor supported by displacement springs at its ends. A lumped-parameter rotor model containing 22 rigid bodies will be employed to calculate* the first three natural frequencies and their associated mode shapes. The calculated natural frequencies are presented in Figure 2.17 as a function of the normalized support stiffness $\hat{k} = k/(48EI/L^3)$, where the factor $(48EI/L^3)$ is the stiffness at the center of the rotor for pinned-end supports. Figure 2.18 illustrates dynamically normalized mode shapes for $\hat{k} = 0.1$, 1.0, 10.0, and 100. The mode shapes demonstrate the spatial distribution of the rigid bodies

---

*Results include shear-deflection effects.

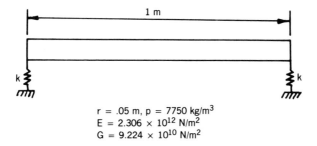

r = .05 m, p = 7750 kg/m^3
E = 2.306 × 10^{12} N/m^2
G = 9.224 × 10^{10} N/m^2

Figure 2.16 Example rotor 1.

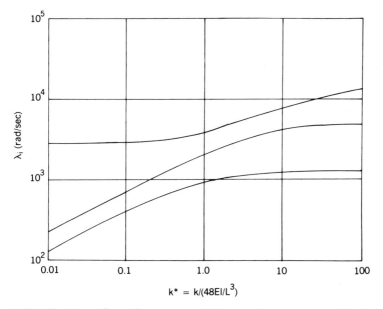

k* = k/(48EI/L^3)

Figure 2.17 First three (lowest) rotor natural frequencies versus normalized support stiffness.

used in modeling the rotor. The mode shapes of Figure 2.18(a)* demonstrate that at $\hat{k} = 0.1$, the first two modes are essentially rigid-body modes of the rotor bouncing and rocking (as a rigid body) on the comparatively soft end springs. These modes are commonly referred to as stick modes. The mode shape for the third natural frequency is the free-free bending mode, essentially unaffected by the soft end supports. An examination of the remaining frames in Figure 2.18 demonstrates that increasing $\hat{k}$ to 100 yields modes which correspond to the results to be expected for pinned-end boundary

*The results of Figure 2.18(a) coincide with those obtained for $\hat{k} = 0.01$.

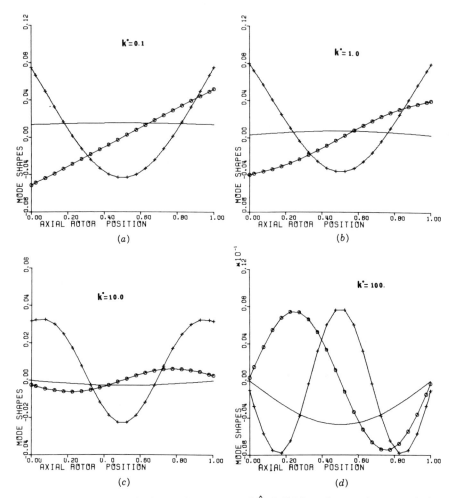

Figure 2.18 Rotor mode shapes for a range of $\hat{k}$: Solid line, first mode; open circles, second mode; crosses, third mode.

conditions. Figure 2.17 illustrates the corresponding increase in natural frequencies. At $\hat{k} = 100$, the first two natural frequencies have approximately reached their asymptotic values corresponding to pinned-end natural frequencies, while the third natural frequency continues to increase.

The support stiffness of a rotor is frequently at the discretion of the designer, and from Figure 2.17 can be used (within limits) to control critical-speed locations. An examination of this figure demonstrates that a comparatively wide, resonance-free, operating speed range can be obtained by selecting either the small value $\hat{k} = 0.01$, or the larger values $\hat{k} = 10$ or 100. For the softly supported rotor, one could propose to accelerate through the first two low-frequency modes, and operate below the third natural frequency. Conversely, one could choose to employ a stiff bearing support, and operate below the first natural frequency. Since the third free-free natural frequency is higher than the first pinned-end mode, designers are occasionally tempted (or forced) to select the softly-supported-rotor configuration represented by $k = 0.01$. However, we learned in Section 1.5 that reducing the natural frequency also lowers the onset speed of instability, and, as a general rule, softly supported rotors should be approached with considerable caution from a stability viewpoint.

Example Rotor 2

Figure 2.19(a) illustrates the dimensions of example rotor 2. We wish to calculate the first three natural frequencies, forward critical speeds, and backward critical speeds. A lumped-parameter model employing 11 rigid bodies will be used. The location of these bodies and the mass to be allocated to each are indicated in Figure 2.19(b). The rigid bodies are connected by the beam elements of Figure 2.19(c). Although they were not presented, the critical speeds for example rotor 1 do not differ substantially from their associated natural frequencies because the rotor is comparatively slender,

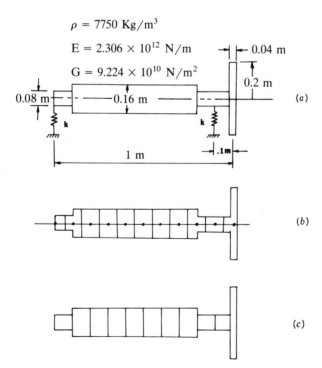

$\rho = 7750 \text{ Kg/m}^3$

$E = 2.306 \times 10^{12} \text{ N/m}$

$G = 9.224 \times 10^{10} \text{ N/m}^2$

0.04 m

0.2 m

0.08 m

0.16 m

(a)

1 m

.1m

(b)

(c)

Figure 2.19 Example rotor 2. (a) Dimensions, (b) inertia allocation, and (c) stiffness allocation.

TABLE 2.1 Natural Frequencies λ_i, Forward Critical Speeds ω_i, and Backward Critical Speeds ω_{-1} in rpm for a Range of Support Stiffness.

$k \ (N/m)$	5×10^6	5×10^7	5×10^8
λ_1	2,199	6,791	14,274
λ_2	3,593	9,796	18,289
λ_3	22,090	23,205	30,739
ω_1	2,210	6,841	15,020
ω_2	3,778	10,610	23,902
ω_3	41,668	44,974	55,053
ω_{-1}	2,195	6,779	13,565
ω_{-2}	3,525	9,464	16,837
ω_{-3}	18,640	19,810	28,691

and there are no disks present to develop appreciable gyroscopic coupling. By contrast, example rotor 2 has a large end disk which has the potential for substantial gyroscopic coupling.

Table 2.1 provides natural frequencies and forward and backward critical speeds for a range of support stiffness. The results of this table indicate that gyroscopic coupling has a minimal influence on the first (lowest) natural frequency, somewhat more influence on the second, and a substantial influence on the third. Figure 2.20 illustrates accompanying mode shapes for $k = 5 \times 10^8$ N/m. The frames of this figure explain why the comparatively flat first mode is not much modified by gyroscopic coupling. The substantial disk-end slopes for the second and third modes explains the considerable influence gyroscopic coupling has on these modes. Many rotor designs rely on gyroscopic coupling to elevate rotor critical speeds (particularly those associated with the second-natural-frequency mode) out of the operating range.

Modeling Alternatives

Both of the preceding rotors were represented by lumped-parameter models. One of the principal questions which arises with this approach is, "how many lumps are enough?" The answer depends entirely on the number and shape of modes to be calculated. Sufficient lumped parameters should be provided to adequately approximate the curvature of the rotor modes being modeled. For example, the three modes of Figure 2.18(a) require fewer lumped parameters than the three modes of Figure 2.18(d), since the curvature of the latter is substantially greater. In the author's experience, an inspection of the mode-shape plot is generally sufficient to determine whether enough coordinates have been provided.

A finite-element model for rotor number 1 would require four elements according to Ruhl and Booker's (1972) suggestion that "the number of

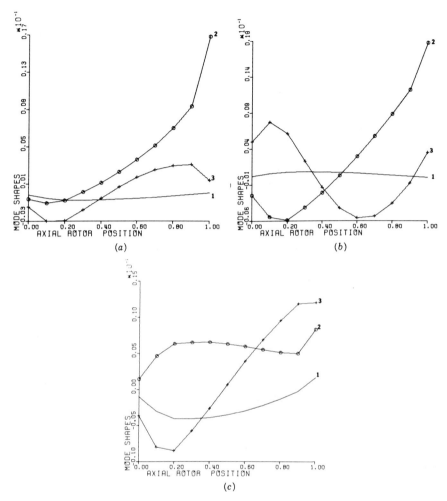

Figure 2.20 Mode shapes for example rotor 2 for (a) natural frequencies, (b) forward critical speeds, and (c) backward critical speeds. Solid line, first mode; open circles, second mode; crosses, third mode.

elements exceeds by one the highest critical to be encountered." This rule would require four elements plus a rigid end disk to model example rotor 2.

2.8 SUMMARY AND EXTENSIONS

The contents of this chapter have provided an introduction to various approaches for modeling flexible rotors, viz., lumped-parameter and finite-element models. Eigenanalysis procedures based on transfer-matrix and general-stiffness-matrix formulations have also been reviewed. Considering the range of possible modeling and eigenanalysis approaches, one might inquire as to the "best" approach. This is a reasonable question if one can agree on the criteria to be used in evaluating the various approaches. Suppose we agree that the following criteria are suitable:

(a) simplicity and ease of understanding,

(b) availability,

(c) computer time and storage requirements,

(d) relative freedom from numerical difficulty, and

(e) breadth of application.

For simplicity and ease of understanding, the Myklestad-Prohl method is probably unbeatable; also, computer programs based on these approaches are widely available or easily developed. Lumped-parameter, stiffness-matrix-based models are also reasonably simple, assuming that one understands the stiffness-matrix definition, with the finite-element approach requiring more mathematical sophistication. A comparison between computer requirements for the various approaches is difficult, since one can use transfer-matrix procedures to calculate only as many eigenvalues and eigenvectors as are required, while the Jacobi method for general-stiffness-based models will yield all eigenvalues and eigenvectors simultaneously. Restricting our attention to stiffness formulations, the finite-element models generally require less computer time, since a substantially smaller model results. In the past, one of the strong arguments for the transfer-matrix approach has been small computer storage requirements. However, the dramatic increases which have been made in storage capability has largely eliminated this advantage. More generally, the advances in computer capability are such that none of the methods are expensive, and, in an absolute sense, there is not a great deal of computer money to be saved in moving from the least to most efficient computer procedure.

For breadth of application, the finite-element method is probably superior, primarily because it represents a very general approach for structural-dynamics analysis of case structures. For many applications, case structure motion is negligible because the case is both massive and comparatively rigid. The opposite situation holds in the analysis of jet engines or liquid rocket engines where a minimum-weight case structure is as flexible as the rotor. Figure 2.21 [Childs (1976a)] illustrates a dual-rotor/case jet-engine configuration, and Figure 2.22 illustrates the first two planar modes for this system. The flexibility of the case is clearly comparable to that of the rotors. Although these modes were actually calculated from a transfer-matrix procedure, a finite-element analysis for the case becomes mandatory for more complicated case structures.

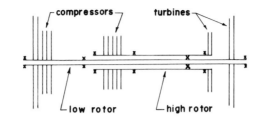

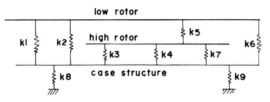

Figure 2.21 Dual-rotor/case system configuration from Childs (1976a).

The contents of this chapter have not provided an exhaustive review of flexible-rotor analysis, but have instead covered the more general and recurrent structural-dynamic modeling and analysis procedures. Among the topics which were not covered are the effects of disk flexibility on critical speeds and natural frequencies. Specifically, we have assumed in the preceding that a turbine wheel would be treated as a rigid disk. Chivens and Nelson (1975) examined the effect of disk flexibility and concluded that this effect lowers the natural frequency of shaft-disk systems, but not the critical speeds. Dopkin and Shoup (1969) also considered this problem, and their analysis yields a transfer-matrix approach including disk flexibility.

The structural-dynamics analysis of "real" rotors can present many practical problems which are not covered in this chapter. For example, built-up rotors which use through-bolts to compress a rotating assembly of disks present serious conceptual problems for simple beam theory. Bolted flange connections are sometimes used in rotating machines and pose comparable problems. Some rotor sections look more like thick shell elements than beams. Tapered beam finite-element models have been developed by Rouch and Kao (1979) and Greenhill et al. (1985) which can be used for linear or gradual changes in cross-section. Stephenson and Rouch (1990) have used standard "brick" finite elements to analyze abrupt changes in cross-sections of rotors, and this more general finite-element approach is increasingly used for the analysis of assymetric housing structures. For serious rotor applications, modal testing to verify structural dynamics predictions is advised.

In proceeding through the balance of this text, we will not require the contents of the present chapter, until we reach Chapter 7. The intervening chapters on hydrodynamic bearings, squeeze-film dampers, seals, etc., pro-

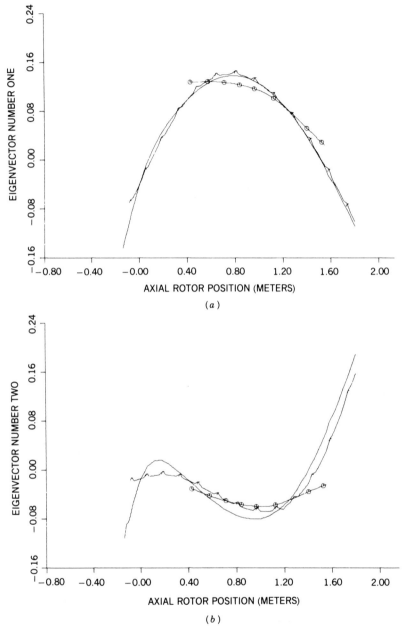

Figure 2.22   Mode shapes for dual-rotor/case system of Figure 2.21. Solid line, low rotor; open circles, high rotor; crosses, case structure [Childs (1976a)].

vide the background necessary for modeling hydrodynamic and aerodynamic forces acting on our structural-dynamic model.

REFERENCES: CHAPTER 2

Biezeno, C., and Grammel, R. (1959), *Engineering Dynamics*, Vol. III of *Steam Turbines*, D. Van Nostrand Co., New York.

Childs, D. (1976a), "A Modal Transient Rotordynamic Model for Dual-Rotor Jet Engine Systems," *Journal of Engineering for Industry*, 876–882.

Childs, D. (1976b), "A Modal Transient Simulation Model for Flexible Asymmetric Rotors," *Journal of Engineering for Industry*, 312–319.

Chivens, D., and Nelson, H. (1975), "The Natural Frequencies of a Rotating, Flexible Shaft-Disk System," *Journal of Engineering for Industry*, 881–886.

Dopkin, J., and Shoup, T. (1969), "Rotor Resonant Speed Reduction Caused by Flexibility of Disks, *Journal of Engineering Industry*, 1328–1331.

Greenhill, L., Bickferd, W., and Nelson, H. (1985), "A Conical Beam Finite Element for Rotordynamic Analysis," *Journal of Vibrations, Acoustics, Stress and Reliability in Design*, **107**, 421–427.

Meirovitch, L. (1975), *Elements of Vibration Analysis*, McGraw-Hill, New York.

Myklestad, N. (1944), "A New Method for Calculating Natural Modes of Uncoupled Bending Vibrations of Airplane Wings and Other Types of Beams," *Journal of Aeronautical Science*, **11**, 153–162.

Nelson, H. (1980), "A Finite Rotating Shaft Element Using Timoshenko Beam Theory," *Journal of Mechanical Design*, **2**(4), 793–803.

Nelson, H., and McVaugh, J. (1976), "The Dynamics of Rotor-Bearing Systems Using Finite Elements," *Journal of Engineering for Industry*, 593–600.

Pestel, E., and Leckie, F. (1973), *Matrix Methods in Elastomechanics*, McGraw-Hill, New York.

Prohl, M. (1945), "A General Method for Calculating Critical Speeds of Flexible Rotors," *Journal of Applied Mechanics*, A142–A148.

Rouch, K., and Kao, J. (1979), "A Tapered Beam Finite Element for Rotordynamics Analysis," *Journal of Sound and Vibration*, **66**(1), 119–140.

Ruhl, R., and Booker, J. (1972), "A Finite Element Model for Distributed Parameter Turborotor Systems," *Journal of Engineering for Industry*, 126–132.

Stephenson, R., and Rouch, K. (1990), "Rotor Modeling Considerations with Shaft Diameter Changes," in J. Kim and W. Yang (Eds.), *Dynamics of Rotating Machinery*, Hemisphere Publishing.

Timoshenko, S. (1955), *Vibration Problems in Engineering* (3rd ed.), D. Van Nostrand Co., New York.

3

ROTORDYNAMIC INTRODUCTION TO HYDRODYNAMIC BEARINGS AND SQUEEZE-FILM DAMPERS

3.1 INTRODUCTION

The general area of rotordynamics has historically involved a combination of the separate areas of (a) vibrations or structural dynamics and (b) hydrodynamic-bearing analysis. The preceding chapter provides an introduction to relevant vibrations material, while the present chapter provides a comparable introduction to the nomenclature, governing equations, analysis procedures, and models for hydrodynamic journal bearings and squeeze-film dampers. The material in this chapter focuses on the relatively small portion of lubrication theory, analysis, and results which apply to rotordynamic analysis. The contents and organization assume no prior knowledge of lubrication theory and only a limited understanding of fluid mechanics.

The derivation of an appropriate version of the Reynolds equation is provided in the next section. This equation defines the pressure field within a bearing as a function of its motion. The pressure field may be integrated to define the forces developed by a bearing, and acting on a rotor as a function of the rotor's position and velocity vectors at the bearing location. For plain journal bearings, this integration yields closed-form impedance vectors, which completely define the bearing's nonlinear static and dynamic characteristics. The development of these impedance descriptions is the subject of Section 3.3.

The author's initial view of hydrodynamic bearings was that their static and small-amplitude-motion characteristics were more peculiar than their nonlinear transient characteristics. Specifically, when one applies a static fixed-direction load to a bearing, the consequent static displacement vector has a component normal to the direction of the applied load. Further, when small motion is postulated about an equilibrium position, complete nonsymmetric stiffness and damping matrices are obtained, which "cross-couple" the rotor motion in orthogonal X-Z and Y-Z rotor planes. Static-equilibrium characteristics and the derivation of stiffness and damping coefficients for plain journal bearings are reviewed in Section 3.4.

The cross-coupling terms in a bearing-stiffness matrix are similar to the terms arising in Section 1.5 due to internal rotor damping and have a similar unpleasant consequence; specifically, they destabilize rotors. Section 3.5 considers the stability characteristics of both rigid- and Jeffcott-rotor models supported in plain journal bearings, defining both the onset speed of instability and the precessional or whirl frequency associated with the unstable motion. Synchronous response characteristics of these rotor models due to imbalance are also examined. Finally, transient solutions based on nonlinear bearing impedances are presented for a rigid rotor on plain journal bearings.

Unstable motion of a rotor supported in bearings *about an equilibrium position* is normally called "oil whirl," with a precessional frequency equal to the rotor's natural frequency. Vertical rotors in plain journal bearings have no static-equilibrium position and are subject to an unstable precessional motion that "tracks" at one-half running speed, which is also called oil whirl. At running speeds above twice the rotor critical speed, this subsynchronous motion "locks in" to the rotor natural frequency and is called "oil whip." Section 3.6 explains the difference between these two phenomena.

Various bearing configurations have better stability characteristics than the plain journal bearing. Additional bearing configurations including multilobe, step, and tilting-pad bearings are introduced in Section 3.7, and their stability properties reviewed.

The Reynolds equation derived in Section 3.2 applies for an isoviscous, Newtonian, incompressible fluid operating in the laminar regime. In fact, many high-performance bearings operate in the turbulent regime, and gas is used as a lubricant fluid in some applications. Further, most liquid bearings are subject to cavitation. Section 3.8 reviews the procedures and models which have been developed to account for these complications in developing numerical solutions for bearings.

The finite-difference and finite-element solution techniques for bearing solutions are introduced in some detail in Section 3.9. The two techniques are commonly used for the calculation of bearing equilibrium characteristics and stiffness and damping coefficients.

Hydrodynamic bearings such as journal bearings are used to support rotors. Squeeze-film dampers are geometrically identical to plain journal bearings but are employed to provide additional rotor damping at the bearings to either assist in stabilizing an otherwise unstable rotor, or to improve the synchronous response characteristics of rotors supported in ball bearings. Models for squeeze-film dampers are discussed in Section 3.10.

3.2   REYNOLDS EQUATION

Our interest in this section is the derivation of a governing equation for the pressure field within a bearing. The pressure field will then be integrated to determine the bearing forces acting on a rotor.

General Development

Figure 3.1 illustrates the cross section of a "plain" journal bearing. The clearance-to-radius ratio C_r/R of this type of bearing is generally on the order of 0.001. Hence, the fluid-film thickness is very small in comparison to the bearing length and circumference. We may accordingly neglect the curvature of the fluid-film surface and consider a fluid film between two plane bearing and journal segments as illustrated in Figure 3.2. Each segment has velocity components in the x and y directions, but has zero velocity components along the z axis of the bearing. The situation illustrated in Figure 3.2 rarely arises in practice, since the bearing portion of a journal bearing is generally motionless; i.e., generally, $V_1 = U_1 = 0$. However, we will retain this generality for the purposes of this derivation.

The Reynolds equation is a greatly simplified version of the Navier-Stokes (NS) equations based on physical arguments related to the smallness of C_r/R and the Reynolds number. The present discussion concerning these simplifications is taken from Szeri (1980).

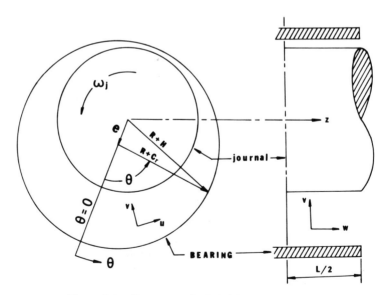

Figure 3.1   Geometry of a "plain" journal bearing.

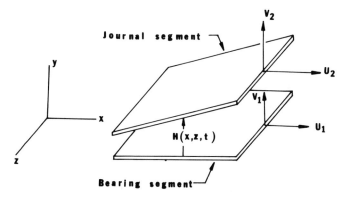

Figure 3.2   Plane journal and bearing segments.

Using Cartesian coordinates and omitting body forces, the NS equations can be stated

$$\frac{\partial u}{\partial t} + u\frac{\partial u}{\partial x} + v\frac{\partial u}{\partial y} + w\frac{\partial u}{\partial z} = -\frac{1}{\rho}\frac{\partial p}{\partial x} + v\left(\frac{\partial^2 u}{\partial x^2} + \frac{\partial^2 u}{\partial y^2} + \frac{\partial^2 u}{\partial z^2}\right),$$

$$\frac{\partial v}{\partial t} + u\frac{\partial v}{\partial x} + v\frac{\partial v}{\partial y} + w\frac{\partial v}{\partial z} = -\frac{1}{\rho}\frac{\partial p}{\partial y} + v\left(\frac{\partial^2 v}{\partial x^2} + \frac{\partial^2 v}{\partial y^2} + \frac{\partial^2 v}{\partial z^2}\right),$$

$$\frac{\partial w}{\partial t} + u\frac{\partial w}{\partial x} + v\frac{\partial w}{\partial y} + w\frac{\partial w}{\partial z} = -\frac{1}{\rho}\frac{\partial p}{\partial z} + v\left(\frac{\partial^2 w}{\partial x^2} + \frac{\partial^2 w}{\partial y^2} + \frac{\partial^2 w}{\partial z^2}\right), \quad (3.1)$$

where ρ and v are, respectively, the fluid density and kinematic viscosity. Further, p is the pressure, and u, v, w are the fluid velocity components in the x, y, z directions. The flow-field model is completed by the incompressible-fluid continuity equation

$$\frac{\partial u}{\partial x} + \frac{\partial v}{\partial y} + \frac{\partial w}{\partial z} = 0. \quad (3.2)$$

The spatial coordinates are naturally nondimensionalized via

$$\bar{x} = \frac{x}{R}, \qquad \bar{y} = \frac{y}{C_r}, \qquad \bar{z} = \frac{z}{R}. \quad (3.3)$$

The circumferential and axial velocity components are also readily nondimensionalized in terms of the rotor surface velocity $R\omega$, where ω is the shaft running speed; viz.,

$$\bar{u} = \frac{u}{R\omega}, \qquad \bar{w} = \frac{w}{R\omega}. \qquad (3.4a)$$

Substituting these definitions into the continuity equation gives

$$\frac{\partial \bar{u}}{\partial \bar{x}} + \left(\frac{R}{C_r}\right) \frac{1}{R\omega} \frac{\partial v}{\partial \bar{y}} + \frac{\partial \bar{w}}{\partial z} = 0.$$

The terms in this equation will all be of order unity providing the following nondimensionalization is employed for v:

$$\bar{v} = \left(\frac{R}{C_r}\right) \frac{v}{R\omega}. \qquad (3.4b)$$

The nondimensionalization is completed by

$$\bar{p} = \mathcal{R}e\left(\frac{C_r}{R}\right) \frac{p}{\rho(R\omega)^2}, \qquad \mathcal{R}e = \frac{C_r(R\omega)}{\nu}, \qquad \bar{t} = \omega t. \qquad (3.5)$$

Substituting the nondimensionalized variables of Eqs. (3.3)–(3.5) into Eq. (3.1) yields

$$\mathcal{R}e \frac{C_r}{R} \left[\frac{\partial \bar{u}}{\partial \bar{t}} + \left(\bar{u} \frac{\partial \bar{u}}{\partial \bar{x}} + \bar{v} \frac{\partial \bar{u}}{\partial \bar{y}} + \bar{w} \frac{\partial \bar{u}}{\partial z} \right) \right]$$

$$= -\frac{\partial \bar{p}}{\partial \bar{x}} + \frac{\partial^2 \bar{u}}{\partial \bar{y}^2} + \left(\frac{C_r}{R}\right)^2 \left(\frac{\partial^2 \bar{u}}{\partial \bar{x}^2} + \frac{\partial^2 \bar{u}}{\partial \bar{z}^2} \right),$$

$$\left(\frac{C_r}{R}\right)^2 \left[\frac{C_r^2 \Omega}{\nu} \frac{\partial \bar{v}}{\partial \bar{t}} + \mathcal{R}e \frac{C_r}{R} \left(\bar{u} \frac{\partial \bar{v}}{\partial \bar{x}} + \bar{v} \frac{\partial \bar{v}}{\partial \bar{y}} + \bar{w} \frac{\partial \bar{v}}{\partial z} \right) \right.$$

$$\left. - \frac{\partial^2 \bar{v}}{\partial \bar{y}^2} - \left(\frac{C_r}{R}\right)^2 \left(\frac{\partial^2 \bar{v}}{\partial \bar{x}^2} + \frac{\partial^2 \bar{v}}{\partial \bar{z}^2} \right) \right] = -\frac{\partial \bar{p}}{\partial \bar{y}},$$

$$\mathcal{R}e \frac{C_r}{R} \left[\frac{\partial \bar{w}}{\partial \bar{t}} + \left(\bar{u} \frac{\partial \bar{w}}{\partial \bar{x}} + \bar{v} \frac{\partial \bar{w}}{\partial \bar{y}} + \bar{w} \frac{\partial \bar{w}}{\partial z} \right) \right]$$

$$= -\frac{\partial \bar{p}}{\partial \bar{z}} + \frac{\partial^2 \bar{w}^2}{\partial \bar{y}^2} + \left(\frac{C_r}{R}\right)^2 \left(\frac{\partial^2 \bar{w}}{\partial \bar{x}^2} + \frac{\partial^2 \bar{w}}{\partial \bar{z}^2} \right).$$

Neglecting second-order terms in (C_r/R) gives

$$\mathscr{R}e\left(\frac{C_r}{R}\right)\left[\frac{\partial \bar{u}}{\partial \bar{t}} + \left(\bar{u}\frac{\partial \bar{u}}{\partial \bar{x}} + \bar{v}\frac{\partial \bar{v}}{\partial \bar{y}} + \bar{w}\frac{\partial \bar{u}}{\partial \bar{z}}\right)\right] = -\frac{\partial \bar{p}}{\partial \bar{x}} + \frac{\partial^2 \bar{u}}{\partial \bar{y}^2},$$

$$0 = -\frac{\partial \bar{p}}{\partial \bar{y}},$$

$$\mathscr{R}e\left(\frac{C_r}{R}\right)\left[\frac{\partial \bar{w}}{\partial \bar{t}} + \left(\bar{u}\frac{\partial \bar{w}}{\partial \bar{x}} + \bar{v}\frac{\partial \bar{w}}{\partial \bar{y}} + \bar{w}\frac{\partial \bar{w}}{\partial \bar{z}}\right)\right] = -\frac{\partial \bar{p}}{\partial \bar{z}} + \frac{\partial^2 \bar{w}^2}{\partial \bar{y}^2}. \quad (3.6)$$

Hence, because of the small (C_r/R) ratio, the pressure gradient across the film is entirely negligible.

The normal practice in analyzing hydrodynamic bearings is to neglect both the temporal and convective acceleration terms on the left of Eq. (3.8). These terms are arguably small provided that $\mathscr{R}e(C_r/R)$ is less than one. Given that C_r/R is generally on the order of 0.001, the inertia terms become significant if $\mathscr{R}e \geq 1000$. As we shall see in Chapter 4, for liquid seals both C_r/R and $\mathscr{R}e$ are larger, and the convective and temporal acceleration terms must be retained. For the analysis of bearings, these terms can be neglected and the dimensional governing equations are

$$\frac{\partial^2 u}{\partial y^2} = \frac{1}{\mu}\frac{\partial p}{\partial x}, \qquad \frac{\partial^2 w}{\partial y^2} = \frac{1}{\mu}\frac{\partial p}{\partial z}, \qquad (3.7)$$

where μ is the absolute viscosity.

Having arrived at the equilibrium equation (3.7) and continuity equation (3.2), we will now complete the development of the Reynolds equation following the text by Pinkus and Sternlicht (1961). First, we will integrate Eqs. (3.7) with respect to y, employing the boundary conditions of Figure 3.2, viz.,

$$y = 0: \quad u = U_1, \quad v = V_1, \quad w = 0,$$
$$y = H: \quad u = U_2, \quad v = V_2, \quad w = 0.$$

Assuming constant viscosity, Eq. (3.7) can be integrated in terms of these boundary conditions to yield the following velocity-field definitions:

$$u = \frac{1}{2\mu}\frac{\partial p}{\partial x}y(y - H) + \left[\left(\frac{H - y}{H}\right)U_1 + \frac{y}{H}U_2\right],$$

$$w = \frac{1}{2\mu}\frac{\partial p}{\partial z}y(y - H). \qquad (3.8)$$

Note that u is the sum of (a) flow due to the circumferential pressure gradient $\partial p / \partial x$, and (b) shear flow due to the no-slip boundary conditions. Note further that the pressure gradients $\partial p / \partial x$, $\partial p / \partial z$ are not functions of y.

The governing equations (3.2) and (3.8) involve the dependent variables u, v, w, and p. Since H is in general a function of x, z, and t, the independent variables are x, y, z, t. We propose to combine and modify Eqs. (3.2) and (3.8) to obtain a single equation in the pressure variable p with independent variables x, z, t.

Substitution from Eqs. (3.8) into (3.2) yields

$$\frac{\partial v}{\partial y} = -\frac{\partial}{\partial x}\left[\frac{1}{2\mu}\frac{\partial p}{\partial x}y(y - H)\right] - \frac{\partial}{\partial z}\left[\frac{1}{2\mu}\frac{\partial p}{\partial z}y(y - H)\right]$$
$$- \frac{\partial}{\partial x}\left[\left(\frac{H - y}{H}\right)U_1 + \frac{y}{H}U_2\right].$$

Integrating with respect to y from $y = 0$ to $y = H$ yields

$$V_2 - V_1 = -\int_0^H \frac{\partial}{\partial x}\left[\frac{1}{2\mu}\frac{\partial p}{\partial x}y(y - H)\right]dy - \int_0^H \frac{1}{2\mu}\frac{\partial}{\partial z}\left[\frac{\partial p}{\partial z}y(y - H)\right]dy$$
$$- \int_0^H \frac{\partial}{\partial x}\left[\left(\frac{H - y}{H}\right)U_1 + \frac{y}{H}U_2\right]dy. \tag{3.9}$$

This step effectively eliminates y as an independent variable. Applying Leibniz's rule for the differentiation of integrals,

$$\frac{d}{d\alpha}\int_0^{u(\alpha)}f(x, \alpha)\,dx = \int_0^{u(\alpha)}\frac{df}{d\alpha}(x, \alpha)\,dx + f(u, \alpha)\frac{du}{d\alpha},$$

to the integrals of Eq. (3.9) gives

$$V_2 - V_1 = -\frac{\partial}{\partial x}\int_0^H \frac{1}{2\mu}\frac{\partial p}{\partial x}y(y - H)\,dy - \frac{\partial}{\partial z}\int_0^H \frac{1}{2\mu}\frac{\partial p}{\partial z}y(y - H)\,dy$$
$$- \frac{\partial}{\partial x}\int_0^H\left[\left(\frac{H - y}{H}\right)U_1 + \frac{y}{H}U_2\right]dy + U_2\frac{\partial H}{\partial x}.$$

Since μ, $\partial p / \partial x$, and $\partial p / \partial z$ are independent of y, the indicated integrations may be readily completed to yield the following laminar-flow Reynolds equation for an isoviscous incompressible fluid:

$$\frac{\partial}{\partial x}\left(\frac{H^3}{\mu}\frac{\partial p}{\partial x}\right) + \frac{\partial}{\partial z}\left(\frac{H^3}{\mu}\frac{\partial p}{\partial z}\right) = 12(V_2 - V_1) + 6(U_1 - U_2)\frac{\partial H}{\partial x}$$

$$+ 6H\frac{\partial}{\partial x}(U_1 + U_2). \tag{3.10}$$

The first, second, and third terms on the right-hand side of this equation are generally characterized as "squeeze," "wedge," and "stretch," respectively. With respect to the bearing configuration of Figures 3.1 and 3.2, the squeeze and wedge terms result from relative radial and tangential velocities, respectively. The results of the following journal-bearing discussion demonstrates that the stretch term is negligible.

Reynolds Equation in Circumferential Coordinates

The objective of the present development is the restatement of the Reynolds equation (3.10) in terms of circumferential coordinates, which are more suitable for the analysis of journal bearings. First, the circumferential variable x is to be replaced by $R\theta$. The angle θ is illustrated in Figure 3.1 and is a counterclockwise rotation originating at the line of centers. Note that the circumferential displacement $R\theta$ corresponds to rotation about the center of the bearing, *not* the center of the journal. The replacement of x by $R\theta$ in Eq. (3.10) yields

$$\frac{1}{R^2}\frac{\partial}{\partial \theta}\left(\frac{H^3}{\mu}\frac{\partial p}{\partial \theta}\right) + \frac{\partial}{\partial z}\left(\frac{H^3}{\mu}\frac{\partial p}{\partial z}\right) = 12V_2 - \frac{6U_2}{R}\frac{\partial H}{\partial \theta} + \frac{6H}{R}\frac{\partial U_2}{\partial \theta}, \tag{3.11}$$

where, for the purposes of this development, the bearing is assumed to be stationary; i.e., $U_1 = V_1 = 0$ in Eq. (3.10).

To complete the conversion of this equation, the variables V_2, U_2, and H must be redefined in terms of θ. We assume that the journal and bearing remain parallel to the z axis illustrated in Figure 3.1. Hence, H is a function of θ only. From Figure 3.1, one observes

$$(R + H)^2 - e^2 - 2(R + C_r)e\cos\theta = (R + C_r)^2. \tag{3.12}$$

Expanding this result, and discarding higher-order terms in e, H/R, and C_r/R yields

$$H = C_r(1 + \varepsilon\cos\theta) = C_r h, \tag{3.13}$$

where ε and h are, respectively, the eccentricity ratio and normalized film

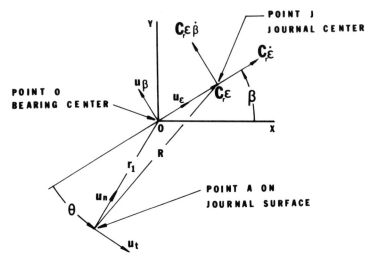

Figure 3.3 Kinematics for the definition of U_2 and V_2 of Eq. (3.11) in polar coordinates [Childs et al. (1977)].

thickness defined by

$$\varepsilon = e/C_r, \qquad h = H/C_r = 1 + \varepsilon \cos \theta. \qquad (3.14)$$

An eccentricity ratio of zero implies that the journal is centered within the bearing, while an eccentricity ratio of one implies that the journal and bearing surfaces are in contact. Figure 3.3 illustrates the additional kinematics required to define U_2 and V_2 in terms of θ. Note that the journal-center velocity is defined by

$$V_J = u_\varepsilon C_r \dot{\varepsilon} + u_\beta C_r \varepsilon \dot{\beta} = \left(C_r \dot{\varepsilon} \cos \theta + C_r \varepsilon \dot{\beta} \sin \theta \right) u_n$$

$$+ \left(C_r \dot{\varepsilon} \sin \theta - C_r \varepsilon \dot{\beta} \cos \theta \right) u_t,$$

where β is the attitude angle of $C_r \varepsilon$ relative to the x axis. The journal rotates about its center with an angular velocity ω_j. The definition of the velocity V_A of point A (an arbitrary point on the journal surface) is required. The variables U_2, V_2 of Eq. (3.11) are, respectively, the u_t and u_n components of V_A. The basic kinematic relationship between the velocity of points A and J is

$$V_A = V_J + K\omega_j \times r_{JA}, \qquad (3.15)$$

where r_{JA} is the vector from point J to point A, defined by

$$r_{JA} = -(r_1 + C_r\varepsilon \cos \theta)u_n - C_r\varepsilon \sin \theta u_t, \tag{3.16}$$

In this relationship, r_1 is the magnitude of the line segment OA, and is defined from Figure 3.3 by

$$r_1^2 + 2r_1 C_r\varepsilon \cos \theta + (C_r\varepsilon)^2 = R^2.$$

Dropping second- and higher-order terms in $C_r\varepsilon$ yields

$$r_1 = R - C_r\varepsilon \cos \theta.$$

Substitution of this result into Eqs. (3.15) and (3.16) yields

$$V_A = V_2 u_n + U_2 u_t,$$

where

$$V_2 = C_r\dot{\varepsilon} \cos \theta + C_r\varepsilon\left(\dot{\beta} - \omega_j\right)\sin \theta,$$

$$U_2 = R\omega_j + C_r\dot{\varepsilon} \sin \theta - C_r\varepsilon\dot{\beta} \cos \theta. \tag{3.17}$$

Substituting these definitions into Eq. (3.11), and discarding second-order terms in ε, $\dot{\varepsilon}$, and H/R yields the following restatement of the Reynolds equation in circumferential coordinates:

$$\frac{\partial}{\partial\theta}\left(\frac{h^3}{\mu}\frac{\partial p}{\partial\theta}\right) + R^2\frac{\partial}{\partial z}\left(\frac{h^3}{\mu}\frac{\partial p}{\partial z}\right) = \frac{12R^2}{C_r^3}\left[C_r\dot{\varepsilon} \cos \theta + C_r\varepsilon\left(\dot{\beta} - \bar{\omega}\right)\sin \theta\right], \tag{3.18}$$

where $\bar{\omega} = \omega_j/2$. The normalized film thickness definition of Eq. (3.14) is also used in obtaining Eq. (3.18). The "stretch" effect represented by the term $(6H/R)(\partial U_2/\partial\theta)$ of Eq. (3.11) is demonstrably second order [from the U_2 definition of Eq. (3.17)], and makes no contribution to Eq. (3.18).

Equation (3.18) may also be stated

$$\frac{\partial}{\partial\theta}\left(\frac{h^3}{\mu}\frac{\partial p}{\partial\theta}\right) + R^2\frac{\partial}{\partial z}\left(\frac{h^3}{\mu}\frac{\partial p}{\partial z}\right) = 12\left(\frac{R}{C_r}\right)^2\left(\bar{\omega}\frac{\partial h}{\partial\theta} + \frac{\partial h}{\partial t}\right). \tag{3.19}$$

The correctness of this version is demonstrated by expanding the right-hand terms as

$$\frac{\partial h}{\partial \theta} = -C_r \varepsilon \sin \theta, \qquad \frac{\partial h}{\partial t} = C_r \dot{\varepsilon} \cos \theta - C_r \varepsilon \sin \theta \dot{\theta}.$$

Equation (3.18) is then obtained from Eq. (3.19) if $\dot{\theta} = -\dot{\beta}$. The correctness of this latter requirement is confirmed by noting that the angular position of a fixed point on the bearing may be located with respect to the X axis by $\Theta = \beta + \pi + \theta$. Since $\dot{\Theta} = 0$, $\dot{\theta} = -\dot{\beta}$.

3.3 IMPEDANCE DESCRIPTIONS FOR PLAIN JOURNAL BEARINGS

Since inertia terms are dropped in deriving the Reynolds equation, it may be expressed without any basic change in format in coordinate systems which are accelerating or rotating relative to the reference system of derivation. This situation obviously differs from the relationships of Newtonian mechanics which are valid only in a fixed, inertial-reference system. For the purposes of this discussion, the X, Y, Z coordinate system of Figure 3.4 is defined to be inertial, and the following constant-viscosity statement of Eq. (3.19) in this coordinate system will be employed:

$$\frac{\partial}{\partial \theta}\left(h^3 \frac{\partial p}{\partial \theta}\right) + R^2 \frac{\partial}{\partial z}\left(h^3 \frac{\partial p}{\partial z}\right) = \frac{12\mu R^2}{C_r^3}\left[C_r \dot{\varepsilon} \cos \theta + C_r \varepsilon (\dot{\beta} - \bar{\omega})\sin \theta\right].$$

(3.20)

The kinematic variables of this equation are illustrated in Figures 3.3 and 3.4.

The impedance descriptions to be derived and stated in this section define the reaction forces developed by a plain journal bearing as a consequence of the journal motion, and are taken from Childs et al. (1977). An impedance

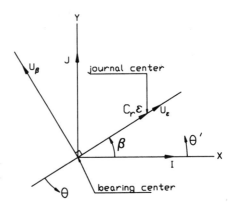

Figure 3.4 Plain journal-bearing kinematic variables, with X, Y, Z inertial reference system [Childs et al. (1977)].

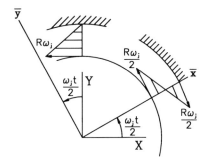

Figure 3.5 $\bar{x}, \bar{y}, \bar{z}$ coordinate system which is rotating at $\bar{\omega} = \omega_j/2$ relative to the XYZ system.

vector provides a complete static and dynamic description of a journal bearing. The impedance vector is an outgrowth of mobility descriptions developed by Booker (1965a, 1971) and Blok (1965) for the analysis of dynamically loaded journal bearings in internal-combustion engines. The mobility description defines the journal velocity vector as a function of its load and position vectors.

The physical interpretation provided by Booker and Blok for the terms on the right-hand side of Eq. (3.20) contributes significantly to an appreciation of the development of either mobility or impedance vectors and goes as follows. The velocity of the journal center relative to the stationary X, Y, Z system is

$$\mathbf{V}_J = \frac{dC_r\mathbf{\varepsilon}}{dt}\bigg|X, Y, Z = \mathbf{u}_\varepsilon C_r\dot{\varepsilon} + \mathbf{u}_\beta C_r\varepsilon\dot{\beta}.$$

The time rate of change of the vector $C_r\mathbf{\varepsilon}$ with respect to a coordinate system that has an angular velocity of $\mathbf{K}\bar{\omega}$ relative to the X, Y, Z system is denoted by $\mathbf{V}_S$, and is related to $\mathbf{V}_J$ by

$$\mathbf{V}_S = \mathbf{V}_J - \mathbf{K}\bar{\omega} \times C_r\mathbf{\varepsilon} = \mathbf{u}_\varepsilon C_r\dot{\varepsilon} + \mathbf{u}_\beta C_r\varepsilon\left(\dot{\beta} - \bar{\omega}\right). \tag{3.21}$$

Hence, the terms $C_r\dot{\varepsilon}, C_r\varepsilon(\dot{\beta} - \bar{\omega})$ of Eq. (3.20) are the velocity components of the journal center with respect to a coordinate system that is rotating at an angular velocity of $\mathbf{K}\bar{\omega}$ relative to the stationary X, Y, Z system.

Figure 3.5 illustrates the tangential velocity distribution which would be observed by observers in the stationary X, Y, Z and rotating $\bar{x}, \bar{y}, \bar{z}$ systems. An observer in the stationary system sees zero velocity on the stator surface and $R\omega_j$ on the rotor surface, while an observer in the $\bar{x}, \bar{y}, \bar{z}$ system sees $R\omega_j/2$ on the rotor and $-(R\omega_j/2)$ on the stator. Hence, in the rotating coordinate system, the average tangential velocity (across the film) is zero.[*]

Booker and Blok note that the journal center's motion would always appear to be in a state of pure squeezing to an observer in this type of

---

[*] Black (1969), Chapter 4, used this insight in his initial analysis of annular seals.

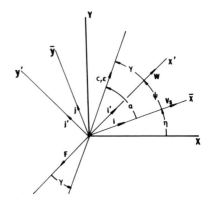

Figure 3.6 Kinematic variables for impedance definition [Childs et al. (1977)].

rotating system, and $\mathbf{V}_S$ is accordingly denoted as the journal's pure-squeeze-velocity vector. The practical consequence of this observation is that a suitable coordinate transformation enables one to completely define a bearing in terms of its pure-squeeze-velocity characteristics. In other words, a bearing's angular velocity ω_j does not enter its impedance definition.

Figure 3.6 illustrates two additional coordinate systems which will be useful in our impedance definitions. The $\bar{x}, \bar{y}, \bar{z}$ system is fixed relative to $\mathbf{V}_S$, while the x', y', z' system is fixed relative to the bearing reaction force $\mathbf{F}$. The force $\mathbf{F}$ is applied from the fluid film to the journal (or rotor), and is defined by the impedance vector as a function of journal motion.

Given that the velocity and position of the journal center relative to the stationary coordinate system are defined by

$$C_r\mathbf{\varepsilon} = \mathbf{I}X + \mathbf{J}Y = C_r\varepsilon(\mathbf{I}\cos\beta + \mathbf{J}\sin\beta),$$

$$\mathbf{V}_J = \mathbf{I}\dot{X} + \mathbf{J}\dot{Y} = \mathbf{u}_\varepsilon C\dot{\varepsilon} + \mathbf{u}_\beta C\varepsilon\dot{\beta}, \tag{3.22}$$

the pure-squeeze-velocity vector is defined in the stationary X, Y, Z system by

$$\mathbf{V}_S = \mathbf{V}_J - (\mathbf{K}\bar{\omega} \times C_r\mathbf{\varepsilon}) = \mathbf{I}(\dot{X} + \bar{\omega}Y) + \mathbf{J}(\dot{Y} - \bar{\omega}X). \tag{3.23}$$

Alternatively, from Figure 3.6,

$$\mathbf{V}_S = V_S(\mathbf{u}_\varepsilon \cos\alpha - \mathbf{u}_\beta \sin\alpha) = V_S(\mathbf{I}\cos\eta + \mathbf{J}\sin\eta). \tag{3.24}$$

Substitution from Eqs. (3.24) and (3.21) yields the following restatement of the Reynolds equation (3.20):

$$\frac{\partial}{\partial\theta}\left(h^3\frac{\partial p}{\partial\theta}\right) + R^2\frac{\partial}{\partial z}\left(h^3\frac{\partial p}{\partial z}\right) = \frac{12\mu R^2 V_S}{C_r^3}\cos(\alpha + \theta). \tag{3.25}$$

The Ocvirk (Short) Bearing Solution

The Ocvirk (1952) solution to Eq. (3.25) is obtained by neglecting the first term on the left, and solving for $p(\theta, z)$ with the boundary conditions $p(\theta, L/2) = p(\theta, -L/2) = 0$, to obtain

$$p(\theta, z) = -3\mu L^2 \left[1 - (2z/L)^2\right] V_S \cos(\alpha + \theta)/2C_r^3 h^3. \quad (3.26)$$

Integrating axially yields the following average pressure definition:

$$P(\theta) = \frac{1}{L} \int_{-(L/2)}^{L/2} p(\theta, z)\, dz = -\mu L^2 V_S \cos(\alpha + \theta)/C_r^3 h^3. \quad (3.27)$$

The pressure is seen to be positive between angles θ_1, θ_2 defined by

$$\theta_1 = \frac{\pi}{2} - \alpha, \qquad \theta_2 = \frac{3\pi}{2} - \alpha.$$

In words, the pressure is positive over a region of π radians centered about $\mathbf{V}_S$.

The Ocvirk approximation generally applies for short bearings having an $L/D \le 0.5$ and an eccentricity ratio less than 0.7. The validity of this approximation becomes more apparent when the Reynolds equation is restated in terms of the following dimensionless variables:

$$z^* = z/L, \qquad p^* = p/12\mu\omega_j \left(\frac{L}{C_r}\right)^2,$$

$$V_s^* = V_s/C_r\omega_j$$

to be

$$4\left(\frac{L}{D}\right)^2 \frac{\partial}{\partial\theta}\left(h^3 \frac{\partial p^*}{\partial\theta}\right) + \frac{\partial}{\partial z^*}\left(h^3 \frac{\partial p^*}{\partial z^*}\right) = V_s^* \cos(\alpha + \theta). \quad (3.28)$$

Clearly, the first term becomes negligible for small L/D ratios. In physical terms, the pressure-gradient contribution to the circumferential velocity component of Eq. (3.8) is neglected.

Equation (3.28) also demonstrates that the first term on the left becomes dominant for an infinitely long bearing, which leads to the following Sommerfeld-solution development.

Sommerfeld (Long) Bearing Solution

The Sommerfeld solution to the Reynolds equation is obtained by dropping the second term on the left in Eq. (3.20), and integrating with respect to θ to obtain

$$P(\theta) = -6\mu R^2 V_S(\cos \alpha \cos \theta - b \sin \alpha \sin \theta)(2 + \varepsilon \cos \theta)/C_r^3 h^2, \quad (3.29)$$

where

$$b = 2/(2 + \varepsilon^2).$$

From Eq. (3.29), the positive pressure sector lies between the angles θ_1, θ_2, defined by

$$b \tan \theta_1 = (\tan \alpha)^{-1}, \qquad \theta_2 = \theta_1 + \pi. \quad (3.30)$$

This solution is taken from Booker (1965a), who credits it to Gross (1962). The boundary conditions used to obtain Eq. (3.29) are (a) periodicity with respect to θ; i.e., $P(\theta) = P(\theta + 2\pi)$, and (b) the requirement that the positive pressure segment extend over π radians; i.e., $P(\theta_1) = P(\theta_1 + \pi) = 0$.

Impedance Descriptions

The forces acting on the journal can be obtained by integrating these pressure distributions. A complete film "2π" bearing is obtained by integrating over $[0, 2\pi]$, while a ruptured-film "π" bearing is obtained by integrating only the positive portion of the film, i.e., $[\theta_1, \theta_2]$. The π bearing provides an approximate model for cavitation, since the assumption is made that film rupture prevents the development of negative pressures, and tends to be the more generally applicable model.

Integration of the pressure distribution from the preceding section yields the following definition for the force components parallel and normal to the eccentricity vector:

$$F_\varepsilon = RL \int P(\theta)\cos \theta \, d\theta,$$

$$F_\beta = RL \int P(\theta)\sin \theta \, d\theta, \quad (3.31)$$

which can be stated

$$F_\varepsilon = -V_S 2\mu L(R/C_r)^3 W_\varepsilon(\varepsilon, \alpha),$$

$$F_\beta = -V_S 2\mu L(R/C_r)^3 W_\beta(\varepsilon, \alpha). \quad (3.32)$$

The quantities W_ε, W_β are the desired components of the impedance vector $\mathbf{W}$, whose vector character is emphasized by the following restatement of Eq. (3.32):

$$\mathbf{u}_\varepsilon F_\varepsilon + \mathbf{u}_\beta F_\beta = -V_S 2\mu L (R/C_r)^3 (\mathbf{u}_\varepsilon W_\varepsilon + \mathbf{u}_\beta W_\beta). \tag{3.33}$$

For computational purposes, the following definition of **W** in terms of its components parallel and normal to the squeeze-velocity vectors V_S is convenient:

$$W_{\bar{x}} = W_\varepsilon \cos\alpha - W_\beta \sin\alpha,$$
$$W_{\bar{y}} = W_\varepsilon \sin\alpha + W_\beta \cos\alpha. \tag{3.34}$$

Figure 3.7 provides impedance plots for the Ocvirk π and 2π solutions in terms of these components. A similar definition of the components of **W** in the X, Y system is readily stated in terms of the angle β.

The impedance descriptions for the Ocvirk (short) bearing model are obtained by substitution from Eq. (3.27) into Eqs. (3.31) and (3.32) to obtain

$$W_\varepsilon = 2(J_3^{20} \cos\alpha - J_3^{11} \sin\alpha)(L/D)^2,$$
$$W_\beta = 2(J_3^{11} \cos\alpha - J_3^{20} \sin\alpha)(L/D)^2, \tag{3.35}$$

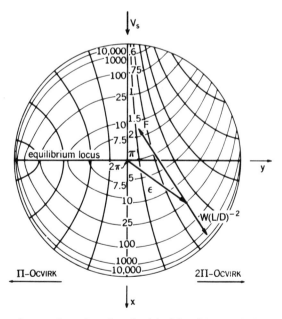

Figure 3.7 Impedance plots for the Ocvirk (short) π and 2π bearing models; $L/D = 1$ [Childs et al. (1977)].

where*

$$J_m^{jk} = \int \sin^j \theta \cos^k \theta h^{-m} \, d\theta.$$ (3.36)

The solutions for both the π and 2π bearings are defined by this relationship. They differ only in bounds of integration used, viz., $[0, 2\pi]$ for the 2π bearing and $[\theta_1, \theta_2]$ for the π bearings. Hence, for the π bearing the integrals are a function of both α and ε, while for the 2π bearing they depend only on ε. For small ε, the π bearing impedance reduces to

$$W_{\bar{x}} = 2(\pi/2 + 4\varepsilon \cos \alpha)(L/D)^2,$$

$$W_{\bar{y}} = 2(2\varepsilon \sin \alpha)(L/D)^2.$$ (3.37)

For the Sommerfeld (long) bearing, the following impedance components are obtained by substitution from Eq. (3.29) into Eqs. (3.31) and (3.32):

$$W_\varepsilon = 3(B_{11} \cos \alpha - B_{21} \sin \alpha),$$

$$W_\beta = 3(B_{21} \cos \alpha - B_{22} \sin \alpha),$$ (3.38)

where

$$B_{11} = 2J_2^{02} + \varepsilon J_2^{03}, \qquad B_{21} = 2J_2^{11} + \varepsilon J_2^{12}, \qquad B_{22} = b(2J_2^{20} + \varepsilon J_2^{21}).$$ (3.39)

As with the short-bearing solution, both the π and 2π impedance vectors are defined by these results and differ only in the integration bounds used to evaluate the integrals of Eq. (3.39). For small ε, the π bearing solution reduces to

$$W_{\bar{x}} = 3(\pi + 4\varepsilon \cos \alpha), \qquad W_{\bar{y}} = 3(2\varepsilon \sin \alpha),$$ (3.40)

which resembles the short-bearing result of Eq. (3.37).

---

*Booker (1965b) provides a convenient summary of these integrals; however, the following alternative definition for J_1^{00} is useful:

$$J_1^{00} = 2(1 - \varepsilon^2)^{-1/2} \left\{ n\pi + \tan^{-1}\left[(1 - \varepsilon)^{1/2}(1 + \varepsilon)^{-1/2} \tan\left(\frac{\theta}{2}\right) \right] \right\}(2n - 1),$$

$$\pi \le \theta \le (2n + 1)\pi.$$

Moes Finite-Length Cavitated Impedance

The following two asymptotic solutions for plain journal bearings have proven to be useful in the mobility analysis of transient bearing phenomena:

(a) the Ocvirk (short) bearing solution for small eccentricity ratios, and
(b) the Sommerfeld (long) bearing solution for large eccentricity ratios.

Individually, these asymptotic solutions have a limited value which is consistent with their restricted range of application in both L/D (length to diameter) and eccentricity ratios. Fortunately, the fact that their ranges of application do not coincide means that the two (vector) solutions can be combined in such a way that an approximate solution is obtained which is valid for general finite-length bearings at both large and small eccentricity ratios.

This method has been used in the development of analytic definitions for finite-length mobilities and is based on the observed fact that a vectoral sum of the Ocvirk and Sommerfeld mobilities provides an excellent approximation for the actual (numeric) mobility vector for all eccentricity and L/D ratios. Hence, Moes used a weighted sum of the asymptotic solutions cited above to obtain an accurate finite-length analytic mobility vector.

This technique was used to directly derive cavitated and 2π finite-length impedance descriptions as presented in Childs et al. (1977). The cavitated impedance vector is particularly useful, and is reviewed below. A definition for the vector **W** is required and will be obtained by stating W and γ in terms of ε and α. The angle γ is defined in terms of ε and α by

$$\gamma = \left[\left(1 - a(1 - b^2)^{-1/2}\right) \right] \left[\tan^{-1} A - \frac{\pi}{2} b \bigg/ |b| + \sin^{-1} b \right] + \alpha - \sin^{-1} b,$$

$$\text{(3.41)}$$

$$A = 4(1 + 2.12B)(1 - b^2)^{1/2}/3(1 + 3.60B)b$$
$$B = (1 - \varepsilon^2)(L/D)^{-2}$$
$$a = \varepsilon \cos \alpha, \qquad b = \varepsilon \sin \alpha.$$

The amplitude W can now be expressed by

$$W = \left[0.15(E^2 + G^2)^{1/2}(1 - a')^{3/2} \right]^{-1}, \qquad \text{(3.42)}$$

$$E = 1 + 2.12Q, \qquad G = 3b'(1 + 3.6Q)/4(1 - a'),$$
$$Q = (1 - a')(L/D)^{-2},$$
$$a' = \varepsilon \cos \gamma, \qquad b' = \varepsilon \sin \gamma.$$

From Figure 3.6, the impedance components parallel and perpendicular to the squeeze velocity $\mathbf{V}_S$ are

$$W_{\bar{x}} = W \cos \psi, \qquad W_{\bar{y}} = W \sin \psi, \qquad \psi = \alpha - \gamma. \qquad (3.43)$$

There have been numerous finite-length bearing models developed in addition to the Moes model above. Barrett et al. (1978) proposed a description based on the short-bearing model with an end-leakage correction which is particularly attractive for $L/D < 1.25$.

3.4 STATIC CHARACTERISTICS, AND STIFFNESS AND DAMPING COEFFICIENTS FOR PLAIN JOURNAL BEARINGS

The definition of an equilibrium position for the journal of a bearing due to a fixed-direction applied load is the initial objective of this section. The second objective is the development of linear stiffness and damping coefficients for small motion of the journal about its equilibrium position.

Equilibrium Definition

The kinematic requirement for equilibrium is that the journal-bearing velocity relative to the stationary X, Y, Z system, $\mathbf{V}_J$, vanish; i.e., $C\dot{\varepsilon} = C\varepsilon\dot{\beta} = 0$. From Eq. (3.23), this yields the squeeze-velocity definition $\mathbf{V}_{S0} = -\mathbf{K}\bar{\omega} \times C\boldsymbol{\varepsilon} = -C\varepsilon\bar{\omega}\mathbf{u}_\beta$, which physically means that the $\mathbf{u}_\varepsilon$ squeeze-velocity component is zero. Hence, from Figure 3.6, $\alpha = \alpha_0 = \pi/2$ at equilibrium. For the impedance plots of Figure 3.7, $\alpha = \pi/2$ is the line $\bar{x} = 0$, i.e., the $\bar{y}$ axis. This statement can be better appreciated by noting that $\mathbf{V}_S$ is parallel to the $\bar{x}$ axis, and only perpendicular to the eccentricity vector on the $\bar{y}$ axis.

To define the journal equilibrium position, we assume (without loss of generality) that the applied static load is directed along the $-Y$ axis. Hence, the reaction load $\mathbf{F}_0$ is directed along the $+Y$ axis, and from Figures 3.6 and 3.8(a), the equilibrium position is defined by the eccentricity magnitude ε_0 and attitude angle γ_0. The direct solution for ε_0 is inherently nonlinear, and its solution is parametrized in terms of the Sommerfeld number definition:*

$$S \doteq \mu \left(\frac{\omega_j}{2\pi} \right) \left(\frac{R}{C_r} \right)^2 \frac{DL}{F_0} = 1/\pi\varepsilon_0 W_0 > 0,$$

$$W_0 = W(\varepsilon_0, \pi/2). \qquad (3.44)$$

---

*Equation (3.44) is the "American" definition for S. Europeans normally use $S = \varepsilon_0 W_0/2 = F_0(C_r/R)^2/2\mu\bar{\omega}LD$. Childs et al. (1977) provide an $S - \varepsilon_0$ plot using this definition.

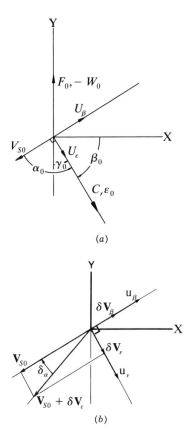

(a)

(b)

Figure 3.8 Equilibrium conditions for a plain journal bearing [Childs et al. (1977)].

This result is obtained by substituting the static force and squeeze-velocity magnitudes into the impedance definition of Eq. (3.32). Figure 3.9 illustrates the form of Eq. (3.44) for the Moes cavitated impedance of Eqs. (3.41) and (3.42). The curves of this figure were obtained by calculating S as a function of ε_0 and $W(\varepsilon_0, \pi/2)$. For a given applied-load magnitude F_0 and bearing, this figure can be used to determine the steady-state eccentricity magnitude ε_0. Specifically, one calculates S in terms of F_0 and the bearing parameters via Eq. (3.44) and then uses Figure 3.9 to determine ε_0.

The equilibrium attitude angle γ_0 is defined from Figure 3.8(a) by

$$\tan \gamma_0 = -W_\beta(\varepsilon_0, \pi/2)/W_\varepsilon(\varepsilon_0, \pi/2), \qquad (3.45)$$

and completes the definition for the journal's equilibrium locus. Figure 3.10 illustrates the form of this locus for the Moes cavitating impedance, demonstrating that a downward load yields horizontal as well as vertical displacements. Further, for light loads (large S and small ε_0), horizontal displacements considerably exceed vertical displacements. The results of

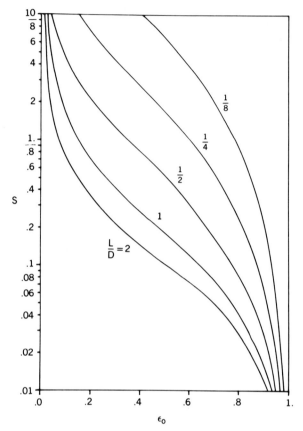

Figure 3.9 Sommerfeld number solution $S = S(\varepsilon_0)$ from Eq. (3.44) for the Moes cavitating impedance of Eqs. (3.41) and (3.42).

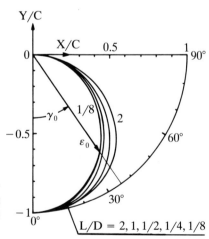

Figure 3.10 Bearing equilibrium locus definition of Eq. (3.45) for the Moes cavitating impedance of Eqs. (3.41) and (3.42) [Childs et al. (1977)].

Figure 3.10 are bizarre when compared to conventional structural results, and as demonstrated in the following section, yield comparably peculiar stiffness and damping coefficients.

The results of Figures 3.9 and 3.10 coincide with previous finite-length calculations [Pinkus and Sternlicht (1961)] and provide a static verification for the Moes cavitated impedance.

Stiffness and Damping Coefficients

The development of linear stiffness and damping coefficients for small motion about a journal's equilibrium position is the objective of our present developments. Stiffness and damping coefficients are required for both synchronous response calculations and linear stability analysis. The desired coefficients yield the following definition for the bearing reaction force:

$$
\begin{Bmatrix} F_X \\ F_Y \end{Bmatrix} = -[K_{I,J}]\begin{Bmatrix} X \\ Y \end{Bmatrix} - [C_{I,J}]\begin{Bmatrix} \dot{X} \\ \dot{Y} \end{Bmatrix},
\tag{3.46}
$$

where $[K_{I,J}]$ and $[C_{I,J}]$ are matrices of stiffness and damping coefficients.

At equilibrium, the $\mathbf{u}_\varepsilon, \mathbf{u}_\beta$ reference system has a fixed position relative to the stationary X, Y, Z system. Hence, stiffness and damping coefficients will be defined in terms of the $\mathbf{u}_\varepsilon, \mathbf{u}_\beta$ reference, and the similarity transformation

$$
[K_{I,J}] = [R_0]^T[K_{\varepsilon,\beta}][R_0], \qquad [C_{I,J}] = [R_0]^T[C_{\varepsilon,\beta}][R_0],
$$

$$
[R_0] = \begin{bmatrix} c\beta_0 & s\beta_0 \\ -s\beta_0 & c\beta_0 \end{bmatrix} = \begin{bmatrix} s\gamma_0 & -c\gamma_0 \\ c\gamma_0 & s\gamma_0 \end{bmatrix}
\tag{3.47}
$$

will then be used to obtain their desired definition with respect to the X, Y axes. In Eq. (3.47), $c\beta_0 = \cos\beta_0$, $s\beta_0 = \sin\beta_0$, $s\gamma_0 = \sin\gamma_0$, $c\gamma_0 = \cos\gamma_0$.

A Taylor-series expansion of the bearing reaction force about the equilibrium position yields

$$
F_\varepsilon(\varepsilon, \beta, \dot{\varepsilon}, \dot{\beta}) = F_\varepsilon(\varepsilon_0, \beta_0, 0, 0) + \frac{1}{C_r}\frac{\partial F_\varepsilon}{\partial \varepsilon}C_r\, d\varepsilon + \frac{1}{C_r\varepsilon}\frac{\partial F_\varepsilon}{\partial \beta}C_r\varepsilon\, d\beta
$$

$$
+ \frac{1}{C_r}\frac{\partial F_\varepsilon}{\partial \dot{\varepsilon}}C_r\, d\dot{\varepsilon} + \frac{1}{C_r\varepsilon}\frac{\partial F_\varepsilon}{\partial \dot{\beta}}C_r\varepsilon\, d\dot{\beta},
$$

$$
F_\beta(\varepsilon, \beta, \dot{\varepsilon}, \dot{\beta}) = F_\beta(\varepsilon_0, \beta_0, 0, 0) + \frac{1}{C_r}\frac{\partial F_\beta}{\partial \varepsilon}C_r\, d\varepsilon + \frac{1}{C_r\varepsilon}\frac{\partial F_\beta}{\partial \beta}C_r\varepsilon\, d\beta
$$

$$
+ \frac{1}{C_r}\frac{\partial F_\beta}{\partial \dot{\varepsilon}}C_r\, d\dot{\varepsilon} + \frac{1}{C_r\varepsilon}\frac{\partial F_\beta}{\partial \dot{\beta}}C_r\varepsilon\, d\dot{\beta}.
\tag{3.48}
$$

where second- and higher-order differential terms have been dropped. By definition, the components of $[K_{\varepsilon,\beta}]$ and $[C_{\varepsilon,\beta}]$ are seen to be

$$K_{\varepsilon\varepsilon} = -\frac{1}{C_r}\frac{\partial F_\varepsilon}{\partial \varepsilon} \qquad K_{\varepsilon\beta} = -\frac{1}{C_r\varepsilon}\frac{\partial F_\varepsilon}{\partial \beta},$$

$$K_{\beta\varepsilon} = -\frac{1}{C_r}\frac{\partial F_\beta}{\partial \varepsilon}, \qquad K_{\beta\beta} = -\frac{1}{C_r\varepsilon}\frac{\partial F_\beta}{\partial \beta},$$

$$C_{\varepsilon\varepsilon} = -\frac{1}{C_r}\frac{\partial F_\varepsilon}{\partial \dot{\varepsilon}}, \qquad C_{\varepsilon\beta} = -\frac{1}{C_r\varepsilon}\frac{\partial F_\varepsilon}{\partial \dot{\beta}},$$

$$C_{\beta\varepsilon} = -\frac{1}{C_r}\frac{\partial F_\beta}{\partial \dot{\varepsilon}}, \qquad C_{\beta\beta} = -\frac{1}{C_r\varepsilon}\frac{\partial F_\beta}{\partial \dot{\beta}}, \qquad (3.49)$$

with the partial derivatives evaluated at the equilibrium position.

The coefficients $K_{\varepsilon\varepsilon}, K_{\beta\varepsilon}$ may be obtained directly from the following restatement of Eq. (3.33):

$$\mathbf{F} = -2\mu L\left(\frac{R}{C_r}\right)^3 V_s\mathbf{W}. \qquad (3.50)$$

The partial derivative of this relationship with respect to ε yields

$$-\frac{1}{C_r}\frac{\partial \mathbf{F}}{\partial \varepsilon} = 2\mu L\left(\frac{R}{C_r}\right)^3\left\{\frac{1}{C_r}\frac{\partial V_s}{\partial \varepsilon}\mathbf{W} + \frac{V_s}{C_r}\frac{\partial \mathbf{W}}{\partial \varepsilon}\right\}. \qquad (3.51)$$

As noted previously, at equilibrium $V_s = V_{s0} = C_r\varepsilon_0\bar{\omega}$; hence from Eqs. (3.49)–(3.51),

$$K_{\varepsilon\varepsilon} = \frac{F_0}{C_r W_0}\left(\frac{W_\varepsilon}{\varepsilon_0} + \frac{\partial W_\varepsilon}{\partial \varepsilon}\right),$$

$$K_{\beta\varepsilon} = \frac{F_0}{C_r W_0}\left(\frac{W_\beta}{\varepsilon_0} + \frac{\partial W_\beta}{\partial \varepsilon}\right). \qquad (3.52)$$

The coefficients $K_{\varepsilon\beta}, K_{\beta\beta}$ are obtained by considering the consequences of a perturbation $\delta\beta_0$ of the equilibrium angle β_0, with ε_0, α_0, and the reaction-force magnitude held constant. The rotation $\delta\beta_0$ yields

$$F_\varepsilon = F_{\varepsilon 0} c(\delta\beta_0) + F_{\beta 0} s(\delta\beta_0) \cong F_{\varepsilon 0} + F_{\beta 0}\delta\beta_0,$$

$$F_\beta = -F_{\varepsilon 0} s(\delta\beta_0) + F_{\beta 0} c(\delta\beta_0) \cong F_{\beta 0} - F_{\varepsilon 0}\delta\beta_0.$$

By comparison to Eq. (3.48), the desired coefficients are

$$K_{\varepsilon\beta} = \frac{-F_{\beta 0}}{C_r \varepsilon_0} = -\frac{F_0}{C_r \varepsilon_0} \frac{W_{\beta 0}}{W_0},$$

$$K_{\beta\beta} = \frac{F_{\varepsilon 0}}{C_r \varepsilon_0} = \frac{F_0}{C_r \varepsilon_0} \frac{W_{\varepsilon 0}}{W_0}. \tag{3.53}$$

Figure 3.8(b) illustrates the two perturbed velocity components $\delta V_\varepsilon = C_r \delta\dot\varepsilon$, $\delta V_\beta = C_r\varepsilon\delta\dot\beta$ and assists in the derivation of damping coefficients. The direct consequence of the change δV_ε (with $\delta V_\beta = 0$ and constant ε_0, β_0) is the perturbation $\delta\alpha$. From Figure 3.8(b),

$$\frac{d\alpha}{dV_\varepsilon} = \frac{-1}{V_{S0}}.$$

Hence, from Eq. (3.50)

$$-\frac{\partial \mathbf{F}}{\partial V_\varepsilon} = 2\mu L\left(\frac{R}{C_r}\right)^3 V_S \frac{\partial \mathbf{W}}{\partial\alpha}\frac{d\alpha}{dV_\varepsilon} = \frac{-F_0}{W_0 V_{S0}}\frac{\partial \mathbf{W}}{\partial\alpha}.$$

By comparison to Eqs. (3.48) and (3.49),

$$C_{\varepsilon\varepsilon} = \frac{-F_0}{W_0 V_{S0}}\frac{\partial W_\varepsilon}{\partial\alpha},$$

$$C_{\beta\varepsilon} = \frac{-F_0}{W_0 V_{S0}}\frac{\partial W_\beta}{\partial\alpha}. \tag{3.54}$$

The perturbed velocity component δV_β is seen from Figure 3.8(b) to yield directly $\delta V_s = -\delta V_\beta$; hence $dV_s/dV_\beta = -1$, and from Eq. (3.50)

$$-\frac{\partial \mathbf{F}}{\partial V_\beta} = 2\mu L\left(\frac{R}{C_r}\right)^3 \mathbf{W}\frac{dV_S}{dV_\beta} = -2\mu L\left(\frac{R}{C_r}\right)^3 \mathbf{W}_0.$$

By comparison to Eqs. (3.48) and (3.49), the remaining damping coefficients are

$$C_{\varepsilon\beta} = \frac{-F_0}{W_0 V_{S0}} W_{\varepsilon 0},$$

$$C_{\beta\beta} = \frac{-F_0}{W_0 V_{S0}} W_{\beta 0}. \qquad (3.55)$$

From Eq. (3.45) the rotordynamic coefficients have the alternative form

$$K_{\varepsilon\varepsilon} = \frac{F_0}{C_r} \left(\frac{c\gamma_0}{\varepsilon_0} - s\gamma_0 \frac{\partial\gamma}{\partial\varepsilon} + \frac{c\gamma_0}{W_0} \frac{\partial W}{\partial\varepsilon} \right),$$

$$K_{\beta\varepsilon} = -\frac{F_0}{C_r} \left(\frac{s\gamma_0}{\varepsilon_0} + c\gamma_0 \frac{\partial\gamma}{\partial\varepsilon} + \frac{s\gamma_0}{W_0} \frac{\partial W}{\partial\varepsilon} \right),$$

$$K_{\varepsilon\beta} = \frac{F_0}{C_r} \frac{s\gamma_0}{\varepsilon_0},$$

$$K_{\beta\beta} = \frac{F_0}{C_r} \frac{c\gamma_0}{\varepsilon_0},$$

$$C_{\varepsilon\varepsilon} = \frac{-F_0}{C_r \bar{\omega} \varepsilon_0} \left(\frac{c\gamma_0}{W_0} \frac{\partial W}{\partial\alpha} - s\gamma_0 \frac{\partial\gamma}{\partial\alpha} \right),$$

$$C_{\beta\varepsilon} = \frac{F_0}{C_r \omega \varepsilon_0} \left(\frac{s\gamma_0}{W_0} \frac{\partial W}{\partial\alpha} + c\gamma_0 \frac{\partial\gamma}{\partial\alpha} \right),$$

$$C_{\varepsilon\beta} = \frac{-F_0}{C_r \bar{\omega}} \left(\frac{c\gamma_0}{\varepsilon_0} \right),$$

$$C_{\beta\beta} = \frac{F_0}{C_r \bar{\omega}} \left(\frac{s\gamma_0}{\varepsilon_0} \right). \qquad (3.56)$$

For convenience, the following normalized coefficient definitions are employed:

$$k_{ij} = (C_r/F_0) K_{ij}, \qquad c_{ij} = (2C_r \bar{\omega}/F_0) C_{ij}. \qquad (3.57)$$

From this relationship and the transformation of Eq. (3.47), the following complete set of dimensionless stiffness and damping coefficients is obtained:

$$k_{XX} = \frac{c\gamma_0}{\varepsilon} - s\gamma_0 \left(\frac{\partial \gamma_0}{\partial \varepsilon} \right)_\alpha,$$

$$k_{XY} = \frac{s\gamma_0}{\varepsilon} + c\gamma_0 \left(\frac{\partial \gamma_0}{\partial \varepsilon} \right)_\alpha,$$

$$k_{YX} = \frac{-s\gamma_0}{\varepsilon} - \frac{s\gamma_0}{W} \left(\frac{\partial W}{\partial \varepsilon} \right)_\alpha,$$

$$k_{YY} = \frac{c\gamma_0}{\varepsilon} + \frac{c\gamma_0}{W} \left(\frac{\partial W}{\partial \varepsilon} \right)_\alpha,$$

$$c_{XX} = \frac{2s\gamma_0}{\varepsilon} \left(\frac{\partial \gamma_0}{\partial \alpha} \right)_\varepsilon,$$

$$c_{XY} = \frac{-2c\gamma_0}{\varepsilon} \left(\frac{\partial \gamma_0}{\partial \alpha} \right)_\varepsilon,$$

$$c_{YX} = \frac{2}{\varepsilon} \left\{ c\gamma_0 + \frac{s\gamma_0}{W} \left(\frac{\partial W}{\partial \alpha} \right)_\varepsilon \right\},$$

$$c_{YY} = \frac{2}{\varepsilon} \left\{ s\gamma_0 - \frac{c\gamma_0}{W} \left(\frac{\partial W}{\partial \alpha} \right)_\varepsilon \right\}. \tag{3.58}$$

Note that these coefficients are to be evaluated at an equilibrium position, and that they apply for any impedance definition, e.g., short, long, finite length, etc. The stiffness and damping coefficients which result from applying these relationships to the finite-length impedance of Eqs. (3.41) and (3.42) are illustrated in Figure 3.11[*] and provided in Childs et al. (1977). The results basically coincide with those of Barrett et al. (1978) except at large and small eccentricities, where their numerical differentiation approach encounters difficulties.

Stiffness and Damping Coefficients for Short Bearings

The stiffness and damping coefficients for short bearings may be derived from the formulas cited above. The following form for stating these coefficients is adopted from Lund (1966), who employs the modified Sommerfeld number

$$\sigma = \pi S \left(\frac{L}{D} \right)^2 = \frac{\mu \omega}{8} \left(\frac{L}{C_r} \right)^2 \frac{DL}{F_0}. \tag{3.59}$$

---

[*]In Figure 3.11, a_{ij} denotes stiffness coefficients and b_{ij} denotes damping coefficients.

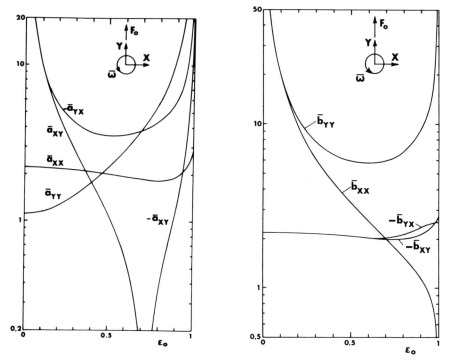

Figure 3.11 Stiffness and damping coefficients for the finite-length impedance of Eqs. (3.41) and (3.42) for $L/D = 1$ [Childs et al. (1977)].

At equilibrium, σ and the attitude angle γ_0 are defined by

$$\sigma = \left(1 - \varepsilon_0^2\right)^2 / \varepsilon_0 \left[16\varepsilon_0^2 + \pi^2\left(1 - \varepsilon_0^2\right)\right]^{1/2},$$

$$\gamma_0 = \tan^{-1}\left[\pi\left(1 - \varepsilon_0\right)^{1/2} / 4\varepsilon_0\right]. \tag{3.60}$$

The dimensionless reaction forces at equilibrium are

$$f_{\varepsilon 0} = \frac{F_{\varepsilon 0}}{F_0} = \sigma 4\varepsilon_0^2 \left(1 - \varepsilon_0^2\right)^{-2},$$

$$f_{\beta 0} = \frac{F_{\beta 0}}{F_0} = \sigma \pi \varepsilon_0 \left(1 - \varepsilon_0^2\right)^{-3/2}.$$

In terms of these parameters, the stiffness and damping coefficients may be stated:

$$k_{XX} = f_{\varepsilon 0}\left[(f_{\varepsilon 0})^2 + 1 - \varepsilon_0^2\right]/\varepsilon_0(1 - \varepsilon_0^2),$$

$$k_{XY} = -f_{\beta 0}\left[(f_{\varepsilon 0})^2 - (1 - \varepsilon_0^2)\right]/\varepsilon_0(1 - \varepsilon_0^2),$$

$$k_{YX} = -f_{\beta 0}\left[(f_{\varepsilon 0})^2 + 1 + 2\varepsilon_0^2\right]/\varepsilon_0(1 - \varepsilon_0^2),$$

$$k_{YY} = f_{\varepsilon 0}\left[(f_{\beta 0})^2 + 1 + 2\varepsilon_0^2\right]/\varepsilon_0(1 - \varepsilon_0^2),$$

$$c_{XX} = 2f_{\beta 0}\left[(2 + \varepsilon_0^2)(f_{\beta 0})^2 - 1 + \varepsilon_0^2\right]/\varepsilon_0(1 - \varepsilon_0^2),$$

$$c_{XY} = -2f_{\varepsilon 0}\left[(2 + \varepsilon_0^2)(f_{\beta 0})^2 - 1 + \varepsilon_0^2\right]/\varepsilon_0(1 - \varepsilon_0^2),$$

$$c_{YX} = c_{XY},$$

$$c_{YY} = 2f_{\beta 0}\left[(2 + \varepsilon_0^2)(f_{\varepsilon 0})^2 + 1 - \varepsilon_0^2\right]/\varepsilon_0(1 - \varepsilon_0^2). \tag{3.61}$$

Figures 3.12 and 3.13 illustrate these coefficients as a function of eccentricity ratio ε_0 and modified Sommerfeld number σ. Stiffness and damping coeffi-

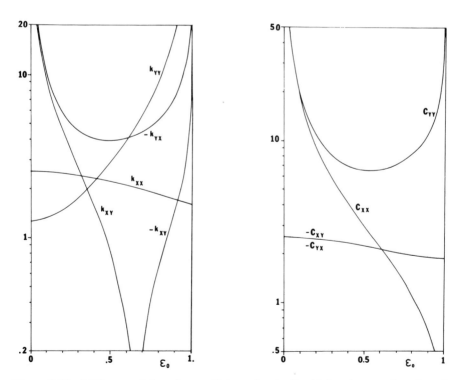

Figure 3.12 Stiffness and damping coefficients for the Ocvirk bearing as a function of ε_0, from Eqs. (3.61).

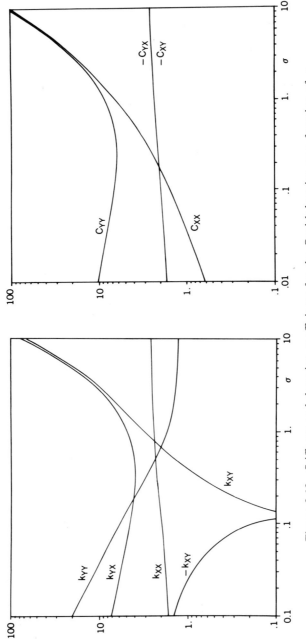

Figure 3.13 Stiffness and damping coefficients for the Ocvirk bearing as a function of σ, from Eqs. (3.60) and (3.61).

cients for other bearing types are commonly provided as a function of either ε_0 or S.

As illustrated in Figure 3.9, the equilibrium position for a bearing journal approaches zero as the speed and Sommerfeld number increase. For this limiting condition, Eq. (3.61) shows

$$k_{XY}(0) = -k_{YX}(0) = \bar{k} \rightarrow \sigma\pi,$$

$$c_{XX}(0) = c_{YY}(0) = \bar{C} \rightarrow 2\sigma\pi.$$

Hence,

$$\frac{k_{XY}(0)}{c_{XX}(0)} \rightarrow \frac{1}{2}$$

and

$$\Omega_w = \frac{K_{XY}(0)}{\omega C_{XX}(0)} \rightarrow \frac{1}{2}. \tag{3.62}$$

The ratio Ω_w is called the whirl-frequency ratio and is an inherent stability parameter for fixed-arc journal bearings. A further discussion of this parameter is undertaken in the following section.

Returning to Eq. (3.61), at small ε_0

$$K_{XX}(0) = 4/\pi, \qquad K_{YY}(0) = 8/\pi.$$

The dissimilar direct stiffnesses at the centered position are to be expected given the equilibrium loci of Figure 3.10. Note in Figure 3.10 that the static vertical load yields a markedly larger displacement in the X direction than the Y direction. A quite different result will be obtained for the annular seals of Chapter 4, where the direct stiffnesses are equal at the centered position. The results are different for bearings and seals because the mechanisms are different. Bearings develop direct stiffnesses through viscous hydrodynamic effects and cavitation. Seals develop direct stiffnesses from a hydrostatic mechanism.

3.5   SYNCHRONOUS-RESPONSE, STABILITY, AND TRANSIENT-ANALYSIS RESULTS FOR SIMPLE ROTOR MODELS SUPPORTED BY PLAIN JOURNAL BEARINGS

The development of the preceding section yielded (a) impedance descriptions, (b) the requirements for bearing equilibrium, and (c) bearing stiffness and damping coefficients for small motion about an equilibrium position. The bearing impedance is a nonlinear description of the bearing reaction force as a function of bearing motion and is appropriate for transient simulations of rotors supported in plain journal bearings. The stiffness and damping coefficients provide an appropriate bearing model for synchronous-response and stability calculations for small motion about equilibrium. The stability, synchronous-response, and transient-response characteristics of rotors are examined in this section for comparatively simple rotor models, viz., (a) a rigid rotor on journal bearings and (b) the Jeffcott rotor on journal bearings. Stability and synchronous-response analysis is covered initially, with transient response results concluding the section.

Stability Analysis of a Rigid Rotor on Plain Journal Bearings

Figure 3.14 illustrates a vertical rigid rotor of mass $2m$, acted on by a side load of $2F_0$, and supported by two identical plain journal bearings. The constant-speed differential equations of motion for this rotor are

$$m\ddot{R}_X = F_X + ma\omega^2 \cos \omega t,$$

$$m\ddot{R}_Y = F_Y + ma\omega^2 \sin \omega t - F_0, \qquad (3.63)$$

where m and a are the rotor's mass and imbalance-vector magnitude. Further, R_X and R_Y are the coordinates of the rotor mass center, and F_X, F_Y are the bearing-reaction components.

We will return to these equations at the conclusion of this section concerning nonlinear transient motion; however, our present interest concerns "small" motion about an equilibrium position R_{X0}, R_{Y0} defined by

$$R_X = R_{X0} + X, \qquad R_Y = R_{Y0} + Y.$$

Substituting these definitions into Eq. (3.63) and expanding the bearing-reac-

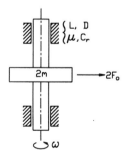

Figure 3.14 A vertical rigid rotor supported by plain journal bearings.

tion forces about the equilibrium position yields

$$m \begin{Bmatrix} \ddot{X} \\ \ddot{Y} \end{Bmatrix} + [C_{I,J}] \begin{Bmatrix} \dot{X} \\ \dot{Y} \end{Bmatrix} + [K_{I,J}] \begin{Bmatrix} X \\ Y \end{Bmatrix} = ma\omega^2 \begin{Bmatrix} \cos \omega t \\ \sin \omega t \end{Bmatrix}.$$

Introducing the dimensionless variables

$$\overline{X} = X/C_r, \qquad \overline{Y} = Y/C_r, \qquad \tau = \omega t$$

yields the differential equations

$$p^2 \begin{Bmatrix} \overline{X}'' \\ \overline{Y}'' \end{Bmatrix} + [c_{I,J}] \begin{Bmatrix} \overline{X}' \\ \overline{Y}' \end{Bmatrix} + [k_{I,J}] \begin{Bmatrix} \overline{X} \\ \overline{Y} \end{Bmatrix} = p^2 \left(\frac{a}{C_r} \right) \begin{Bmatrix} \cos \tau \\ \sin \tau \end{Bmatrix}, \qquad p^2 = \frac{C_r m \omega^2}{F_0},$$

$$\tag{3.64}$$

where C is the bearing clearance, and the prime denotes differentiation with respect to τ. Note that this equation uses the nondimensional rotordynamic coefficients defined by Eq. (3.57).

The characteristic equation for Eq. (3.64) is

$$\hat{s}^4 p^4 + \hat{s}^3 p^2 (c_{XX} + c_{YY}) + \hat{s}^2 \left[p^2 (k_{XX} + k_{YY}) + c_{XX} c_{YY} - c_{XY} c_{YX} \right]$$
$$+ \hat{s} (c_{XX} k_{YY} + c_{YY} k_{XX} - k_{YX} c_{XY} - k_{XY} c_{YX})$$
$$+ (k_{XX} k_{YY} - k_{XY} k_{YX}) = 0, \tag{3.65}$$

where $\hat{s} = s/\omega$ is the dimensionless Laplace operator. For a given eccentricity ratio, the onset speed of instability ω_s can be directly calculated from this equation by increasing the running speed ω (and hence p^2), until the real part of one of the four roots to Eq. (3.65) changes from negative to positive; i.e., $\hat{s}_i = 0 + j\Omega_s$. The running speed at which this occurs is by definition ω_s, and the imaginary portion of the root defines the precessional (or whirl[*]) frequency ratio of the incipient rotor instability. Specifically, at ω_s, the homogeneous solution to Eq. (3.64) corresponding to the unstable root may be stated:

$$\overline{X} = A e^{j\Omega_s \tau} = A e^{j\Omega_s \omega_s t},$$
$$\overline{Y} = B e^{j\Omega_s \tau} = B e^{j\Omega_s \omega_s t}, \tag{3.66}$$

and consists of forward precessional motion at the frequency $\Omega_s \omega_s$.

---

[*]Relationships between $\Omega_s(\varepsilon_0)$ and Ω_w, defined by Eq. (3.62), which are both called the whirl-frequency ratio, will be discussed subsequently.

Lund (1965) demonstrated that ω_s and Ω_s can be more directly obtained by simply substituting Eq. (3.66) into the homogeneous version of Eq. (3.64) to obtain

$$(p_s \Omega_s)^2 = D = (c_{XX} k_{YY} + c_{YY} k_{XX} - c_{YX} k_{XY} - c_{XY} k_{YX}) / (c_{XX} + c_{YY}),$$
$$\Omega_s^2 = [(D - k_{XX})(D - k_{YY}) - k_{XY} k_{YX}] / (c_{XX} c_{YY} - c_{XY} c_{YX}).$$
$$(3.67)$$

For a given value of eccentricity ε_0, one evaluates the first equation to obtain D, and then the second equation to obtain $\Omega_s(\varepsilon_0)$. Back substitution then yields $p_s^2 = D / \Omega_s^2$, which defines ω_s.

Figure 3.15 is taken from Lund (1966) and illustrates p_s^2 versus eccentricity ratios for various L/D ratios. The results demonstrate that the rotor is completely stable for eccentricity ratios greater than approximately 0.75, for all L/D ratios, but is comparatively insensitive to eccentricity ratios for $\varepsilon_0 \leq 0.5$. In examining this figure, one should recall the results of Figure 3.9; specifically, for a given load (as specified by the Sommerfeld number) the eccentricity is very much a function of the L/D ratio. Figure 3.16 illustrates

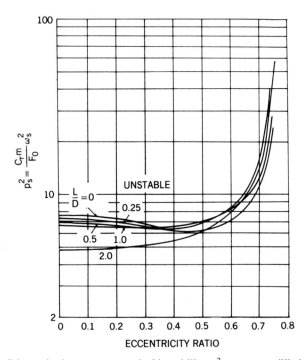

Figure 3.15   Dimensionless onset speed of instability p_s^2 versus equilibrium eccentricity ε_0 and L/D ratio [Lund (1966)].

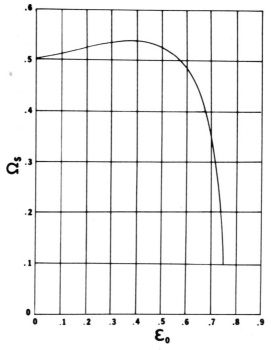

Figure 3.16   Whirl-frequency ratio Ω_s versus equilibrium eccentricity ratio ε_0 for the short-π bearing [Lund (1966)].

Ω_s versus ε_0 for the π short-bearing coefficients of Section 3.4. Note from Figure 3.9 that an increase in speed also increases S, reduces the equilibrium eccentricity ε_0, and eventually destabilizes the rotor. From Figure 3.16, the whirl-frequency ratio is approximately 0.5 at small ε_0. In practice, the whirl ratio of unstable rotors supported by plain journal bearings is somewhat less than 0.5.

Stability and Synchronous Response of a Jeffcott-Rotor Model in Short Journal Bearings

The influence of rotor flexibility on the stability and synchronous response characteristics of a rotor supported in plain journal bearings is of interest in this section. The model of Figure 3.17 will be employed to examine this topic and consists of a Jeffcott flexible-rotor model on short π journal bearings. The mass of the rotor is $2m$, and each shaft section has a stiffness of k. The constant-running-speed equations of motion for the rotor may be stated:

$$2m\ddot{R}_X = -2k(R_X - B_X) + 2ma\omega^2 \cos \omega t,$$
$$2m\ddot{R}_Y = -2k(R_Y - B_Y) + 2ma\omega^2 \sin \omega t - 2W, \qquad (3.68)$$

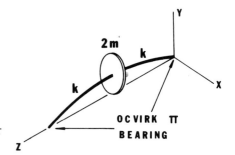

Figure 3.17 A horizontal Jeffcott rotor supported in Ocvirk π bearings.

and the equilibrium requirements at each bearing are

$$F_X = k(R_X - B_X),$$
$$F_Y = k(R_Y - B_Y). \tag{3.69}$$

In these equations, R_X, R_Y are the coordinates of the rotor mass, and B_X, B_Y are the coordinates of the rotor at bearing locations. Further, F_X and F_Y are components of the nonlinear bearing reaction forces.

Our current interest is with small motion about a rotor and journal equilibrium position as defined by

$$R_X = R_{X0} + X, \qquad R_Y = R_{Y0} + Y,$$
$$B_X = B_{X0} + x, \qquad B_Y = B_{Y0} + y.$$

The appropriate governing equations of motion are

$$m\ddot{X} + k(X - x) = ma\omega^2 \cos \omega t,$$
$$m\ddot{Y} + k(Y - y) = ma\omega^2 \sin \omega t,$$
$$[C_{I,J}]\left\{ \begin{matrix} \dot{x} \\ \dot{y} \end{matrix} \right\} + [K_{I,J}]\left\{ \begin{matrix} x \\ y \end{matrix} \right\} = -k\left\{ \begin{matrix} x - X \\ y - Y \end{matrix} \right\}. \tag{3.70}$$

We are interested in the steady-state solution to these differential equations as running speed ω is increased and anticipate from our earlier experience with the rigid-rotor model that the flexible-rotor model will also eventually become unstable.

To examine stability, the dimensionless variables

$$\bar{X} = X/C_r, \qquad \bar{Y} = Y/C_r, \qquad \bar{x} = x/C_r, \qquad \bar{y} = y/C_r, \qquad \tau = \omega\tau$$

are introduced to yield the following homogeneous equations:

$$\left(\frac{\delta}{C_r}\right)p^2\bar{X}'' + (\bar{X} - \bar{x}) = 0,$$

$$\left(\frac{\delta}{C_r}\right)p^2\bar{Y}'' + (\bar{Y} - \bar{y}) = 0,$$

$$\left(\frac{\delta}{C_r}\right)[c_{I,J}]\left\{\begin{matrix}\bar{x}'\\ \bar{y}'\end{matrix}\right\} + \left(\frac{\delta}{C_r}\right)[k_{I,J}]\left\{\begin{matrix}\bar{x}\\ \bar{y}\end{matrix}\right\} = \left\{\begin{matrix}\bar{X} - \bar{x}\\ \bar{Y} - \bar{y}\end{matrix}\right\}, \qquad (3.71)$$

where $\delta = W/k$ is the static deflection of the rotor, $p = \omega\sqrt{C_r/g}$, and the prime denotes differentiation with respect to τ. These equations have the following sixth-order characteristic equations:

$$\hat{s}^6 p^4 (\delta/C_r)^2 (c_{XX}c_{YY} - c_{YX}c_{XY}) + \hat{s}^5 p^2 (\delta/C_r)$$
$$\times \left[c_{YY} + c_{XX} + (\delta/C_r)(c_{YY}k_{XX} + c_{XX}k_{YY} - c_{YX}k_{XY} - c_{XY}k_{XY})\right]$$
$$+ \hat{s}^4 p^2 \left\{p^2 + (\delta/C_r)\left[p^2(k_{XX} + k_{YY} + 2(c_{XX}c_{YY} - c_{XY}c_{YX}))\right]\right.$$
$$\left. + (\delta/C_r)^2 p^2 (k_{XX}k_{YY} - k_{XY}k_{YX})\right\}$$
$$+ \hat{s}^3 p^2 \left[(c_{YY} + c_{XX}) + 2(\delta/C_r)\right.$$
$$\times (c_{YY}k_{XX} + c_{XX}k_{YY} - c_{YX}k_{XY} - c_{XY}k_{YX})\right]$$
$$+ \hat{s}^2 \left[p^2(k_{XX} + k_{YY}) + (c_{XX}c_{YY} - c_{YX}c_{XY})\right.$$
$$\left. + 2(\delta/C_r)p(k_{XX}k_{YY} - k_{XY}k_{YX})\right]$$
$$+ \hat{s}(c_{YY}k_{XX} + c_{XX}k_{YY} - c_{YX}k_{XY} - c_{XY}k_{YX})$$
$$+ (k_{XX}k_{YY} - k_{YX}k_{XY}) = 0, \qquad (3.72)$$

where $\hat{s} = s/\omega$ is the dimensionless Laplace operator. For $\delta = 0$, Eq. (3.72) is seen to reduce to Eq. (3.65), previously derived for a rigid rotor on hydrodynamic bearings.

As noted above, the system model of Eq. (3.71) becomes unstable when the real part of one of the roots to Eq. (3.72) becomes positive, and the roots may be directly calculated to determine ω_s and Ω_s. However, again following Lund (1965), the desired results may be more quickly and instructively derived by substituting the assumed solutions,

$$\bar{X} = Ae^{j\Omega_s\tau}, \qquad \bar{Y} = Be^{j\Omega_s\tau},$$
$$\bar{x} = ae^{j\Omega_s\tau}, \qquad \bar{y} = be^{j\Omega_s\tau}, \qquad (3.73)$$

into Eq. (3.71) to obtain

$$\frac{(p_{sf}\Omega_s)^2}{1 - (p_{sf}\Omega_s)^2(\delta/C_r)}$$
$$= D = (c_{XX}k_{YY} + c_{YY}k_{XX} - c_{YX}k_{XY} - c_{XY}k_{YX})/(c_{XX} + c_{YY}),$$
$$\Omega_s^2 = [(D - k_{XX})(D - k_{YY}) - k_{XY}k_{YX}]/(c_{XX}c_{YY} - c_{XY}c_{YX}). \quad (3.74)$$

Note that the whirl-frequency ratio Ω_s depends only on the bearing coefficients and is completely independent of rotor flexibility. Hence, the Ω_s versus ε_0 illustration of Figure 3.16 for a rigid rotor also applies for a flexible rotor. More specifically, the Ω_s versus ε_0 relationship defines an innate characteristic of the bearing, independent of rotor flexibility. From the first of Eq. (3.74), the flexible-rotor dimensionless onset speed of instability p_{sf} may be stated

$$p_{sf}^2 = D/\Omega_s^2[1 + D(\delta/C_r)] = p_s^2/[1 + D(\delta/C_r)], \quad (3.75)$$

which shows that rotor flexibility reduces the onset speed of instability by the factor $[1 + D(\delta/C_r)]^{1/2}$. Figure 3.18 illustrates the dependency of rotor stability on shaft flexibility. The dependence of ω_{sf} on rotor flexibility will be considered further after the following discussion of rotor synchronous response.

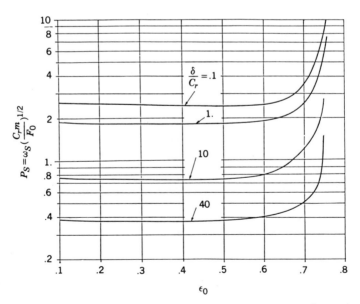

Figure 3.18 Flexible-rotor onset speed of instability versus δ/C_r [Lund (1965)].

The synchronous solution to Eq. (3.70) is obtained by substituting the assumed solution format

$$
\begin{Bmatrix} X \\ Y \\ x \\ y \end{Bmatrix} = \begin{Bmatrix} X_c \\ Y_c \\ x_c \\ y_c \end{Bmatrix} \cos \omega t + \begin{Bmatrix} X_s \\ Y_s \\ x_s \\ y_s \end{Bmatrix} \sin \omega t, \tag{3.76}
$$

equating coefficients of $\cos \omega t$ and $\sin \omega t$, and solving the resultant simultaneous equations for the two constant vectors on the right-hand side of Eq. (3.76). For a given running speed, the analysis procedure to be followed consists of the following steps:

(a) The equilibrium position *for the bearing* is calculated in terms of the equilibrium eccentricity ratio ε_0 and equilibrium attitude angle γ_0. This step is accomplished by first calculating the Sommerfeld number from Eq. (3.44) as

$$
S = \mu \left(\frac{\omega}{2\pi} \right) \left(\frac{R}{C_r} \right)^2 \frac{DL}{W},
$$

and then calculating ε_0 and γ_0 from Eqs. (3.59) and (3.60).
(b) The bearing-stiffness and damping matrices $[c_{I,J}]$ and $[k_{I,J}]$ are determined from Eq. (3.61) as a function of ε_0, and the synchronous solution to Eq. (3.70) is calculated.
(c) Equation (3.75) is checked to see if the system is stable.

The example problem to be solved is defined by the parameters

$$
\begin{aligned}
\mu &= 0.035 \text{ N s/m}^2, & D &= 0.1 \text{ m}, \\
W &= 1335 \text{ N}, & C_r &= 7.5 \times 10^{-2} \text{ mm}, \\
L &= 0.05 \text{ m}, & a &= C_r.
\end{aligned} \tag{3.77}
$$

The rotor flexibility is parametrized in terms of δ/C_r, where δ is the rotor midspan static deflection; i.e., $\delta = W/k$. The rigid-rotor onset speed of instability for the data of Eq. (3.77) is $\omega_s = 9530$ rpm.

Figure 3.19 illustrates the solution to Eq. (3.70), with Figure 3.19(a) illustrating ε_0 and γ_0 as a function of running speed. The results of this figure reflect the increase in S with increasing running speed, and the consequent decrease in ε_0 and asymptotic motion of γ_0 towards 90°. The bearing equilibrium position is simply mapping out the locus of Figure 3.10 with increasing running speed. In other words, the rotor is rising in its bearings with increasing speed. Figure 3.19(b) illustrates the principal major

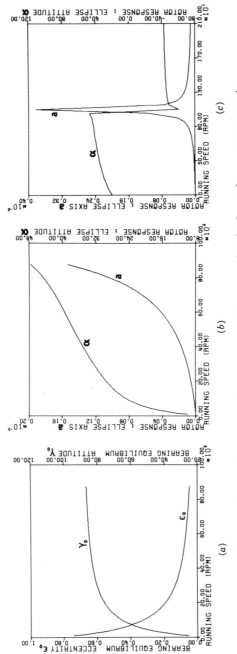

Figure 3.19 (a) Bearing equilibrium eccentricity ε_0 and attitude angle γ_0 versus running speed for the data of Eqs. (3.77). (b) Rotor response for the example problem of Eqs. (3.77) for $\delta/C_r = 0.1$; elliptic major axis amplitude and attitude angle versus running speed; $\omega_{sf} = 8730$ rpm $= 914$ rad/sec; whirl frequency $= 459$ rad/sec. (c) Rotor response for the example problem of Eqs. (3.77) for $\delta/C_r = 10.0$; elliptic major axis amplitude and attitude angle versus running speed; $\omega_s = 2090$ rpm $= 219$ rad/sec; whirl frequency $= 111$ rad/sec.

170

axis of the rotor's elliptical orbit* and its attitude angle as a function of running speed for $\delta/C_r = 0.1$. For this δ/C_r value, the flexible-rotor onset speed of instability $\omega_{sf} = 8730$ rpm, as compared to the rigid-rotor onset speed of instability $\omega_s = 9530$ rpm. Figure 3.19(c) illustrates these same variables for $\delta/C_r = 10.0$. The response solutions are presented out to the onset speed of instability, $\omega_{sf} = 2090$ rpm and demonstrate that rotor flexibility introduces a rotor critical speed while lowering the onset speed of instability. Note that the critical speed is lightly damped, as evidenced by the sharp peak in amplitude and associated change in phase as it is traversed.

One would expect the rotor critical speed to be approximately defined by a relationship of the form

$$\omega_1 \cong \sqrt{k_e/m}\,, \qquad \frac{1}{k_e} = \frac{1}{k} + \frac{1}{\overline{K}},$$

where $\overline{K}$ is, in some sense, an average bearing stiffness. Since the direct bearing stiffnesses are known to be functions of running speed, the rotor critical speed would also be expected to show speed dependence. In fact, the bearing stiffness generally exceeds the rotor stiffness by such a large magnitude that the critical speed is only slightly less than the rigid-bearing critical speed; i.e.,

$$\omega_1 \cong \lambda = \sqrt{k/m}\,. \tag{3.78}$$

This statement can be confirmed by an inspection of the roots to Eq. (3.72), which are complex-conjugate pairs of the form

$$s_i = \sigma_i \pm j\omega_i, \qquad i = 1, 2, 3.$$

Table 3.1 contains roots calculated for $\delta/C_r = 10$ [$\lambda = 114.4$ rad/sec for the data of Eq. (3.77)] and demonstrates that the roots ω_1, ω_2 are relatively constant as running speed is increased, and only slightly smaller than λ. Bearing asymmetry accounts for the small differences between ω_1 and ω_2. The rotor model becomes unstable when σ_1 changes sign between 215 and 220 rd/sec.

The root ω_3 increases with increasing ω, with the ratio ω_3/ω approaching an asymptote of approximately $\frac{1}{2}$. It is the bearing motion, not the rotor, which is primarily defined by the complex-conjugate roots $s_3 = \sigma_3 \pm j\omega_3$. However, the resonance associated with these roots is not excited by imbalance[†], since ω_3 is always less than ω. Further, the motion has appreciable damping.

---

*These variables are illustrated in Figure A.1 of Appendix A.
[†]This motion can be excited in a model rotor by providing an external half-running-speed excitation.

TABLE 3.1 Roots From Eq. (3.72) for $\delta/C_r = 10$, Which Yields $\lambda = 114.4$ rd/sec for the Data of Eq. (3.77). From Eq. (3.75), $\omega_{sf} = 2090$ rpm $= 219$ rd/sec.

ω (rd/sec)	σ_1	ω_1	σ_2	ω_2	σ_3	ω_3
10.47	−0.637	114.3	−0.063	114.4	−10.78	8.31
41.90	−1.141	114.1	−0.259	114.3	−23.93	25.40
83.80	−1.404	113.8	−0.425	114.3	−31.80	46.30
126.00	−1.590	113.2	−0.492	114.3	−35.55	67.20
168.00	−1.470	112.3	−0.501	114.3	−37.82	88.60
209.00	−0.339	111.5	−0.488	114.3	−40.24	110.00
215.00	−0.149	111.4	−0.485	114.3	−40.57	113.00
220.00	0.038	111.5	−0.483	114.3	−40.89	115.00

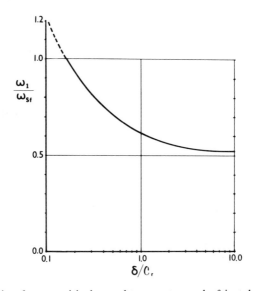

Figure 3.20 Ratio of rotor critical speed to onset speed of instability for increasing rotor flexibility.

Figure 3.20 illustrates the ratio ω_1/ω_{sf} versus δ/C_r. The curve is dashed for $\omega_1/\omega_{sf} > 1$ to illustrate that the rotor becomes unstable before the critical speed is encountered. Hence, increasing the rotor stiffness relative to the bearing eventually eliminates the rotor critical speed as demonstrated by Figure 3.19(b). As δ/C_r is increased in Figure 3.20, the asymptote $\omega_1/\omega_{sf} \cong 0.5$ is approached, which from Eq. (3.78) yields

$$\omega_{sf} \cong 2\lambda. \tag{3.79}$$

An explanation for this limiting behavior is provided by our earlier discussion of the complex roots to Eq. (3.72). At the rotor instability, σ_1 is zero, and $\omega_1 \cong \lambda$; hence, the assumed solution of Eq. (3.73) yields

$$\overline{X} = Ae^{j\Omega_s \omega t} = Ae^{j\omega_1 t} \cong Ae^{j\lambda t} \cdots,$$

$$\overline{Y} = Be^{j\Omega_s \omega t} = Be^{j\omega_1 t} \cong Be^{j\lambda t} \cdots, \tag{3.80}$$

etc., and the onset speed of instability is approximately defined by

$$\omega = \omega_{sf} = \lambda/\Omega_s. \tag{3.81}$$

As discussed earlier, the whirl-frequency ratio Ω_s is a function of the eccentricity ratio ε_0, and at small eccentricities is approximately $\frac{1}{2}$, which explains Eq. (3.79).

The following message provided by Eq. (3.81) is simple but easily overlooked. The onset speed of instability for a *flexible* rotor supported by plain journal bearings can only be elevated by increasing the rotor's stiffness. Various additional bearing types are introduced in Section 3.6 which have the capacity for substantially elevating the *rigid-rotor* onset speed of instability; however, when dealing with "very" flexible rotors, *only bearings with Ω_s appreciably smaller than $\frac{1}{2}$ can be expected to materially elevate ω_{sf}.*

Transient Motion of a Rigid Rotor on Journal Bearings

The linear stability and synchronous-response procedures outlined above are generally adequate for the analysis of rotors supported in hydrodynamic bearings. Hence, stiffness and damping coefficients generally provide an appropriate model for hydrodynamic bearings in rotordynamic analysis. However, situations occasionally arise which require a transient nonlinear solution, and the contents of this section demonstrate the utility of impedance descriptions as a nonlinear model for journal bearings.

The rigid-rotor example of Eq. (3.63) will be considered in a horizontal position; i.e., $F_0 = W$, the rotor weight. The bearing reaction-force components F_X, F_Y are to be defined in terms of the bearing position (R_X, R_Y) and velocity ($\dot{R}_X, \dot{R}_Y$) vector components and the spin velocity ω. The solution to this problem for the short and long bearing impedances is summarized in the following steps:

(a) From Eq. (3.22) one calculates

$$\varepsilon = \left(R_X^2 + R_Y^2 \right)^{1/2}/C_r.$$

The 2π bearing impedances can now be calculated.

(b) From Eqs. (3.22)–(3.24) and Figure 3.6,

$$\sin \beta = R_Y/C_r \varepsilon, \qquad \cos \beta = R_X/C_r \varepsilon, \qquad \bar{\omega} = \omega/2,$$

$$\sin \eta = \left(\dot{R}_Y + \bar{\omega} R_X\right)/V_s, \qquad \cos \eta = \left(\dot{R}_X - \bar{\omega} R_Y\right)/V_s,$$

$$\sin \alpha = \sin(\beta - \eta) = \sin \beta \cos \eta - \cos \beta \sin \eta,$$

$$\cos \alpha = \cos(\beta - \eta) = \cos \beta \cos \eta + \sin \beta \sin \eta. \qquad (3.82)$$

The short and long π bearing integrals can now be evaluated in terms of ε, $\sin \alpha$, and $\cos \alpha$.

(c) The short- and long-bearing impedances can be evaluated in terms of the W_ε, W_β component definitions of Eqs. (3.35) and (3.38), respectively. The reaction-force components F_ε, F_β are then defined by Eq. (3.32), and the stationary reaction components are defined via the coordinate transformation

$$F_X = F_\varepsilon \cos \beta - F_\beta \sin \beta,$$

$$F_Y = F_\varepsilon \sin \beta + F_\beta \cos \beta. \qquad (3.83)$$

Alternatively, Eq. (3.34) can be used to obtain (explicitly) the impedance definition $W_{\bar{x}}(\varepsilon, \alpha), W_{\bar{y}}(\varepsilon, \alpha)$ in the "squeeze-velocity-vector" oriented $\bar{x}, \bar{y}$ system, which yields

$$F_X = F_{\bar{x}} \cos \eta - F_{\bar{y}} \sin \eta,$$

$$F_Y = F_{\bar{x}} \sin \eta + F_{\bar{y}} \cos \eta. \qquad (3.84)$$

The small ε definitions of Eqs. (3.37) and (3.40) were used here for $\varepsilon < 0.01$ to avoid numerical difficulties associated with J_1^{00}.

The procedure for using the Moes finite-length impedance description involves the definition of ε, β, η, and α as outlined in steps (a) and (b) above. Equation (3.41) is then used to successively calculate a, b, and γ, with Eq. (3.42) used to successively calculate a', b', and W. Finally, the impedance components of Eq. (3.43) are used to calculate the reaction components $F_{\bar{x}}, F_{\bar{y}}$, and Eq. (3.84) yields the desired stationary-reference reaction definition. The Moes finite-element impedance is well behaved and does not require a separate "small ε" definition, comparable to Eqs. (3.37) and (3.40).

Transient solutions are obtained by numerical integration of Eq. (3.63) for the basic data of Eq. (3.77). The Ocvirk π impedance and the cavitating Moes impedance were used to model the bearings. Figure 3.21 illustrates transient solutions with zero imbalance for both the Ocvirk and Moes

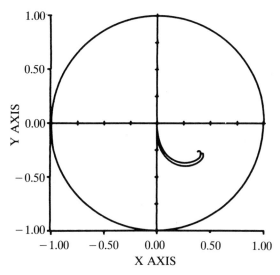

Figure 3.21 Transient solutions obtained by integrating Eq. (3.63) for the π Ocvirk and cavitated Moes impedance; zero imbalance; $\omega_j = 500$ rpm. The Ocvirk impedance yields a smaller ε_0.

impedances at $\omega = 950$ rpm. The solutions presented are converging toward the steady-state equilibrium position, with the Ocvirk solution yielding eccentricities which are too small (or clearances which are too large) when compared to the finite-length-impedance result. These results would be expected, since the $L/D = \frac{1}{2}$ ratio is nearing the end of the valid range for the Ocvirk model, particularly at eccentricity ratios greater than 0.5.

Figure 3.22 illustrates the solution for the same running conditions with $a = 100C_r = 7.5 \times 10^{-3}$ m. The initial conditions for this figure are the equilibrium results obtained in the integration of Figure 3.21. Due to the large imbalance magnitude, the orbits are also large and are not centered about the equilibrium position. Figure 3.23 illustrates comparable solutions for the reduced imbalance $a = 10C_r = 7.5 \times 10^{-2}$ m. Small orbits about the equilibrium position are seen to result from this imbalance magnitude. The type of motion illustrated in this figure could reasonably be modeled by stiffness and damping coefficients, instead of the more complicated nonlinear impedance descriptions.

Figure 3.24 illustrates the short-bearing solution response from zero position and velocity initial conditions at a running speed of $\omega = 9550$ rpm = 1000 rad/sec. The imbalance in this case is $a = C_r = 7.5 \times 10^{-3}$ m. The solution demonstrates the nature of the unstable rotor motion. A limit cycle solution results with the dominant motion at approximately one-half the running speed.

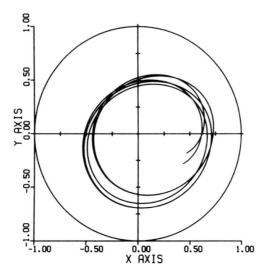

Figure 3.22 Transient solution obtained by integrating Eq. (3.63) for the π Ocvirk and cavitated Moes impedances; $a = 100C = 7.5 \times 10^{-2}$ m; $\omega_j = 500$ rpm.

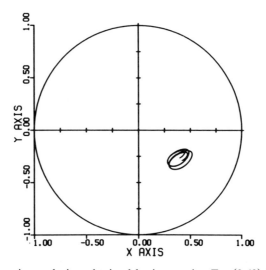

Figure 3.23 Transient solution obtained by integrating Eq. (3.63) for the π Ocvirk and the Moes cavitated impedance; $a = 10C_r = 7.5 \times 10^{-2}$ m; $\omega_j = 500$ rpm.

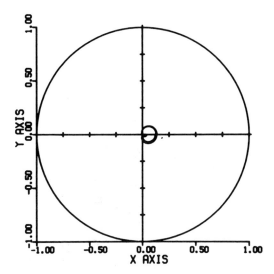

Figure 3.24 Transient solution obtained by integrating Eq. (3.63) for the Moes cavitated impedance; $a = C_r = 7.5 \times 10^{-5}$ m; $\omega = 9550$ rpm.

3.6 "OIL WHIP" AND "OIL WHIRL"

The weight of the Jeffcott-rotor example of Figure 3.17 yields finite bearing equilibrium positions. A stability analysis for small motion about the equilibrium position demonstrates the possibility of a linear instability at running speeds above $\omega_s \cong 2\omega_n$, the onset speed of instability. The unstable motion consists of precessional orbits at the rotor-bearing-system natural frequency and is called, "oil whirl." Placing the rotor of Figure 3.17 in a vertical position removes the static load from the bearings, eliminates the static equilibrium position, and introduces the possibility for a quite different type of subsynchronous orbital motion whose frequency "tracks" at approximately one-half running speed and is also called "oil whirl."

Figure 3.25 is taken from Muszynska (1986) and demonstrates a "waterfall" plot for the response of a vertical flexible rotor as the running speed is increased. The plot shows response spectra at a set of increasing speeds. A unity-slope line across this figure shows the synchronous response due to imbalance. Observe that increasing running speed yields a peak synchronous-response amplitude at the rotor critical speed with decreasing amplitudes as the running speed is increased further. The line whose slope is 2 in Figure 3.25 identifies spectral peaks at one-half running speed. The one-half-running-speed spectral contribution grows steadily with increasing running speed until the running speed is about twice the critical speed. For further increases in running speed, the subsynchronous response is at the natural frequency, not one-half the running speed, and grows progressively.

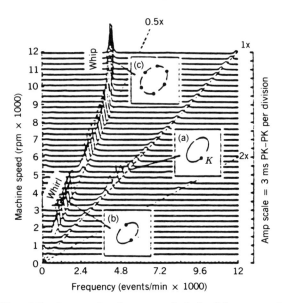

Figure 3.25 Waterfall plot results for a vertical flexible rotor, demonstrating a progression from oil whirl to oil whip [Muszynska, 1986]. Reproduced with permission of Academic Press Ltd.

As illustrated in Figure 3.25, subsynchronous motion at about one-half running speed is called "oil whirl" and is relatively benign, while subsynchronous motion at the natural frequency is called oil whip,* and can be highly destructive.

Muszynska (1986), in a series of publications, is largely responsible for explaining the occurrence and transition from oil whirl to oil whip. However, the discussion of this section is taken from an appealingly simple analysis provided by Crandall (1990). Returning to a vertical-rotor version of the model of Figure 3.17, we will assume that the rotor is perfectly balanced, and that centered circular orbits within the bearings are sufficiently small to preclude cavitation. Further, the orbits of the journal center within the bearing is at the frequency Ω with an amplitude $e = C_r \varepsilon$. Hence, the velocity of the journal center is, in the notation of Figures 3.3 and 3.4,

$$V_j = u_\beta C_r \varepsilon \dot{\beta} = u_\beta C_r \varepsilon \Omega, \qquad (3.85)$$

---

*The designations whip and whirl of this section differ from the terminology used in most of this book where an unstable rotor, which is precessing *at its natural frequency*, is referred to as whirling, not whipping. The term whipping is used only with respect to motion driven by dry friction, Section 1.10, and unloaded rotors on plain journal bearings.

and from Eq. (3.21), the squeeze velocity is

$$V_s = -\mathbf{u}_\beta C_r \varepsilon \left(\frac{\omega}{2} - \Omega \right). \tag{3.86}$$

For $\Omega < \omega/2$, V_s is 180° out of phase with V_j, and from Figure 3.6, $\alpha = \pi/2$. Without cavitation, there is no radial-bearing reaction component, and the bearing reaction force is from Eq. (3.32):

$$F_b = -V_s 2\mu L (R/C_r)^3 W_\beta(\varepsilon, \pi/2) \mathbf{u}_\beta$$

$$= C_r \varepsilon \left(\frac{\omega}{2} - \Omega \right) 2\mu L \left(\frac{R}{C_r} \right)^3 W_\beta(\varepsilon, \pi/2) \mathbf{u}_\beta.$$

Hence, for $\Omega < \omega/2$, the bearing reaction force is aligned with V_j and is destabilizing. The impedance of this equation could be uncavitated versions of the Ocvirk, Sommerfeld, or Moes impedances. Crandall used the following Sommerfeld solution due to Robertson (1933):

$$F_b = 12\pi\mu L \left(\frac{R}{C_r} \right)^3 \frac{\Omega C_r \varepsilon (\overline{\omega}/2\overline{\Omega} - 1)}{(1 + \varepsilon^2/2)\sqrt{1 - \varepsilon^2}}, \tag{3.87}$$

where the disk natural frequency $\omega_n = \sqrt{k/m}$ has been used to normalize ω and Ω as

$$\overline{\omega} = \omega/\omega_n, \qquad \overline{\Omega} = \Omega/\omega_n.$$

The disk drag force F_d is logically a function of the disk's velocity $\dot{R} = \rho\Omega$. Crandall used the following power-law definition

$$F_d = 2C_d k C_r (\overline{\rho}\overline{\Omega})^\gamma, \qquad \overline{\rho} = \rho/C_r, \tag{3.88}$$

where C_d and γ are empirical parameters. Equations (3.87) and (3.88) provide physical definitions for the external forces F_b and F_d acting on the rotor. These physical definitions plus alternative definitions from equilibrium requirements will be used to define the rotor response in terms of ε, ρ, δ of Figure 3.26 and $\overline{\Omega}$ as function of $\overline{\omega}$.

Figure 3.26 illustrates the equilibrium orbit orientation of the bearing and the rotor. The bearing-orbit radius is $C_r \varepsilon$, the disk-orbit radius is ρ, and the elastic deflection of the shaft is δ. Equilibrium of the bearing reaction force with the load transmitted through the shaft (i.e., $2k\delta$), requires that δ and F_b be colinear. Hence, $\rho = C_r \varepsilon + \delta$.

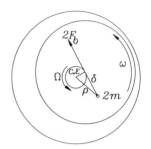

Figure 3.26 Bearing journal, precessing at Ω while rotating at ω. The bearing-reaction force F_b is aligned with the rotor elastic deflection vector δ.

Figure 3.27 illustrates the balance of forces corresponding to a limit cycle with a precessional frequency Ω. The bearing reactions $2F_b$ are balanced by the shaft force $2k\delta$. The drag force acts in a direction opposite to the disk velocity $\dot{R} = \rho\Omega$. Summing forces which are parallel to F_b yields

$$2F_b = 2k\delta = 2m\rho\Omega^2\left(\frac{\delta}{\rho}\right) + F_d\left(\frac{C_r\varepsilon}{\rho}\right), \tag{3.89a}$$

while summation of forces perpendicular to F_b provides

$$F_d\left(\frac{\delta}{\rho}\right) = 2m\rho\Omega^2\left(\frac{C_r\varepsilon}{\rho}\right). \tag{3.89b}$$

Equations (3.89) provide the desired alternative governing equations for F_d and F_b. Direct alternate expressions for F_d and F_b will now be developed.

From Figure 3.27, similar triangles yield

$$\frac{m\rho\Omega^2}{k\delta} = \frac{\delta}{\rho} \Rightarrow \frac{\delta}{\rho} = \frac{\Omega}{\omega_n} = \overline{\Omega}. \tag{3.90}$$

Substitution from Eq. (3.90) into Eq. (3.89b) yields

$$F_d = 2k\overline{\Omega}C_r\varepsilon, \tag{3.91}$$

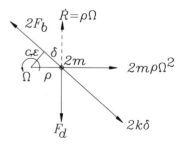

Figure 3.27 Equilibrium orientation of precessing bearing and disk centers.

which is the desired alternative definition for F_d. From $\rho^2 = (C_r \varepsilon)^2 + \delta^2$, Eq. (3.90) gives

$$\rho = \frac{C_r \varepsilon}{\sqrt{1 - \overline{\Omega}^2}}, \qquad \delta = \frac{C_r \varepsilon \overline{\Omega}}{\sqrt{1 - \overline{\Omega}^2}}. \qquad (3.92)$$

Substitution from Eqs. (3.90)–(3.92) into Eq. (3.89a) provides

$$F_b = \frac{k \overline{\Omega} C_r \varepsilon}{\sqrt{1 - \overline{\Omega}^2}}, \qquad (3.93)$$

which is the desired alternate definition for F_b.

Equating the definitions for F_b given by Eqs. (3.87) and (3.93) gives

$$\left(\frac{\overline{\omega}}{2\overline{\Omega}} - 1 \right) \sqrt{1 - \overline{\Omega}^2} = \left(\frac{C_r}{R} \right)^3 \frac{\sqrt{mk}}{12\pi\mu L} (1 + \varepsilon^2/2)\sqrt{1 - \varepsilon^2}. \qquad (3.94)$$

The right-hand side of this equation is approximately zero because of the C_r/R ratio and is a very weak function of ε. *Equation (3.94) says that the precessional frequency Ω will be approximately equal to either $\omega/2$ (oil whirl) or ω_n (oil whip).*

Equating the definitions for F_d in Eqs. (3.88) and (3.91) gives

$$\varepsilon = C_d (\overline{\rho} \overline{\Omega})^{\gamma} / \overline{\Omega}.$$

Substitution from the first of Eq. (3.92) eliminates $\overline{\rho}$, yielding

$$\varepsilon = \frac{C_d}{\overline{\Omega}} \left(\frac{\varepsilon \overline{\Omega}}{\sqrt{1 - \overline{\Omega}^2}} \right)^{\gamma} \qquad (3.95a)$$

or

$$\varepsilon = \frac{(1 - \overline{\Omega}^2)^{\gamma/2(\gamma - 1)}}{\overline{\Omega} C_d^{1/(\gamma - 1)}}. \qquad (3.95b)$$

Equations (3.94) and (3.95b) define ε and $\overline{\Omega}$ as a function of $\overline{\omega}$. Equations (3.92) provide separate definitions for ρ and δ as functions of ε and $\overline{\Omega}$.

To yield oil whip, ε must increase as $\overline{\Omega}$ approaches 1. To achieve this desired result, γ must be less than 1 in Eq. (3.95b), since $\gamma > 1$ causes ε to approach zero as $\overline{\Omega}$ approaches 1. The limiting case of $\gamma = 1$ also fails to yield either oil whirl ($\Omega \cong \omega/2$) or oil whip ($\Omega \cong \omega_n$), since for $\gamma = 1$,

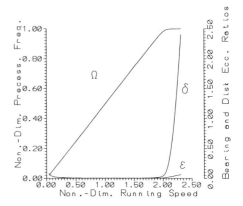

Figure 3.28 Bearing and disk eccentricity ratios and nondimensional precession frequency $\overline{\Omega}$ versus nondimensional running speed $\overline{\omega}$.

Eq. (3.95a) gives

$$\overline{\Omega} = \sqrt{1 - C_d^2}\,.$$

For $\gamma = \frac{1}{2}$, Eq. (3.95b) reduces to

$$\varepsilon = C_d^2/\overline{\Omega}(1 - \overline{\Omega}^2)^{1/2}. \tag{3.96}$$

Figure 3.28* illustrates a solution to Eqs. (3.94) and (3.96) for $\overline{\Omega}$, ε, and $\overline{\rho}$ versus $\overline{\omega}$ for $C_d = 0.04$, and

$$\left(\frac{C_r}{R}\right)^3 \frac{\sqrt{mk}}{12\pi\mu L} = 0.004.$$

Observe that $\overline{\Omega}$ stays at about $0.5\overline{\omega}$ until $\overline{\omega}$ is slightly below 2.0 (oil whirl) and then asymptotically approaches 1.0 for further increases in $\overline{\omega}$ (oil whip). The bearing eccentricity ratio ε starts at about 0.03, gets smaller, and then grows (modestly) as $\overline{\omega}$ exceeds 2.0. Note that this solution for ε is consistent with the initial assumptions of the analysis. The disk radius $\overline{\rho}$ follows a similar pattern but increases sharply as $\overline{\omega}$ passes 2.0.

Putting aside the beauty of Crandall's analysis in explaining the observed phenomena, oil whip and oil whirl have the following practical consequences. Many vertical pumps operate with plain journal bearings and demonstrate low-level oil-whirl motion without any significant problems. However, oil whip due to a running speed at or in excess of about twice the natural frequency can cause excessive vibrations and damage. Deliberate introduction of side loads or imbalance or changing to another bearing configuration can help alleviate the problem.

---

*Not all combinations of $\gamma < 1$ and C_d yield smooth transitions. For example, $\gamma = \frac{2}{3}$, $C_d = 0.40$ yielded a divergent solution as $\overline{\omega}$ approached 2.0.

3.7   ADDITIONAL BEARING CONFIGURATIONS

The results of Section 3.5 demonstrate that unstable subsynchronous motion is a major problem associated with high-speed rotors supported in plain journal bearings. This motion is characterized by large-amplitude limit cycle orbits at approximately one-half running speed. For a flexible rotor, this unstable motion initiates at an onset speed of instability which is approximately twice the critical speed, and its amplitude increases sharply (and in some cases dangerously) as the running speed increases. The replacement of plain journal bearings with other bearing types has proven to be the most direct approach for elevating the onset speed of instability and alleviating this problem. A variety of additional commonly occurring bearing types with improved stability characteristics are discussed in the balance of this subsection. However, no attempt is made to exhaustively cover all bearing types.

Multilobe Bearings

The results of Section 3.5 demonstrate that the onset speed of instability for rotors supported in plain journal bearings can be elevated by increasing the bearing's operating eccentricity. The multilobe journal bearings of Figure 3.29 are designed to take advantage of this characteristic by "preloading" the bearing segments. For example, the elliptic bearing has its upper and lower partial arc segments displaced $C_r \varepsilon_p$ from the bearing center, where ε_p is the bearing preload. Hence, when $\varepsilon_p = 0$, the bearing becomes cylindrical, and when $\varepsilon_p = 1$, the journal is in contact with the lobes. As a consequence of this design, the effective operating eccentricity of the loaded bearing segment is always higher in a multilobe design than the comparable eccentricity for a plain journal bearing. Lobed bearings are stiffer than plain journal bearings and consequently tend to be more stable, particularly at the centered position where plain journal bearings have lower direct stiffnesses.

Figure 3.30 illustrates the geometry normally used in describing one of the circular arcs of a multilobe bearing. The angles $\bar{\theta}_l$ and $\bar{\theta}_l$ denote the leading and trailing edges of the arc (with respect to the journal rotation). The angular coordinate α,

$$\alpha = \pi(\bar{\theta} - \bar{\theta}_l)\big/(\bar{\theta}_t - \bar{\theta}_l), \tag{3.97}$$

serves as a local circumferential coordinate. With the journal centered in the bearing, pad preload generally causes the clearance between pad and journal to first decrease and then increase with increasing α. In words, a converging section precedes a diverging section. If the point of minimum clearance occurs at α_p, the pad offset factor is defined by

$$\delta_p = \alpha_p/\pi = \left(\bar{\theta}_p - \bar{\theta}_l\right)\big/\left(\bar{\theta}_t - \bar{\theta}_l\right) \tag{3.98}$$

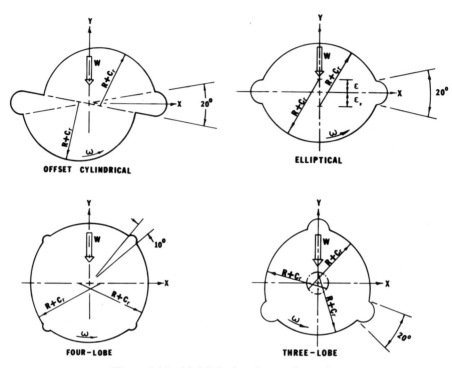

Figure 3.29   Multilobe bearing configurations.

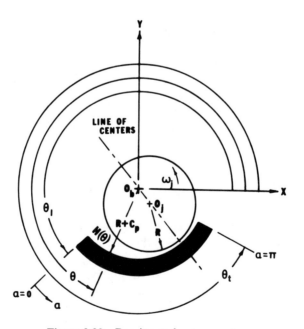

Figure 3.30   Bearing pad arc geometry.

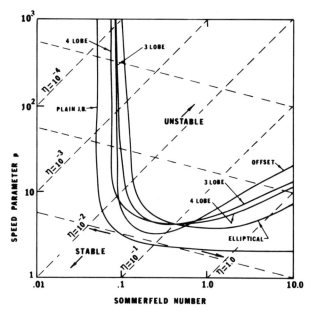

Figure 3.31   Rigid-rotor onset speed of instability for the multilobe bearing configurations of Figure 3.28 [Li et al. (1980)].

If the minimum clearance occurs at the center of the pad, $\delta_p = 0.5$, while a minimum clearance at the trailing edge would yield $\delta_p = 1.0$. Generally, a pad offset of 0.5 is chosen to permit journal rotation in either direction.

Figure 3.31 illustrates the *rigid-rotor* onset speed of instability p_s versus Sommerfeld number for both a plain journal bearing and the multilobe bearings of Figure 3.29. The multilobe results of this figure and a substantial portion of the following discussion is based on work by Li et al. (1980). In comparing these results to our earlier presentation in Figure 3.15, of p_s^2 versus equilibrium eccentricity ratio ε_0, recall that increasing S implies decreasing ε_0. The vertical asymptote for the plain journal bearing in this figure corresponds to $\varepsilon_0 \cong 0.75$. The region to the left of the vertical asymptotes corresponds to heavily loaded bearing conditions for which the onset speed of instability is infinite. Note that at high Sommerfeld numbers where bearings are likely to operate the plain journal bearing is the least stable.

The unit-slope straight lines in this figure correspond to constant bearing parameters defined by

$$\eta = \frac{S}{p} = \frac{\mu L D}{8\pi F_0} \left(\frac{D}{C_r}\right)^2 \left(\frac{F_0}{C_r m}\right)^{1/2}. \tag{3.99}$$

This parameter is independent of running speed; however, the constant η lines may be used to demonstrate the loss of stability with increasing running speed. Suppose, for example, that for a particular bearing $\eta = 0.01$. Then, as $p = \omega\sqrt{Cm/F_0}$ is increased from 1 to 10, Eq. (3.99) requires that S increase from 0.01 to 0.1, and Figure 3.31 demonstrates that the plain journal bearing, four-lobe bearing, and the offset bearing have all exceeded their onset speeds of instability.

The lines in Figure 3.31 having a negative slope of -0.25 illustrate the result of a change in clearance on both p and S. This statement is justified by expressing the explicit dependence of these variables on C_r by

$$S = BC_r^{-2}, \qquad p = AC_r^{1/2},$$

which implies

$$\log S = \log B - 2 \log C_r, \qquad \log p = \log A + \tfrac{1}{2} \log C_r.$$

From these equations, a perturbation in clearance δC_r yields

$$\delta(\log S) = \frac{\partial(\log S)}{\partial C_r} \delta C_r = -\frac{2 \log e}{C_r} \delta C_r,$$

$$\delta(\log p) = \frac{\partial(\log p)}{\partial C_r} \delta C_r = \frac{\log e}{2C_r} \delta C_r,$$

and the required slope in the $S - p$ graph of Figure 3.30 is

$$\frac{\delta(\log p)}{\delta(\log S)} = -\frac{1}{4}.$$

Hence, starting at an arbitrary point (S_0, p_0) on this graph, an increase in clearance moves one to the left and up the lines, while a decrease in clearance moves one to the right and down the lines.

The superior *rigid-rotor* stability performance of multilobe bearings, as compared to plain journal bearings, is most evident for higher Sommerfeld numbers (between 1 and 10) with small clearances. These operating conditions are characteristic of high-speed machinery for which plain journal bearings are impractical. At high Sommerfeld numbers, the offset-bearing geometry of Figure 3.29 is superior, trailed in order by the three-lobe, four-lobe, and elliptic bearings.

Note that the results of Figure 3.31 demonstrating improved stability characteristics of multilobe bearings as opposed to plain journal bearings only applies for the rigid-rotor model. The outcome for flexible rotors is generally less encouraging, as explained below.

The results of Section 3.5 demonstrate that the flexible-rotor onset speed of instability ω_{sf} is approximately related to the rotor-bearing critical speed ω_1 and the bearing whirl-frequency ratio Ω_w by the relationship

$$\omega_{sf} \cong \omega_1/\Omega_w$$

When a flexible rotor-bearing system is unstable due to its bearings, the bearing is generally much stiffer than the rotor. Hence, ω_1 is generally quite insensitive to increases in bearing stiffness accomplished by bearing replacement. The whirl-frequency ratios Ω_w provided in Table 3.2 give a much better indication of the prospects for an elevation of ω_{sf} by bearing replacement. Specifically, bearings for which the limiting value of Ω_w is smaller than one-half would be expected to yield higher onset speeds of instability than plain journal bearings. Table 3.2 is taken from Allaire and Flack (1980) for an example bearing with $\eta = 0.3$. Observe from a whirl-frequency-ratio basis that the elliptical bearing is only slightly better than a plain journal bearing with improved performance in ascending order for the four-lobe, three-lobe, and offset bearings.

Several additional precautionary notes are in order concerning the stability results of Figure 3.31. First, the results predict the onset speed of instability, but do not indicate the degree of instability, i.e., the violence of the motion at onset, and the growth of the unstable motion with increasing speed. While the Reynolds-equation model is appropriate for predicting the onset speed of instability, once this motion initiates it becomes questionable because of the uncertainty of cavitation boundaries during whirling motion.

An additional precaution concerning multilobe bearings is provided by Kirk (1978), whose experimental measurements demonstrate that the geometry of installed bearings may differ significantly from the design geometries, yielding markedly different stability characteristics from those calculated.

TABLE 3.2   Rigid-Rotor Stability Thresholds and Whirl-Frequency Ratios for Various Bearing Types with Bearing Parameter $\eta = 0.3$ [Allaire and Flack (1980)].

Bearing Type	Sommerfeld Number S	Whirl Frequency Ratio Ω_w	Stability Threshold p_s
Elliptical	1.15	0.495	3.82
Four-lobe	1.52	0.475	5.08
Three-lobe	1.86	0.470	6.21
Offset	3.62	0.361	12.07

Axial-Groove Bearing

The "plain" journal bearing discussed so far in this chapter is to some extent a mathematical abstraction, since axial grooves are customarily provided to supply lubricant to the bearing. Axial-groove bearings may employ two, three, or four grooves. They differ from the multilobe bearings considered above in that their bearing segments have zero preloads. As a rule, their onset speeds of instability are comparable to plain journal bearings. However, Kirk (1978) indicates that units operating in this type of bearing may sometimes perform much better than predicted due to installation distortions, which yield unintended bearing-arc preloads.

Pressure-Dam or Step Bearings

The pressure-dam or step bearing of Figure 3.32 is a commonly used industrial bearing. To the author's knowledge, Nicholas (1977) and Nicholas and Allaire (1980) have performed the only analysis of this useful bearing type, and the results and discussion of this subsection is based on their work.

The pressure-dam bearing achieves high operating eccentricities by the following two means:

(a) The relief track cut into the lower pad has the simple and direct effect of reducing the L/D ratio of the load-carrying bearing segment.

(b) The step in the upper pad yields a pressure distribution, which increases the downward load on the journal as the speed and Sommerfeld number increase.

Figure 3.33 compares the rigid-rotor stability characteristics of pressure-dam-bearing configurations (with and without steps and relief tracks), a two-axial-groove bearing, and a plain journal bearing. This figure demonstrates the high operating eccentricity characteristics of pressure-dam bearings and consequent improvement in stability performance. Bearing 5, which combines a properly designed step with a 25% relief track width, shows the best overall performance. Specifically, it has the largest region of infinite p_s, and good stability characteristics at high values for S.

The step angle $\theta_s = 125°$, and clearance ratio* $K' = h_1/h_2 = 3.0$ used in Figure 3.33 are shown in Nicholas (1977) and Nicholas and Allaire (1980) to be near-optimal overall choices from a stability viewpoint. Their additional results demonstrate the advisability of being on the "high" side of these optimal choices. Specifically, stability is not significantly degraded by moderately higher values of these parameters, but may be significantly reduced by the choice of lower values, particularly for K'.

---

*See Figure 3.32(a).

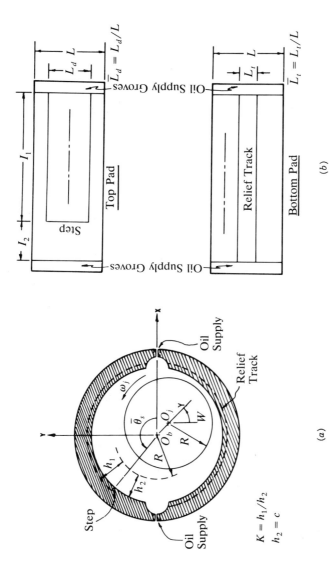

Figure 3.32 The pressure-dam or step bearing; (a) side view, (b) geometries of top and bottom pad [Nicholas (1977), Nicholas and Allaire (1980)]. Reprinted by permission of the Society of Tribologists and Lubrication Engineers. All rights reserved.

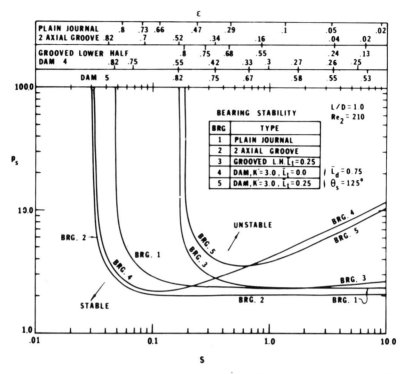

Figure 3.33   Rigid-rotor stability characteristics for five bearings, demonstrating the stability characteristics of a properly designed pressure-dam bearing [Nicholas (1977) and Nicholas and Allaire (1980)]. Two 20° oil inlet grooves located at 0° and 180° are used here for the axial-groove and pressure-dam bearing. Reprinted by permission of the Society of Tribologists and Lubrication Engineers. All rights reserved.

Whirl-frequency ratios have not as yet been published for pressure-dam bearings to provide an indication of their prospects for improving the stability of flexible-rotor systems. However, Leader et al. (1978) have compared experimental results for a flexible rotor on axial-groove bearings and on a properly designed pressure-dam bearing. Their critical-speed and onset-speed-of-instability results are as follows:

	ω_1 (rpm)	ω_{sf} (rpm)	ω_1/ω_{sf}
Axial groove	2750	5750	0.478
Pressure dam	3000	8400	0.357

From the reduced ω_1/ω_{sf} results, one could conclude that pressure-dam bearings are an attractive replacement option for elevating a flexible-rotor onset speed of instability.

Tilting-Pad Bearings

The bearings discussed above are designed to improve the stability character-istics of a journal-type bearing by increasing the operating eccentricity of the bearing or its loaded sector. The tilting-pad bearing design *eliminates* the bearing-stability problem by eliminating the stiffness cross-coupling terms in the bearing. To appreciate this statement, consider the lower pad of the tilting-pad bearing illustrated in Figure 3.34. The pad is pivoted by an axial pin, which will not support a moment; hence, the pad (force) reaction must pass through the pivot point, and a vertical load applied to the journal must also pass through the pivot point; i.e., for equilibrium, the pad pivots to yield a vertical reaction load through the pivot point. The important point to note here is that the reaction is developed without a sidewards displacement of

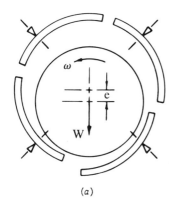

(a)

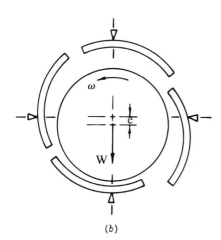

(b)

Figure 3.34 On-pad (a), and between-pad (b) loading for a four-pad tilting-pad bearing.

the loaded journal; i.e., a vertical load yields only a vertical displacement, and the cross-coupling is eliminated.

The situation illustrated in Figure 3.34(b) is called load-on-pad loading since a downward vertical load will pass through the lower pad pivot. The symmetric pad arrangement of Figure 3.34(a) yields a "between-pad" loading situation. Stiffness cross-coupling is only eliminated for these two types of symmetric pad-pivot arrangements. Tilting-pad bearings employing 3, 4, 5, 6, and 12 pads have been reported.

The fact that tilting-pad bearings do not themselves cause rotor instabilities means that they are customarily selected for use with high-speed flexible rotors, which are subject to a variety of instability mechanisms other than oil whip, e.g., labyrinth seals, turbine forces,* etc. Under these circumstances, the tilt-pad bearing design should provide the maximum possible positive stabilizing influence on the rotor to counteract external destabilizing factors. Nicholas et al. (1978) have examined the optimal stability design of a five-pad tilting-pad bearing with respect to pad preload, pad offset, and on-pad and between-pad bearings. *Marked* differences were found in the stabilizing capacity of bearing configurations due to pad preload and pad offset, with negligible differences (at high S) due to on-pad or between-pad loading conditions. For the rotor examined, the "best" design used zero preload and centrally pivoted pads.

Rolling-Element Bearings

Rolling-element bearings are commonly used in aircraft gas-turbine engines because of their durability and low power requirements. In rotordynamics analysis, they are modeled simply as speed-dependent, linear, direct springs. They provide minimal damping, and, in contrast to journal bearings, do not destabilize rotors. Recall, however, from Sections 1.8 and 1.9 that radial clearances at bearing locations can markedly influence rotor motion.

Analysis procedures have been available for some time which predict bearing stiffnesses [Jones (1960) and Harris (1991)]. Generally speaking, computer predictions of bearing stiffnesses provide ballpark estimates, which should be calibrated by test or operating experience. High-speed ball bearings normally have stiffnesses on the order of 0.88×10^8 N/m (0.5×10^6 lb/in.), while roller bearings are approximately 5–10 times stiffer. Axially preloaded bearings are frequently used in high-speed applications, and their stiffnesses depend on both the preload and running speed, as illustrated in the predictions of Figure 3.35. Experience with this bearing has demonstrated that the actual stiffnesses are approximately 50% softer than predicted. However, test results bear out the increase in stiffness with increasing preload, and the loss of stiffness with increasing speed.

*See Section 6.2.

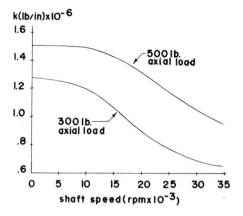

3.8   MODELING COMPLICATIONS IN BEARING ANALYSIS

From a rotordynamics viewpoint, the principal requirements of bearing analysis are the calculations of (a) bearing equilibrium positions as a function of operating conditions and bearing parameters, and (b) stiffness and damping coefficients for synchronous-response and stability calculations. In addition, a general nonlinear bearing model is occasionally required for transient analysis. The procedures which have evolved to meet these requirements must not only account for the different types of bearing geometries discussed above, but must also realistically account for cavitation, turbulence, and fluid compressibility effects. The balance of this section deals with the modeling requirements for cavitation, turbulence, and compressibility effects.

Turbulence Models for Journal Bearings

The generally accepted approach for modeling turbulence in journal bearings was developed by Ng (1964) and Ng and Pan (1965). Their analysis, based on eddy viscosity effects, yields a modified Reynolds equation for an incompressible fluid of the form

$$\frac{1}{R^2}\frac{\partial}{\partial\theta}\left(\frac{H^3}{\mu}G_\theta\frac{\partial p}{\partial\theta}\right) + \frac{\partial}{\partial z}\left(\frac{H^3}{\mu}G_z\frac{\partial p}{\partial z}\right) = \left[C_r\dot\varepsilon\cos\theta + C_r\varepsilon(\dot\beta - \bar\omega)\sin\theta\right].$$

$$(3.100)$$

By comparison to Eq. (3.18), this equation differs from the laminar-flow Reynolds number only by the additional parameters G_θ and G_z. These parameters are functions of the local Reynolds number $\mathcal{R}_\theta = R\omega H/\nu$, as given in Ng and Pan and reduce to the linear "result" of $\frac{1}{12}$ for "small" values of $\mathscr{R}_\theta$. The quotients $\mu/12G_\theta$ and $\mu/12G_z$ may be viewed as effective local viscosities. The "turbulent" Reynolds equation is only moderately more complicated than the laminar Reynolds equation (3.18) and is not markedly more difficult to solve by numerical means. The laminar model is presumed to be valid out to the critical Taylor number $(C_r/R)^{1/2}\mathscr{R}_\theta = 41.3$, with complete turbulence assumed to exist for $\mathscr{R}_\theta$ on the order of 1500.

The turbulent Reynolds equation (3.100) has been verified as an adequate model for plain journal and partial-arc bearings by Orcutt and Arwas (1967) and for tilting-pad bearings by Orcutt (1967). These investigators used Eq. (3.100) to calculate bearing equilibrium eccentricity ε_0 and attitude angle γ_0. They also calculated stiffness and damping coefficients for small motion about the equilibrium position, and verified the accuracy of these results by comparing measured rotor response with calculated rotor response (based on the calculated bearing properties). For the plain journal and partial-arc bearings, these comparisons included calculated and measured onset speeds of instability. The correlation between theory and experiment was generally excellent. For the plain and partial-arc bearing, correlation in the turbulent flow region was equivalent to that attained in the laminar region.

While Ng and Pan (1965) calculate the parameters G_θ, G_z based on law-of-the-wall velocity profiles, other investigators have derived these coefficients based on correlation with shear-stress relationships and basic wall-friction laws. Black (1970a, 1970b) has taken this "bulk-flow" approach, together with Hirs (1973). Hirs's model will be used in Chapter 4 in the analysis of turbulent annular seals.

Compressible Lubricants

One approach to reducing power losses in bearings is to replace an incompressible lubricant with a compressible gas. The direct potential advantage of this approach is an increase in kinematic viscosity, which implies that the top operating speed ω can be increased without pushing the Reynolds number $\mathscr{R}_\theta = R\omega H/\nu$ into the turbulent regime. The appropriate Reynolds equation for a compressible ideal gas under isothermal conditions is

$$\frac{1}{R^2}\frac{\partial}{\partial\theta}\left(\frac{h^3}{12\mu}\bar{p}\frac{\partial\bar{p}}{\partial\theta}\right) + \frac{\partial}{\partial z}\left(\frac{h^3}{12\mu}\bar{p}\frac{\partial\bar{p}}{\partial z}\right) = \frac{1}{2}R\omega\frac{\partial(\bar{p}h)}{\partial\theta} + \frac{\partial(\bar{p}h)}{\partial t}, \quad (3.101)$$

where $\bar{p} = p/p_a$, with p_a the ambient pressure, Pinkus and Sternlicht (1961) and Gross (1962). This equation is obtained from the development of Section 3.2 if the compressible continuity equation is used instead of Eq. (3.2), and the lubricant density is assumed to be proportional to pressure.

Cavitation

Figure 3.36 illustrates the cavitation phenomenon encountered in a plain journal bearing at equilibrium. In the absence of cavitation, negative pressure magnitudes experienced in the diverging section of the film would be comparable to the positive pressures developed in the converging section. In fact, pressure drops only to a limiting cavitation pressure p_{cav}, and is prevented from dropping further by lubricant boiling and/or escape of dissolved gases from the lubricant. Streamers of flow interspersed with vapor result in the cavitation region, and *the flow in this region is not defined by the Reynolds equation* [Floberg (1965)].

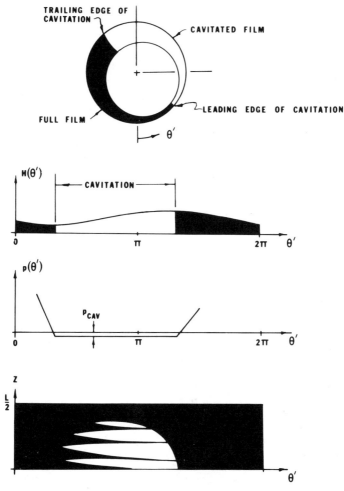

Figure 3.36   Cavitation region in a plain journal bearing.

The determination of the free boundary between the full-film and cavitated regions of a bearing is one of the principal complications involved in numerical bearing calculations. The "obvious" boundary condition employed in defining this free boundary is the requirement $p = p_{cav}$. The additional pressure boundary condition is normally obtained from the requirement of flow continuity *at the leading edge of cavitation*. Flow rate, per unit width, is defined within the full-film region by integrating the velocity components of Eq. (3.8) over the fluid-film height to obtain

$$q_\theta = R\omega_j \left(\frac{H}{2} \right) - \frac{H^3}{12\mu} \frac{1}{R} \frac{\partial p}{\partial \theta},$$

$$q_z = -\frac{H^3}{12\mu} \frac{\partial p}{\partial z}. \tag{3.102}$$

The first term in q_θ is shear flow induced by no-slip boundary conditions at the rotating journal and stationary bearing. The flow rate q_z and the second term in q_θ are induced by the pressure gradient within the bearing. In the full-fluid film preceding the free-boundary condition, the pressure is dropping with increasing θ toward the limiting value of P_{cav}; hence, $\partial p / \partial \theta < 0$, and $q_\theta > R\omega_j(H/2)$. However, within the cavitation region the pressure is constant, $\partial p / \partial \theta = 0$, and $q_\theta \leq R\omega_j(H/2)$. There is accordingly the clear potential for a flow discontinuity on the order of $(H^3/12\mu)(1/R)(\partial p/\partial \theta)$ at the leading edge of cavitation. A similar argument can be made with regard to q_z, since within the cavitation region $\partial p/\partial z$ is also zero. Hence, to ensure flow continuity at the leading edge of cavitation, the following "Reynolds" pressure boundary conditions are imposed:

$$p = p_{cav}; \quad \frac{1}{R} \frac{\partial p}{\partial \theta} = \frac{\partial p}{\partial z} = 0. \tag{3.103}$$

These pressure boundary conditions do not eliminate flow discontinuity at the trailing edge of the cavitation region.

The π Ocvirk and Sommerfeld pressure distributions developed in Section 3.3 violate continuity, since the pressure gradient is not required to vanish at the leading edge of cavitation. The boundary conditions and resultant pressure distributions employed to obtain these π bearing models are customarily designed as "half-Sommerfeld," as opposed to the "full-Sommerfeld" designation for 2π bearing models.

The flow discontinuity involved in the π Ocvirk model can be evaluated at equilibrium ($V_s = C_r\bar{\omega}\varepsilon$, $\alpha = -\pi/2$) from Eqs. (3.102) and (3.26). For $\theta = \pi$, the circumferential flow from the fluid-film into the cavitation boundary is

$$Q_\theta = \int_{-L/2}^{L/2} q_\theta d_z = R\omega_j L \left(\frac{C_r}{2} \right) (1 - \varepsilon) \left[1 + \frac{(L/D)^2}{3(1 - \varepsilon)} \right].$$

However, the flow from the boundary into the cavitation region is only

$$Q_\theta = R_{\omega_j} L \left(\frac{C_r}{2} \right) (1 - \varepsilon).$$

The fact that the error involved is proportional to $(L/D)^2$ provides the basic justification for the short-bearing model. Specifically, for small L/D, the flow induced by the circumferential pressure gradient becomes negligible.

The absence of appropriate integration constants in the Ocvirk bearing model prevents implementation of the Reynolds boundary condition of Eq. (3.103); however, they may be employed to obtain an improved version of the Sommerfeld (long) bearing pressure distribution of Eq. (3.29). Pinkus and Sternlicht (1961) satisfy the boundary conditions

$$p(\theta = 0) = 0, \qquad p = \left. \frac{\partial p}{\partial \theta} \right|_{\bar{\theta}} = 0.$$

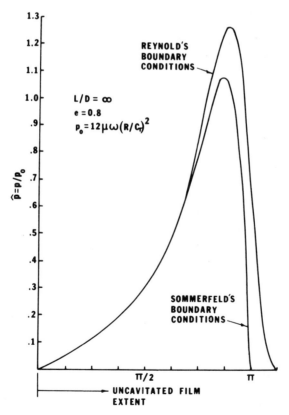

Figure 3.37 Pressure solutions for the half-Sommerfeld and Reynolds boundary conditions for the Sommerfeld (long) bearing model [Pinkus and Sternlicht (1961)].

In words, the full fluid film begins at the maximum clearance, and extends to $\theta = \bar{\theta}$, where both p and $\partial p/\partial\theta$ are zero. A comparison of the two solutions is provided in Figure 3.37, and demonstrates that the Reynolds boundary condition yields a longer uncavitated pressure region and higher peak pressures than the half-Sommerfeld. Consequently, the Reynolds solution yields slightly smaller eccentricities and attitude angles than the half-Sommerfeld solution for the infinitely long Sommerfeld bearing model. The differences between the two solutions increase with increasing L/D and are maximum for the Sommerfeld bearing model.

3.9   FINITE-DIFFERENCE AND FINITE-ELEMENT SOLUTIONS FOR THE REYNOLDS EQUATION

As noted earlier, analytical solutions for bearings have only been developed for the plain journal bearing, largely because of its circumferential symmetry. Closed-form analytical solutions for more complicated bearing geometries, turbulent bearings, and bearings employing compressible fluids are simply not feasible. The numerical procedures which have been developed for the solution of more general bearing configurations and operating conditions are *briefly* introduced in this section.

Finite-Difference Solutions

The finite-difference procedure was the first general numerical method employed for the direct solution of the Reynolds equation to develop equilibrium characteristics and stiffness and damping coefficients. The initial solutions for plain journal bearings were developed by Lund and Sternlicht (1962). Lund is responsible for most of the subsequent refinements of this procedure, and with Thomsen (1978) has provided an excellent review of the finite-difference approach as applied to partial-arc or tilting-pad bearings. The balance of this subsection is largely based on their reference.

From Eq. (3.19), the governing incompressible-fluid Reynolds equation may be stated as follows for a bearing pad (lobe, arc, etc.):

$$\frac{1}{R^2}\frac{\partial}{\partial\bar{\theta}}\left(\frac{H^3}{12\mu}\frac{\partial P}{\partial\bar{\theta}}\right) + \frac{\partial}{\partial z}\left(\frac{H^3}{12\mu}\frac{\partial P}{\partial z}\right) = \bar{\omega}\frac{\partial H}{\partial\bar{\theta}} + \frac{\partial H}{\partial t}, \qquad (3.104)$$

with $\bar{\theta}$ and the additional pad geometry variables illustrated in Figure 3.38. In this figure, the bearing and journal centers are identified by O_b and O_j, respectively, and the pad center of curvature is identified by O_p. The angle γ continues to define the journal attitude relative to the $-Y$ axis, while γ_p defines the journal attitude relative to a vertical line through the pad center of curvature O_p. Note that point O_p is fixed relative to the bearing by the

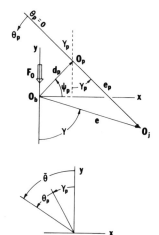

Figure 3.38 Bearing, journal, and pad geometry.

distance d_p and angle ψ_p. Hence, the pad eccentricity e_p and attitude γ_p are defined in terms of the journal eccentricity e and attitude γ by

$$e_p \sin \gamma_p = e \sin \gamma - d_p \cos \psi_p,$$
$$e_p \cos \gamma_p = e \cos \gamma + d_p \sin \psi_p. \tag{3.105}$$

By comparison to Figure 3.1, the angle θ_p would be a natural choice to define the pad clearance function H, since it originates on the line of centers defined by O_p, O_j. However, since forces and dynamic coefficients are required in the X-Y frame, Lund and Thomsen employ $\tilde{\theta}$ instead to obtain

$$H = C_r + e_p \cos \theta_p = C_r + e_p \cos\left(\tilde{\theta} - \gamma_p\right). \tag{3.106}$$

The explicit boundary condition to be satisfied in solving Eq. (3.104) is that the pressure vanish at the pad boundaries. When symmetry about the bearing midplane is accounted for, a pressure solution is only required for half the bearing $(0 \le z \le L/2)$, and the appropriate boundary conditions are

$$\frac{\partial p}{\partial z} = 0 \text{ at } z = 0,$$

$$p = 0 \text{ at } z = L/2 \quad \text{and} \quad \tilde{\theta} = \tilde{\theta}_{lp}, \tilde{\theta}_{tp}, \tag{3.107}$$

where $\tilde{\theta}_{lp}$ and $\tilde{\theta}_{lp}$ identify the leading* and trailing edge of the *pad*. The free-boundary condition to be defined at the trailing edge of cavitation will be discussed later.

*See Figure 3.39.

An additional bearing constraint is provided by the requirement that the static bearing-reaction force vector, defined by the components

$$\begin{Bmatrix} F_{X0} \\ F_{Y0} \end{Bmatrix} = 2\sum_p \int_0^{L/2} \int_{\bar{\theta}_{lp}}^{\bar{\theta}_{tp}} p_0 \begin{Bmatrix} \sin \bar{\theta} \\ -\cos \bar{\theta} \end{Bmatrix} R\, d\bar{\theta}\, dz, \qquad (3.108)$$

be vertical; i.e., $F_{X0} = 0$. In this relationship, p_0 is the static solution to Eq. (3.104); i.e., $\partial H / \partial t = 0$, and the summation is to be taken over all of the bearing pads. For a given running speed, the calculation proceeds from a specified journal eccentricity e_0 and assumed journal attitude γ, with the various pad eccentricities and attitudes defined by Eq. (3.105). The angle γ is calculated iteratively to yield $F_{X0} = 0$ to any desired degree of accuracy. The corresponding F_{Y0} equals the load F_0 and defines the Sommerfeld number

$$S = \mu \left(\frac{\omega}{2\pi} \right) \left(\frac{R}{C_r} \right)^2 \frac{DL}{F_0}. \qquad (3.44)$$

Lund and Sternlicht initially calculated bearing-stiffness coefficients by a numerical differentiation approach, which involved introducing a small perturbation from the equilibrium position, calculating the difference in the bearing-reaction force, and forming the quotients

$$K_{XX} \cong \frac{\delta F_{x0}}{\delta x}, \qquad K_{XY} \cong \frac{\delta F_{x0}}{\delta y}, \qquad \text{etc.}$$

This approach is fundamentally approximate and causes discernible problems at large and small eccentricities. The desired coefficient definitions, corresponding to the above approximate relationships, are

$$K_{XX} = \left(\frac{\partial F_X}{\partial X} \right)_0 = 2\sum_p \int_0^{L/2} \int_{\bar{\theta}_{lp}}^{\bar{\theta}_{tp}} \frac{\partial p_0}{\partial X} \sin \bar{\theta} R\, d\bar{\theta}\, dz,$$

$$K_{XY} = \left(\frac{\partial F_X}{\partial Y} \right)_0 = -2\sum_p \int_0^{L/2} \int_{\bar{\theta}_{lp}}^{\bar{\theta}_{tp}} \frac{\partial p_0}{\partial Y} \cos \bar{\theta} R\, d\bar{\theta}\, dz. \qquad (3.109)$$

Hence, one would prefer to directly develop the partial derivatives $\partial p_0 / \partial X$, $\partial p_0 / \partial Y$, etc., and eliminate the requirement for numerical differentiation. In fact, this is precisely what Lund accomplished by introducing the pressure perturbations

$$p = p_0 + \Delta p; \qquad \Delta p = p_X\, \delta X + p_Y\, \delta Y + p_{\dot{X}}\, \delta \dot{X} + p_{\dot{Y}}\, \delta \dot{Y}, \qquad (3.110)$$

where $p_X = \partial p/\partial X$, etc. The associated clearance function is

$$H = H_0 + \delta X \sin \tilde{\theta} - \delta Y \cos \tilde{\theta}, \tag{3.111}$$

where

$$H_0 = C_r + e_{p0} \cos(\tilde{\theta} - \gamma_{p0}), \tag{3.112}$$

and $\delta X, \delta Y$ are perturbations in the *journal* position. Equation (3.111) is obtained by noting from Figure 3.38 that

$$e \cos \gamma = X_0 + \delta X, \qquad e \sin \gamma = -(Y_0 + \delta Y),$$

and substituting into Eqs. (3.105) and (3.106).

Substitution from Eqs. (3.110) and (3.111) into Eq. (3.104) yields the following first-order perturbation equations:

$$\frac{1}{R^2} \frac{\partial}{\partial \tilde{\theta}} \left(\frac{H_0^3}{12\mu} \frac{\partial p_i}{\partial \tilde{\theta}} \right) + \frac{\partial}{\partial z} \left(\frac{H_0^3}{12\mu} \frac{\partial p_i}{\partial z} \right) = f_i; \qquad i = 0, X, Y, \dot{X}, \dot{Y}, \tag{3.113}$$

where

$$f_0 = \bar{\omega} \frac{\partial H_0}{\partial \tilde{\theta}}, \qquad f_{\dot{X}} = \sin \tilde{\theta}, \qquad f_{\dot{Y}} = \cos \tilde{\theta},$$

$$f_X = \bar{\omega} \left(\cos \tilde{\theta} - \frac{3 \sin \tilde{\theta}}{H_0} \frac{\partial H_0}{\partial \tilde{\theta}} \right) - \frac{H_0^3}{4\mu R^2} \frac{\partial p_0}{\partial \theta} \frac{\partial}{\partial \theta} \left(\frac{\sin \tilde{\theta}}{H_0} \right),$$

$$f_Y = \bar{\omega} \left(\sin \tilde{\theta} + \frac{3 \cos \tilde{\theta}}{H_0} \frac{\partial H_0}{\partial \tilde{\theta}} \right) + \frac{H_0^3}{4\mu R^2} \frac{\partial p_0}{\partial \theta} \frac{\partial}{\partial \theta} \left(\frac{\cos \tilde{\theta}}{H_0} \right).$$

The boundary conditions for the variable set p_i at $z = 0, L/2$ are

$$p_i = 0 \text{ at } z = L/2,$$

$$\frac{\partial p_i}{\partial z} = 0 \text{ at } z = 0. \tag{3.114}$$

As noted in the preceding section, the Reynolds boundary conditions for cavitation in a plain journal bearing require that the pressure vanish at the trailing edge of cavitation and that both the pressure and its gradient vanish at the leading edge. Under static conditions, the trailing edge of cavitation in a plain journal bearing occurs at the maximum clearance.

For a bearing *pad*, the trailing edge is generally taken to be the actual physical boundary defined by a groove. However, if the groove is located in a diverging section of the fluid film, the *pad* edge is set at the maximum film thickness location. In either case, the boundary conditions are

$$p_i = 0 \text{ at } \tilde{\theta} = \tilde{\theta}_{lp}. \tag{3.115}$$

If the pad trailing-edge groove is in a converging section of the film, a similar simple set of boundary conditions applies. However, if the pad trailing-edge groove is in a diverging section of the film, the free-boundary condition due to cavitation must be determined as part of the overall problem solution; i.e., the curve $\tilde{\theta}(z)_{tp}$ must be determined. The Reynolds boundary condition requires that both the pressure and its gradient normal to the curve $\tilde{\theta}(z)_{tp}$ vanish on the leading edge of the cavitation boundary; i.e.,

$$p = \frac{\partial p}{\partial n} = 0 \rightarrow p = \frac{\partial p}{\partial \tilde{\theta}} = \frac{\partial p}{\partial z} = 0 \text{ at } \tilde{\theta} = \tilde{\theta}(z)_{tp}. \tag{3.116}$$

The development of corresponding boundary conditions for the variables p_i proceeds as follows. Fluid-film pressure at the cavitation boundary is required to remain zero under small perturbations of its nominal boundary curve location; i.e., for the perturbation from the point $(\tilde{\theta}_0, z_0)$ to $(\tilde{\theta}_0 + \delta\tilde{\theta}, z_0 + \delta z)$,

$$p(\tilde{\theta}) = 0 = p(\tilde{\theta}_0, z_0) + \left(\frac{\partial p}{\partial \tilde{\theta}}\right)_0 \delta\tilde{\theta} + \left(\frac{\partial p}{\partial z}\right)_0 \delta z$$

$$= p_0(\tilde{\theta}_0, z_0) + \delta p(\tilde{\theta}_0, z_0) + \left(\frac{\partial p_0}{\partial \tilde{\theta}}\right)_0 \delta\tilde{\theta} + \left(\frac{\partial p_0}{\partial z}\right)_0 \delta z. \tag{3.117}$$

From Eq. (3.116), the static pressure and its derivatives must satisfy

$$p_0 = \frac{\partial p_0}{\partial \tilde{\theta}} = \frac{\partial p_0}{\partial z} \text{ at } \tilde{\theta} = \tilde{\theta}(z)_{tp}, \tag{3.118}$$

and from Eq. (3.105), the perturbed pressure δp must also vanish within the cavitation regime and on the boundary which yields

$$p_i = 0 \text{ at } \tilde{\theta} = \tilde{\theta}(z)_{tp}. \tag{3.119}$$

This completes the boundary-condition requirements for the p_i variables.

The finite-difference problem involves solving Eqs. (3.113) for the variables p_i subject to the boundary conditions of Eqs. (3.114), (3.115), (3.118), and (3.119). Figure 3.39 illustrates the finite-difference grid employed by Lund and Thomsen. The finite-difference increments are observed to be $\Delta\tilde{\theta} = (\tilde{\theta}_{tp} - \tilde{\theta}_{lp})/m$; $\Delta z = L/2n$. The difference equations for point (i, j) may be stated

$$b_{j+1}p_{i,j+1} + b_{j-1}p_{i,j-1} + \alpha_j p_{i-1,j} + \alpha_j p_{i,j} + \alpha_j p_{i+1,j} = g_{i,j}, \tag{3.120}$$

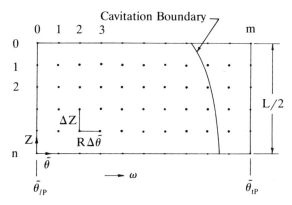

Figure 3.39 Finite-difference grid for solution of a bearing pad.

where p_{ij} is the pressure at the point $(\tilde{\theta}_j = \tilde{\theta}_{lp} + j\,\Delta\theta,\ z_i = L/2 - i\,\Delta z)$, g_{ij} is the right side of Eq. (3.113) evaluated at the point (i, j), and the coefficients are

$$b_{j-1} = H_{j-1/2}^3 / 12\mu(R\,\Delta\tilde{\theta})^2,$$

$$b_{j+1} = H_{j+1/2}^3 / 12\mu(R\,\Delta\tilde{\theta})^2,$$

$$\alpha_j = H_j^3 / 12\mu\,\Delta z^2, \qquad a_j = -(b_{j-1} + b_{j+1} + 2\alpha_j). \qquad (3.121)$$

A recursive procedure developed by Castelli and Pirvics (1967) is most advantageously employed by Lund and Thomsen; this procedure is based on the observation that the difference equation involves only immediately adjacent points. Hence, the following expression holds for adjacent *columns* of pressure variables p_i in the matrix of Figure 3.39:

$$[A_j](p_i)_j + [B_j](p_i)_{j-1} + [C_j](p_i)_{j+1} = (G_i)_j; \qquad j = 1, 2, \ldots, m, \tag{3.122}$$

where the ith entry of $(p_i)_j$ and $(G_i)_j$ are $p_{i,j}$ and $g_{i,j}$, respectively. Further, $[A_j]$ is the following tridiagonal $n \times n$ matrix

$$[A_j] = \begin{bmatrix} a_j & \alpha_j & 0 & \cdots & & & \\ \alpha_j & a_j & \alpha_j & 0 & & & \\ 0 & & & & & & \\ & & & & 0 & 0 & 0 \\ & & & & \alpha_j & a_j & \alpha_j \\ & & & & 0 & 2\alpha_j & a_j \end{bmatrix},$$

whose first and last rows account for the $z = 0$, $L/2$ boundary conditions of Eq. (3.105). The matrices $[B_j], [C_j]$ are diagonal, with jth diagonal entries of b_{j-1} and b_{j+1}, respectively.

Substitution of the recurrence relationship

$$(p)_{j-1} = [D_{j-1}](p_i)_j + (E)_{j-1} \qquad (3.123)$$

into Eq. (3.122) yields

$$[D_j] = -[[A_j] + [B_j][D_{j-1}]]^{-1}[C_j], \qquad j = 1, 2, \ldots, m-1,$$

$$(E)_j = [[A_j] + [B_j][D_{j-1}]]^{-1}((G_i)_j - [B_j](E)_{j-1}). \qquad (3.124)$$

Putting aside cavitation complications for the moment, the recurrence relationship works by proceeding from left to right across the pad successively calculating (and storing) $[D_j]$'s and (E_j)'s. The choice $[D_0] = 0$, $(E)_0 = (p)_0 = 0$ satisfies Eq. (3.123) and is used to initiate the calculations of Eq. (3.124). When these calculations are completed by the calculation of $[D_{m-1}], (E)_{m-1}$, one marches back from right to left calculating pressures from Eq. (3.123), with the starting value $(p_i)_m = 0$.

Cavitation is accounted for in calculating p_0 by setting any calculated negative pressure in column $(p_i)_j$ to zero* before calculating the succeeding $(p_i)_{j-1}$ vector. Lund and Thomsen state that the pressure profile calculated in this manner will satisfy the zero-pressure-gradient requirement at the cavitation interface but suggest a precise determination of the cavitation boundary curve for each i row by interpolation based on the three calculated positive pressures adjacent to the cavitation region. An accurate determination of the cavitation region permits an adjustment of the finite-difference grid to improve the accuracy in a subsequent calculation.

As noted earlier in discussing the static-load definition of Eq. (3.108), the journal angle γ is calculated iteratively to yield a vertical bearing reaction load; i.e., $F_{X0} = 0$. When this condition is satisfied, the static-pressure-field calculation is complete; i.e., p_0 is defined at each of the grid points of Figure 3.39.

The problem is completed by the calculation of $p_X, p_Y, p_{\dot{X}}, p_{\dot{Y}}$ and subsequent integration to obtain the stiffness and damping coefficients. This undertaking is markedly simplified by noting that these variables have precisely the same boundary conditions as p_0; hence, they are also governed by Eq. (3.122), differing only in the entries of $(G_i)_j$. The significance of this observation is that the matrices $[[A_j] + [B_j][D_{j-1}]]^{-1}$ may be stored when calculating the p_0 distribution, reducing the computer-time requirements for

---

*This only applies to p_0 in the cavitated region. Calculated negative values for the derivatives p_X, p_Y, etc. are expected in the full-film region and are not set to zero.

determining $p_X, p_Y, p_{\dot{X}}, p_{\dot{Y}}$ distributions. After these functions have been determined, the stiffness and damping coefficients are readily calculated following the procedure suggested by Eq. (3.109); i.e.,

$$K_{XX} = 2\sum_{p} \int_0^{L/2} \int_{\bar{\theta}_{lp}}^{\bar{\theta}_{tp}} p_X \sin \bar{\theta} R \, d\bar{\theta} \, dz,$$

$$K_{XY} = 2\sum_{p} \int_0^{L/2} \int_{\bar{\theta}_{lp}}^{\bar{\theta}_{tp}} p_Y \cos \bar{\theta} R \, d\bar{\theta} \, dz. \tag{3.125}$$

In summary, the finite-difference calculation procedure involves the following specific steps:

(a) A value of γ is guessed for a fixed journal eccentricity e_0, clearance C_r, viscosity μ, and running speed ω_j.

(b) Pad eccentricities e_p and attitudes γ_p are calculated from Eq. (3.105).

(c) The static-pressure solution is calculated for each pad based on the recurrence relationship of Eqs. (3.123) and (3.124), and the matrices $[[A_j] + [B_j][D_{j-1}]^{-1}]$ developed in this calculation are stored.

(d) The journal reaction forces are calculated from Eq. (3.108) to determine if the requirement $F_{X0} = 0$ is satisfactorily met. If it is not, a new γ is selected, and steps (b) and (c) are repeated until a satisfactory solution is obtained.

(e) The functions $p_X, p_Y, p_{\dot{X}}, p_{\dot{Y}}$ are calculated from Eqs. (3.123) and (3.124) using the previously stored matrices $[[A_j] + [B_j][D_{j-1}]^{-1}]$.

(f) The stiffness and damping coefficients are calculated following the procedure suggested by Eq. (3.125).

The finite-difference procedure works equally well for pivoted-pad, [Lund (1964)] gas [Lund (1968)], or turbulent bearings [Orcutt and Arwas (1967)]. However, the procedure is not well suited to bearings with clearance discontinuities such as the pressure-dam bearing of Section 3.7, for which the finite-element procedure of the following subsection is preferred.

Finite-Element Methods for Bearing Analysis

The FEM (finite-element method) was developed for incompressible lubrication by Reddi (1969). However, the contents of this subsection are largely based on the papers by Booker and Heubner (1972) and Allaire et al. (1977) and the thesis of Nicholas (1977). The simplest possible version of the method, consistent with the preceding material of this chapter, is presented.

The FEM, as applied to the rotordynamics vibration problem, was introduced in Sections 2.5 and 2.6. As applied to vibration analysis, the FEM involves the following characteristic features:

(a) The spatial distribution of the independent variable (or variables) is defined within the element in terms of discrete variables at the elements' boundary by interpolation functions.*

(b) The governing equations can be developed as a necessary condition for the extremization of a functional. This latter condition may not be obvious in the development of Sections 2.5 and 2.6; however, note that Lagrange's equations of motion,

$$\frac{d}{dt}\left(\frac{\partial \mathcal{L}}{\partial \dot{q}_i}\right) - \left(\frac{\partial \mathcal{L}}{\partial q_i}\right) = 0, \qquad i = 1, 2, \ldots, n,$$

follow from minimizing the action integral

$$J = \int_{t_0}^{t_1}(T + V)\, dt,$$

where T and V are the kinetic and potential energies, respectively.

While a variational principle is not customarily invoked to derive governing lubrication equations, Reddi (1969) has developed a functional for which the governing differential equations follow as necessary conditions for an extremum. However, before considering this functional, a thoughtful restatement of the governing equations in general vector form is helpful. Figure 3.40 illustrates a fluid-film segment separating two rigid bearing surfaces. It differs from the model of Figure 3.1 in that the bearing surface velocities $\mathbf{U}_1, \mathbf{U}_2$ have components in the z direction as well as the x. A vector statement of the Reynolds equation for the fluid element of Figure 3.39 is

$$\nabla \cdot \left(\frac{H^3}{12\mu}\nabla p\right) = \nabla \cdot \rho H \mathbf{U} + \rho \frac{\partial H}{\partial t}, \qquad (3.126)$$

where

$$\mathbf{U} = \mathbf{i}(U_{1x} + U_{2x})/2 + \mathbf{k}(U_{1z} + U_{2z})/2.$$

Both flow rates and pressures are employed as variables in the FEM in a manner analogous to the use of both deflections and reaction forces as variables in the beam FEM of Sections 2.5 and 2.6. This is in contrast to the finite-difference method of the preceding subsection, for which nodal pressures are the only variables. To define the flow rates, Eq. (3.8) is integrated to

---

*Equations (2.79) and (2.80) in Section 2.5 and Eqs. (2.95)–(2.96) in Section 2.6 illustrate this feature.

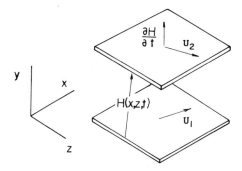

Figure 3.40 General fluid-film geometry for finite-element development.

yield the following average (across the film) fluid velocity components:

$$\tilde{u} = U_x - \frac{H^2}{12\mu}\frac{\partial p}{\partial x}, \qquad \tilde{w} = U_z - \frac{H^2}{12\mu}\frac{\partial p}{\partial z}.$$

The vector version of these equations is

$$\tilde{u} = \mathbf{U} - \frac{H^2}{12\mu}\nabla p, \qquad (3.127)$$

and the corresponding mass flow rate per unit fluid-element width is simply

$$\mathbf{q} = \rho H \mathbf{U} - \rho\frac{H^3}{12\mu}\nabla p. \qquad (3.128)$$

The boundary conditions for a bearing (or an element) are expressed in one of the two following terms:

(a) Known pressures are specified, generally to be zero at the edges of a bearing or bearing pad.
(b) Known flow rates are specified, generally to be zero at a fixed boundary; e.g., in a plain journal bearing, symmetry requires that the flow across the bearing midplane be zero.[*]

Hence, in a bearing analysis, one customarily has pressure boundary conditions specified over a portion of the solution boundary (bearing edge) and flow or velocity boundary conditions specified over the *balance*.

In summary, a variational principle for an incompressible lubricant should yield both the Reynolds equation (3.126) and appropriate pressure and flow boundary conditions. The required variational principle as derived by Reddi

---

[*] In the finite-difference method this boundary condition is enforced by requiring that $\partial p/\partial z = 0$.

(1969) follows. Determine the function $p(x, z)$ which minimizes the functional

$$J = \int_A \left[\left(\frac{\rho H^3}{24\mu} \nabla p - \rho H \mathbf{U} \right) \cdot \nabla p + p\rho \frac{\partial H}{\partial t} \right] dA + \int_{S_q} (\mathbf{q} \cdot \mathbf{n}) p \, ds, \quad (3.129)$$

and is specified on the S_p portion of the boundary of fluid surface A. In the second integral of this functional, $\mathbf{n}$ is the unit normal vector on the edge or boundary of A, and S_q is the balance of the boundary; i.e., the total boundary surface is $S = S_p + S_q$. The first necessary condition that a function $p^*(x, z)$ define a minimum for this function is obtained by defining the family of nearby functions

$$p = p^* + \delta p = p^* + \varepsilon \eta(x, z), \quad (3.130)$$

where ε is a small quantity [Gelfand and Fomin (1963)]. Substituting this definition into Eq. (3.129) and discarding the higher-order terms in ε yields

$$J = J_0 + \delta J,$$

where

$$\delta J = \varepsilon \int_A \left[\mathbf{G} \cdot \nabla \eta + \rho \frac{\partial H}{\partial t} \eta \right] dA + \varepsilon \int_{S_q} \eta (\mathbf{q} \cdot \mathbf{n}) \, ds,$$

$$\mathbf{G} = \frac{\rho H^3}{12\mu} \nabla p - \rho H \mathbf{U}.$$

The first necessary condition that p^* minimize J is that $\delta J = 0$. To express this requirement in terms of p and q, δJ is first restated as

$$\delta J = -\varepsilon \int_A \eta \left[\nabla \cdot \mathbf{G} - \rho \frac{\partial H}{\partial t} \right] dA + \varepsilon \int_A \nabla \cdot \eta \mathbf{G} \, dA + \varepsilon \int_{S_q} (\mathbf{q} \cdot \mathbf{n}) \, ds,$$

and then from Green's theorem, $\int_A \nabla \cdot \eta \mathbf{G} \, dA = \int_S \eta (\mathbf{G} \cdot \mathbf{n}) \, ds$, the desired result is

$$0 = -\varepsilon \int_A \eta \left[\nabla \cdot \left(\frac{\rho H^3}{12\mu} \nabla p - \rho H \mathbf{U} \right) - \rho \frac{\partial H}{\partial t} \right] dA$$

$$+ \varepsilon \int_{S_q} \eta \left[\left(\frac{\rho H^3}{12\mu} \nabla p - \rho H \mathbf{U} + \mathbf{q} \right) \cdot \boldsymbol{\eta} \right] ds$$

$$+ \varepsilon \int_{S_p} \eta \left(\frac{\rho H^3}{12\mu} \nabla p - \rho H \mathbf{U} \right) \cdot \mathbf{n} \, ds.$$

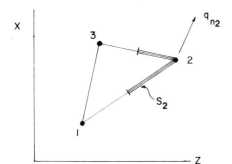

Figure 3.41 Fluid-film finite elements.

Since $\eta(x, z)$ is a general nonzero function on A, the term within the square brackets of the first integral must vanish on A; this requirement is seen to yield the Reynolds equation (3.126). A similar argument is made for the term within square brackets of the second integral, yielding Eq. (3.128) as a free-boundary condition. Finally, since p is specified on S_p, its variation* on this boundary must vanish and the last integral is also zero.

In summary, Reddi's variational principle for an incompressible fluid gives as necessary conditions, the Reynolds equation defined on A and the exit flow Eq. (3.128) on that segment of the boundary S_q for which pressure is not specified. The next step in the development of the FEM is the introduction of interpolation functions.

Initial FEM analyses used the triangular element illustrated in Figure 3.41, with the nodes of the element numbered counterclockwise. The functions $p(x, z), H(x, z), (\partial H/\partial t)(x, z), U_x(x, z), U_z(x, z)$ are to be defined over the element in terms of their values at the nodes by linear[†] interpolation functions. For example, the pressure is defined by

$$p(x, z) = \sum_{i=1}^{3} p_i L_i(x, z) = (L)^T(p),\qquad(3.131)$$

where $p_i = p(x_i, z_i)$, and the linear interpolation functions are defined by

$$L_i(x, z) = (a_i + b_i x + c_i z)/2A, \qquad i = 1, 2, 3,$$

$$a_1 = x_2 z_3 - x_3 z_2, \qquad b_1 = z_2 - z_3, \qquad c_1 = x_3 - x_2,$$
$$a_2 = x_3 z_1 - x_1 z_3, \qquad b_2 = z_3 - z_1, \qquad c_2 = x_1 - x_3,$$
$$a_3 = x_1 z_2 - x_2 z_1, \qquad b_3 = z_1 - z_2, \qquad c_3 = x_2 - x_1,$$

$$2A = \begin{vmatrix} 1 & x_1 & z_1 \\ 1 & x_2 & z_2 \\ 1 & x_3 & z_3 \end{vmatrix}.\qquad(3.132)$$

---

*That is, $\delta p = 0$ implies $\eta(x, z) = 0$ on S_p.

[†]Recall that the beam-shape functions of Sections 2.5 and 2.6 were quadratic.

The *same* interpolation function is employed for all variables, i.e., $H(x, z) = (L)^T(H)$, etc.

For a specified element, substitution of the interpolation-function definitions for $p, U_x, U_z, \partial H/\partial t$ into the extended variational definition

$$\Delta J = \int_A \left\{ \frac{\rho H^3}{24\mu} \left[\left(\frac{\partial p}{\partial x} \right)^2 + \left(\frac{\partial p}{\partial z} \right)^2 \right] - \rho H \left[U_x \frac{\partial p}{\partial x} + U_z \frac{\partial p}{\partial z} \right] + \rho p \frac{\partial H}{\partial t} \right\} dA$$

$$+ \int_{S_q} (q_x n_x + q_y n_y) p \, ds$$

yields

$$\Delta J = -\tfrac{1}{2}(p)^T [Kp](p) - (p)^T (Q - q_n), \qquad (3.133)$$

where

$$(Q) = [KUx](U_x) + [KUz](U_z) + [K\dot{H}](\dot{H})$$

and

$$Kp_{ij} = -\int_A \frac{\rho H^3}{12\mu} \left(\frac{\partial L_i}{\partial x} \frac{\partial L_j}{\partial x} + \frac{\partial L_i}{\partial z} \frac{\partial L_j}{\partial z} \right) dA,$$

$$KUx_{ij} = \int_A \rho H \frac{\partial L_i}{\partial x} L_j \, dA,$$

$$KUz_{ij} = \int_A \rho H \frac{\partial L_i}{\partial z} L_j \, dA,$$

$$K\dot{H}_{ij} = \int_A \rho L_i L_j \, dA,$$

$$q_{ni} = \int_{Sq} \rho H(q_x n_x + q_y n_y) L_i \, dS. \qquad (3.134)$$

Appendix C contains constant-property definitions for the fluidity matrices $[Kp], [KU_x], [KU_z], [K\dot{H}]$. Minimizing the function defined by Eq. (3.133) with respect to the pressures at the nodes of the element; i.e., $\partial \Delta J/\partial p_i = 0$ yields the nodal-flow equation

$$(q_n) = [Kp](p) + [KU_x](U_x) + [KU_z](U_z) + [K\dot{H}](\dot{H}). \quad (3.135)$$

This is the desired finite-element equation, defining the nodal mass flow rate in terms of the nodal pressures, velocities, and squeeze rates. The nodal-flow rate is the net flow exiting from the element over the element boundary adjacent to the node; e.g., q_{n2} is the exit mass flow rate over boundary S_2 in Figure 3.41.

In our previous discussion of finite-element methods for beam vibrations, system mass and stiffness matrices are assembled from element equations by imposing continuity and equilibrium requirements at joints. The triangular squeeze pad (squeezing motion only, zero surface velocity) illustrated in Figure 3.42 is employed by Booker and Heubner to illustrate comparable continuity conditions in lubrication. The following constant-film properties are assumed to hold for the pad:

$$\rho = 36, \qquad \mu = \sqrt{3}/2 \times 10^{-6}, \qquad H = 1 \times 10^{-2}.$$

From Appendix C, Eq. (3.135) yields for the three pads

$$\begin{Bmatrix} q_{n2} \\ q_{n3} \\ q_{n4} \end{Bmatrix}^{\mathrm{I}} = \begin{bmatrix} -2 & -1 & 3 \\ -1 & -2 & 3 \\ 3 & 3 & -6 \end{bmatrix} \begin{Bmatrix} p_2 \\ p_3 \\ p_4 \end{Bmatrix} + \begin{bmatrix} -2 & -1 & -1 \\ -1 & -2 & -1 \\ -1 & -1 & -2 \end{bmatrix} \begin{Bmatrix} \dot{H}_2 \\ \dot{H}_3 \\ \dot{H}_4 \end{Bmatrix},$$

$$\begin{Bmatrix} q_{n3} \\ q_{n1} \\ q_{n4} \end{Bmatrix}^{\mathrm{II}} = \begin{bmatrix} -2 & -1 & 3 \\ -1 & -2 & 3 \\ 3 & 3 & -6 \end{bmatrix} \begin{Bmatrix} p_3 \\ p_1 \\ p_4 \end{Bmatrix} + \begin{bmatrix} -2 & -1 & -1 \\ -1 & -2 & -1 \\ -1 & -1 & -2 \end{bmatrix} \begin{Bmatrix} \dot{H}_3 \\ \dot{H}_1 \\ \dot{H}_4 \end{Bmatrix},$$

$$\begin{Bmatrix} q_{n1} \\ q_{n2} \\ q_{n4} \end{Bmatrix}^{\mathrm{III}} = \begin{bmatrix} -2 & -1 & 3 \\ -1 & -2 & 3 \\ 3 & 3 & -6 \end{bmatrix} \begin{Bmatrix} p_1 \\ p_2 \\ p_4 \end{Bmatrix} + \begin{bmatrix} -2 & -1 & -1 \\ -1 & -2 & -1 \\ -1 & -1 & -2 \end{bmatrix} \begin{Bmatrix} \dot{H}_1 \\ \dot{H}_2 \\ \dot{H}_4 \end{Bmatrix}.$$

While these equations retain *element* flow rate definitions on the left-hand side, *system* pressures and squeeze rates are employed on the right. System flow rates are defined by

$$q_{n1} = q_{n1}^{\mathrm{II}} + q_{n1}^{\mathrm{III}}, \qquad q_{n2} = q_{n2}^{\mathrm{III}} + q_{n2}^{\mathrm{I}},$$
$$q_{n3} = q_{n3}^{\mathrm{I}} + q_{n3}^{\mathrm{II}}, \qquad q_{n4} = q_{n4}^{\mathrm{I}} + q_{n4}^{\mathrm{II}} + q_{n4}^{\mathrm{III}} = 0, \qquad (3.136)$$

and yield the system equation

$$\begin{bmatrix} -4 & -1 & -1 & 6 \\ -1 & -4 & -1 & 6 \\ -1 & -1 & -4 & 6 \\ 6 & 6 & 6 & -18 \end{bmatrix} \begin{Bmatrix} p_1 \\ p_2 \\ p_3 \\ p_4 \end{Bmatrix} = \begin{bmatrix} 4 & 1 & 1 & 2 \\ 1 & 4 & 1 & 2 \\ 1 & 1 & 4 & 2 \\ 2 & 2 & 2 & 6 \end{bmatrix} \begin{Bmatrix} \dot{H}_1 \\ \dot{H}_2 \\ \dot{H}_3 \\ \dot{H}_4 \end{Bmatrix} + \begin{Bmatrix} q_{n1} \\ q_{n2} \\ q_{n3} \\ q_{n4} \end{Bmatrix}.$$
$$(3.137)$$

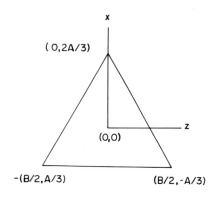

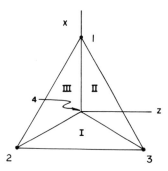

Figure 3.42 Example problem from Booker and Huebner (1972).

The nodal pressures at the boundary (p_1, p_2, p_3) are zero. For $\dot{H}_i = -1$, the unknown pressure p_4 is determined from the last of Eq. (3.137) as $p_4 = \frac{2}{3}$.

Note that the unknown pressure p_4 is defined by the $q_{n4} = 0$ equation. A comparable result holds for pads of journal bearings with m equations for unknown pressures resulting from m requirements that the net nodal-flow rates be zero. This statement is supported by Figure 3.43, which illustrates a finite-element grid on a bearing pad comparable to that discussed in the preceding subsection. For a journal-bearing pad, there is no axial motion; hence $U_{z1} = 0$, and Eq. (3.135) reduces to

$$(q_n) = [Kp](p) + [KU_\theta](U_\theta) + [K\dot{H}](\dot{H}), \tag{3.138}$$

where $U_{\theta i} = R\omega_j/2$. Returning to Figure 3.43, $z = 0$ defines the bearing midplane, while $z = L/2$ defines the axial bearing boundary and $\tilde{\theta} = \tilde{\theta}_{lp}$, $\tilde{\theta} = \tilde{\theta}_{tp}$ define the leading and trailing edges of the bearing. The pressure is zero at $z = L/2$ and at the pad boundary defined by grooves at $\tilde{\theta} = \tilde{\theta}_{lp}$ and $\tilde{\theta} = \tilde{\theta}_{tp}$. Further, the exit flow rates are zero along the lower edge of the pad due to symmetry. The circled nodes correspond to *both* unknown pressures *and* zero nodal flows. Hence, at each of the m nodes for which the pressure is unknown and to be calculated, the flowrate contributions from each

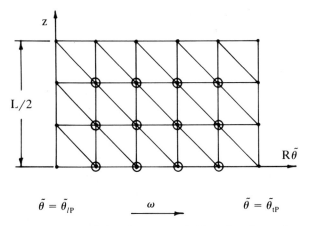

Figure 3.43 Finite-element model for a bearing pad.

element adjacent to the node are summed and the resultant nodal flowrate set equal to zero. This procedure yields a system matrix equation defining the unknown pressures.

As documented by Allaire et al. (1977) a systematic approach for numbering the nodes and elements yields equations identical in form to the finite-difference results of Eq. (3.122), i.e.,

$$[A_j](p_i)_j + [B_j](p_i)_{j-1} + [C_j](p_i)_{j+1} = (G_i)_j,$$

where $(p_i)_j$ is the vector of unknown pressures in a vertical column on the grid and $(p_i)_{j-1}, (p_i)_{j+1}$ are adjacent columns to the left and right, respectively. Hence, the recurrence procedure of Eq. (3.124) may be used to expedite the solution.

The author is not aware of any analytical perturbation procedure in connection with the FEM, comparable to Lunds's of the preceding section. Hence, stiffness and damping coefficients are calculated via a numerical differentiation procedure. In general, cavitation is accounted for by setting negatively calculated pressures to zero.

The FEM is decidedly more complicated than the finite-difference procedure; however, it is also more general. It readily accommodates step changes in clearances typical of the pressure-dam bearing illustrated in Figure 3.32, which present problems for the finite-difference method. It is also quite flexible in following curved boundaries.*

The FEM procedure outlined above for an incompressible fluid is less general than that developed by Reddi and used by Booker and Heubner, who

*See S. M. Rhode in a discussion of Allaire et al. (1977).

also consider surface diffusion, body forces, etc. Reddi and Chu (1970) have also provided a FEM formulation for compressible fluids.

Closing Comments

Generally speaking, test programs for bearings tend to agree adequately with numerical predictions in terms of static (eccentricity and attitude angle versus load) and dynamic (imbalance response, stability calculations) predictions. However, careful test programs tend to be carried out for comparatively small bearings with carefully controlled dimensions and rigid supports for the bearings. Predictions do not always agree so well with results in practice. As previously noted, Kirk's (1978) measurements have shown major differences between assumed and actual dimensions for installed bearings and provides one explanation for discrepancies. Nicholas and Barrett (1985) and Nicholas et al. (1986) have demonstrated that failure to include the influence of bearing support structure in rotordynamic analyses is another major reason for incorrect predictions. Bearing-support flexibility has an obvious influence of reducing critical speeds, but also can markedly reduce effective bearing damping.

3.10 SQUEEZE-FILM DAMPERS

As outlined in Chapter 1, the two most commonly recurring problems in rotordynamics are excessive steady-state synchronous vibration levels, and subharmonic rotor instabilities. Steady-state vibration levels may be reduced by rotor/bearing modifications, which shift the critical speed locations, or by improved balancing. The oil whirl instability of Section 3.5 is normally remedied by changing the bearing design, with the tilting-pad bearing completely eliminating bearings as a mechanism for rotor instability. However, as explained in Chapters 4–6, other destabilizing mechanisms may be large enough to destabilize flexible rotors supported by tilting-pad bearings.

Squeeze-film dampers have been developed and successfully employed to stabilize a variety of otherwise unstable units operating on tilting-pad bearings [Malanowski (1979)]. Most of these applications are of a retrofit nature where a unit is found to be unstable at full operating conditions. However, some high-pressure compressor manufacturers provide squeeze-film dampers at the outset to eliminate stability problems [Shemeld (1986)]. Figure 3.44 illustrates the normal arrangement, with the bearing portion of the tilting-pad bearing attached to the journal element of the damper. The damper is fed oil through a centered circumferential groove, with O-ring or piston-ring end seals used to restrict leakage. The damper surface is supported by a centering* spring which is significantly softer than the tilting-pad bearing. This support centers the rotor during running conditions and prevents the damper

---

*Additional *soft* centering springs are generally provided to compensate for rotor weight.

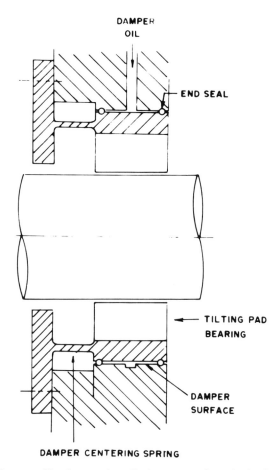

DAMPER
OIL

END SEAL

TILTING PAD
BEARING

DAMPER
SURFACE

DAMPER CENTERING SPRING

Figure 3.44 Squeeze-film damper installation supporting a hydrodynamic bearing.

element from rotating. A squeeze-film damper is seen to resemble a nonrotating plain journal bearing and provides rotor damping at the bearing location as a result of pure squeezing motion of the damper's journal element.

Modern aircraft jet engines use rolling-element bearings because of their durability and low power consumption. Since these bearings provide negligible damping, squeeze-film dampers are generally provided to reduce synchronous response amplitudes, to enhance stability, and to provide a margin of safety for blade-loss conditions. In these applications, the centering spring is generally eliminated, and the cavitated oil film within the damper is used to support the rotor. With this design approach, antirotation tabs or keys are required. Three common end-sealing arrangements for dampers used in aircraft engines are illustrated in Figure 3.45. A single centered oil-supply hole is commonly provided for the radial seals of Figures 3.45(a) and 3.45(b),

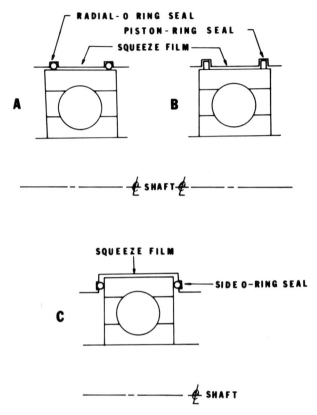

Figure 3.45 Squeeze-film damper arrangements for ball-bearing-supported rotors.

and a circumferential end groove is provided for the end-seal arrangement of Figure 3.45(c).

As with journal bearings, the incompressible-fluid Reynolds equation is generally used to model squeeze-film dampers. However, because of cavitation phenomena, the correlation between theory and experiment is considerably less compelling for dampers than bearings [Hibner and Bansal (1979), Walton et al. (1987), and Zeidan and Vance (1988)]. To be blunt, for most aircraft-gas-turbine applications, test results are strongly in conflict with Reynolds-equation predictions. We will return to this point after a review of damper designs which have historically been modeled by the short, long, or finite-length bearing solutions.

Damper Models

Figure 3.46 illustrates the dynamic pressure distribution *proposed* for the central-groove oil-supply arrangement of Figure 3.44, Cunningham et al.

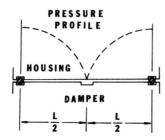

Figure 3.46 Proposed dynamic pressure distribution for the centrally grooved damper of Figure 3.44.

(1975). The pressure fluctuations at the center of the damper are assumed to be zero because of the deep circumferential supply groove, while the pressure gradient is zero at the ends due to the end seals. The damper may be directly modeled by the Ocvirk bearing solution of Eq. (3.26), using the damper length of L. [Imagine the pressure distribution of Eq. (3.26) cut in the middle and placed end to end.]

Tonneson (1976) has extensively tested dampers which are reasonably modeled by the Ocvirk impedance. His experimental apparatus causes a mass, which is spring supported in a squeeze-film damper, to be excited by an imbalance mass rotating at the frequency Ω. Tonneson's linearized model for the damper is

$$F_X = C_X \dot{X}, \qquad F_Y = -C_Y \dot{Y},$$

for small motion about an equilibrium position of the form

$$X = C_r \varepsilon_0 + \delta \cos \Omega t, \qquad Y = \delta \sin \Omega t,$$

where Ω is the imbalance frequency. The coefficients C_X, C_Y are derived by separately considering velocity perturbations in the X and Y directions and then calculating the average damping coefficients over the period $T = 2\pi/\Omega$. For a noncavitating 2π Ocvirk bearing, the damping coefficients can be defined from Eq. (3.32) by

$$C_X(2\pi) = 2\mu L \left(\frac{R}{C_r} \right)^3 W_\varepsilon(\varepsilon_0, 0),$$

$$C_Y(2\pi) = 2\mu L \left(\frac{R}{C_r} \right)^3 W_\beta(\varepsilon_0, 0), \qquad (3.139)$$

where W_ε and W_β are components of the appropriate 2π impedance. The corresponding coefficients for a cavitating π bearing are simply

$$C_X(\pi) = C_X(2\pi)/2, \qquad C_Y(\pi) = C_Y(2\pi)/2.$$

Tonnesen found the correlation between theory and experimental data to be excellent for small concentric orbits but concluded that the motion at large eccentricity ratios is a nonlinear* phenomenon, not appropriately modeled by linear damping coefficients. With centered orbits, experimentally determined values for C_X and C_Y lie between the theoretically predicted values of $[C_X(\pi), C_X(2\pi)]$ and $[C_Y(\pi), C_Y(2\pi)]$, respectively. Increasing the damper's supply pressure reduces cavitation within the damper and moves the experimental results from the cavitated π coefficients toward the uncavitated 2π coefficients. Tonnesen also found an abrupt drop in damping with increasing speed above some limiting value of running speed. This loss of damping effectiveness could be forestalled by increasing damper supply pressure.

In addition to the damper coefficient formulas of Tonnesen given in Eq. (3.127), *for circular orbits*, the stiffness and damping coefficients of Eqs. (3.52) and (3.56) can be used to define "equivalent" linear stiffness and damping coefficients for a cavitated damper. Specifically, appropriate stiffness and damping coefficients are

$$K_d = K_{\varepsilon\varepsilon} = 2\mu L \left(\frac{R}{C_r}\right)^3 \varepsilon_0 \omega_j \left\{ \frac{W_\varepsilon(\varepsilon_0, \alpha_0)}{\varepsilon_0} + \frac{\partial W_\varepsilon(\varepsilon_0, \alpha_0)}{\partial \varepsilon} \right\},$$

$$C_d = C_{\beta\beta} = 2\mu L \left(\frac{R}{C_r}\right)^3 |W_\beta(\varepsilon_0, \alpha_0)|, \tag{3.140}$$

where $\alpha_0 = \pi/2$ for clockwise rotation and $\alpha_0 = -(\pi/2)$ for counterclockwise rotation. This formula for K_d differs from that employed by Cunningham et al. (1975), who used the following "average" definition:

$$K_d \cong -\frac{F_\varepsilon}{C_r \varepsilon_0} = 2\mu L \left(\frac{R}{C_r}\right)^3 \omega_j W_\varepsilon(\varepsilon_0, \alpha_0). \tag{3.141}$$

Equation (3.141) yields the correct orbit amplitude in a synchronous response calculation, while Eq. (3.140) yields the correct local radial stiffness for an orbit of radius $C_r\varepsilon_0$, and should be used in an orbital stability calculation.

The point is emphasized that the above stiffness and damping coefficients apply for a circular orbit of radius $C_r\varepsilon_0$. Dampers have no *equilibrium position* for an applied static load in the sense of a journal bearing. The "stiffness" developed by a damper results from its orbital motion due to a rotating load and could more properly be defined with a cross-coupled damping coefficient. The O-ring sealed damper of Figure 3.45(a) is customar-

---

*Nonlinear simulation results using the impedance definition of Eqs. (3.35) are provided in Childs et al. (1977).

ily described by the Sommerfeld bearing model of Eq. (3.29). This model is appropriate because the end seals eliminate the axial flow in Eq. (3.7). Hence $\partial p / \partial z = 0$, and the Sommerfeld model applies, independent of the damper L/D ratio.

The preceding discussion for dampers described by the Ocvirk bearing model applies here as well. Specifically, Eqs. (3.140) and (3.141) provide linear stiffness and damping coefficients for circular orbits, while the impedance model of Eqs. (3.38)–(3.40) would be employed for nonlinear transient analysis.

Cavitation Complications in Dampers

As noted at the beginning of this section, serious discrepancies have been observed between many of the measurements for dampers and Reynolds-equation-based solutions. Tests conducted at comparatively low speeds, larger clearances, and higher supply pressures [Botman (1976), Feder et al. (1978) and Bansal and Hibner (1978)] show reasonably good agreement with predictions. However, tests at speeds, pressures, and clearances which are representative of aircraft-gas-turbine operating conditions have typically shown *major* and systematic deviations between measurements and theory. Hibner and Bansal made this point forcefully (1979), noting much lower peak-pressure measurements in a forced-orbital-motion rig than predicted. Their damper used both piston-ring seals and butt O-ring seals. Oil was supplied through a central supply hole and discharged through two exit holes which were 180° away from the supply hole. The deviation increased as speed and/or orbit amplitudes increased. Hibner and Bansal observed a frothy discharge of bubbles from their damper and postulated that *gaseous* cavitation was the source of deviation between theory and experiment. Specifically, they argued that air bubbles were drawn into the damper creating a two-phase flow which gave a dramatically different pressure distribution than predicted by the incompressible Reynolds equation.

Zeidan and Vance (1988, 1989) reported similar results for an orbital test rig with a piston-ring damper. Flow was injected in a circumferential feed groove on one side of the damper between a solid and a serrated piston-ring seal and across an exit serrated piston ring. Dynamic pressure measurements were taken and supplemented by high-speed photographic results. A typical low-speed uncavitated dynamic pressure wave is shown in Figure 3.47(a). This figure would correspond to a 2π Reynolds-equation solution. A π Reynolds-equation solution would drop the negative position of this curve to account for cavitation. However, Figure 3.47(b) illustrates the measured pressure conditions under realistic operating conditions. The reduced pressure arises because of gas bubbles, and the positive pressure region extends over $\pi/2$ radians of the orbit (not π), and leads the maximum clearance by approximately $3\pi/2$ radians (not $\pi/2$). Moreover, even though a deep feed groove was used, measurable pressure oscillations in the groove were compa-

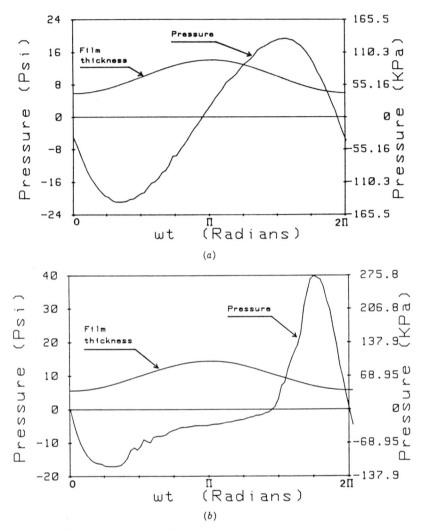

Figure 3.47 Dynamic pressure distribution in a fixed-orbit test rig squeeze-film damper test rig with and without gaseous cavitation [Zeidan and Vance (1988, 1989)]. Reprinted by permission of the Society of Tribologists and Lubrication Engineers. All rights reserved.

rable, albeit smaller, than those in the land. Hence, both the model and the proposed boundary conditions of Figure 3.46 are wrong.

Photographic studies by Zeidan and Vance confirmed Bansal and Kirk's initial suggestion that air was being drawn into the damper and persisted throughout the orbit, invalidating the Reynolds-equation model. The observed erratic nature of the bubbles does not encourage one to believe that a homogeneous two-phase model would yield any significant improvement.

There is at present no adequate model for predicting the performance of squeeze-film dampers under normal operating conditions. Hibner and Bansal's (1979) advice "that viscous dampers either be used within a range of parameters where the incompressible Reynolds equation is valid or specific testing be performed for a damper to determine the degree of deviation from theory" continues to be sound today.

Different, but comparably discouraging results concerning theory versus experiment were provided by Walton et al. (1987). Specifically, results from a rotordynamic rig demonstrated a quite large "added-mass" rotordynamic coefficient for a squeeze-film damper, which is completely unpredicted by the Reynolds equation.

Damper Design

Despite the pessimism of the preceding section, and in the absence of a preferred alternative, most squeeze-film damper designs are based on the Reynolds equation and provide a satisfactory degree of damping. The main practical problem that one can get into with dampers is excessively tight clearances. Recall from Section 1.6 that an optimum value of support damping exists. Hence, if the clearances are too tight the squeeze-film damper provides too much damping and is ineffective.

Cavitation-induced divergence between an incompressible Reynolds-equa-tion-based theory can yield either improved or degraded rotordynamics vibration performance compared to a Reynolds-equation-based prediction. For a given application, an increase in supply pressure provides the only readily available parameter to retard cavitation and improve the correlation between theory and practice.

Elastomeric Dampers

O-ring supports for ball bearings have long been recognized as providing a substantial increase in rotor damping [Smalley et al. (1977)]. In a series of articles, Tecza, Darlow et al. (1979) and Tecza, Smalley et al. (1979) have proposed a damper design based on the damping capacity of elastometers, but using three equally spaced (circumferentially) pads to support a rotor bearing. Their design employs elastomeric buttons within each pad. The preload in the pads is variable and can be set to any desired value. Assuming that satisfactory in-service lifetimes can be demonstrated for this type of design, elastomeric dampers show promise as a simple alternative to squeeze-film dampers.

3.11   SUMMARY AND EXTENSIONS

The reader will doubtlessly agree that a considerable body of information has been compressed into this chapter. The intended lessons of the material presented are as follows:

(a) The Reynolds partial differential equation for an incompressible fluid is the basic governing equation for lubrication and is a greatly simplified version of the Navier-Stokes equation.

(b) Integration of the pressure field defined by the Reynolds equation yields analytical solutions for the plain journal bearing in the form of either nonlinear impedances or linear stiffness and damping coefficient matrices for small motion about an equilibrium position.

(c) The stiffness matrix for a journal bearing is nonsymmetric and causes a rotor supported in plain journal bearing to become unstable above a limiting running speed called the onset speed of instability. The unstable motion associated with journal bearings is called oil whirl and involves large-amplitude subsynchronous motion at the rotor critical speed. Rotor flexibility reduces the onset speed of instability, but does not change its character.

(d) Vertical rotors without side loads are subject to motion consisting of a limit cycle whose frequency "tracks" at approximately one-half running speed which is also called oil whirl. At speeds above twice the rotor's natural frequency, the rotor subsynchronous motion stops tracking running speed and precesses at the natural frequency. The latter motion is called oil whip and can be quite destructive.

(e) There are various bearing types which are more stable than plain journal bearings and which may be used to elevate a rotor's onset speed of instability. Tilting-pad bearings completely eliminate bearings as a source of rotor instability.

(f) Cavitation, compressibility, and turbulence are all phenomena which complicate the modeling and analysis of hydrodynamic bearings; however, solution techniques and models have been developed and are available for their resolution.

(g) Finite-difference and finite-element procedures are generally used for the numerical solution of hydrodynamic bearings. The finite-difference method is simpler and more directly implemented; however, the FEM is more general and readily accounts for curved boundary surfaces and discontinuous changes in film thickness.

(h) Squeeze-film dampers are employed to improve the vibration characteristics of rotors by adding damping to either stabilize otherwise unstable rotors or to reduce synchronous rotor amplitudes. If cavitation is not excessive, dampers may be adequately modeled by the incompressible Reynolds equation. However for most aircraft-gas-turbine applications, high speeds and low supply pressures combine to give excessive gaseous cavitation, which completely invalidates any Reynolds-equation-based solution.

The contents of this and the preceding chapter defining bearing models and structural-dynamic models for rotors provides the basic rotordynamic models for most rotating machines. The contents of Chapters 4–6 complete the modeling requirements for forces acting on rotors due to fluid-structure interaction at seals, turbines, and impellers.

REFERENCES: CHAPTER 3

Allaire, P., and Flack, R. (1980), "Instability Thresholds for Flexible Rotors on Hydrodynamic Bearings," Rotordynamic Instability Problems in High Performance Turbomachinery, NASA Report No. CP2133, proceedings of a workshop held at Texas A & M University, pp. 403–425.

Allaire, P., Nicholas, J., and Gunter, E. (1977), "Systems of Finite Elements for Finite Bearings," *Journal of Lubrication Technology*, 187–197.

Bansal, P., and Hibner, D. (1978), "Experimental and Analytical Investigation of Squeeze-Film Bearing Damper Forces Induced by Offset Circular Whirl Orbits," *Journal of Mechanical Design*, 549–557.

Barrett, L., Allaire, P., and Gunter, E. (1978), "The Dynamic Analysis of Journal Bearings Using a Finite-Length Correction for Short Bearing Theory," in S. Rhode, P. Allaire, and C. Madey (Eds.), *Topics in Fluid Film Bearings and Rotor Bearing System Design and Optimization*, ASME.

Black, H. (1970a), "Empirical Treatment of Hydrodynamic Journal Bearing Performance in the Superlaminar Regime," *Journal of Mechanical Engineering Science*, **12** (2), 116–122.

Black, H. (1970b), "On Journal Bearings with High Axial Flows in the Turbulent Regime," *Journal of Mechanical Engineering Science*, **12** (4), 301–303.

Blok, H. (1965), "Topological Aspects and the Impulse/Whirl Angle Method in the Orbital Hydrodynamics of Dynamically Loaded Journal Bearings," *Lecture Notes* (condensed English version), Delft, The Netherlands.

Booker, J. (1971), "Dynamically Loaded Journal Bearings: Numerical Application of the Mobility Method," *Journal of Lubrication Technology*, 168–176 and 351.

Booker, J. (1965a), "Dynamically Loaded Journal Bearings: Mobility Methods of Solution," *Journal of Basic Engineering*, 537–546.

Booker, J. (1965b), "A Table of the Journal Bearing Integral," *Journal of Basic Engineering*, 533–535.

Booker, J., and Heubner, K. (1972), "Application of Finite Element Methods to Lubrication: An Engineering Approach," *Journal of Lubrication Technology*, 313–323.

Botman, M. (1976), "Experiments on Oil-Film Dampers for Turbomachinery," *Journal of Engineering for Power*, 393–399.

Castelli, V., and Pirvics, J. (1967), "Equilibrium Characteristics of Axial-Groove Gas-Lubricated Bearings," *Journal of Lubrication Technology*, Ser. F, **89** (2), 177–196.

Childs, D., Moes, H., and van Leeuwen, H. (1977), "Journal Bearing Impedance Descriptions for Rotordynamic Applications," *Journal of Lubrication Technology*, 198–219.

Crandall, S. (1990), "From Whirl to Whip in Rotordynamics," in *Transactions, IFToMM 3rd International Conference on Rotordynamics*, Lyon, France, pp. 19–26.

Cunningham, R., Fleming, D., and Gunter, E. (1975), "Design of a Squeeze-Film Damper for a Multimass Flexible Rotor," *Journal of Engineering for Industry*, 1383–1389.

Feder, E., Bansal, P., and Blanco, A. (1978), "Investigation of Squeeze-Film Damper Forces Produced by Circular Centered Orbits," *Journal of Engineering for Power*, 15–21.

Floberg, L. (1965), "On Hydrodynamic Lubrication with Special Reference of Sub-Cavity Pressures and Numbers of Streamers in Cavitation Region," *Acta Polytechnica Scandinavica, Mechanical Engineering Series*, **19**.

Gelfand, I., and Fomin, S. (1963), *Calculus of Variations*, Prentice-Hall, Englewood Cliffs, NJ.

Gross, W. (1962), *Gas Film Lubrication*, Wiley, New York.

Harris, T. (1991), *Rolling Bearing Analysis* (3rd ed.), Wiley, New York.

Hibner, D., and Bansal, P. (1979), "Effects of Fluid Compressibility on Viscous Damper Characteristics," *Proceedings of the Conference on the Stability and Dynamic Response of Rotors with Squeeze-Film Bearings*, University of Virginia.

Hirs, G. (1973), "A Bulk-Flow Theory for Turbulence in Lubricant Films," *Journal of Lubrication Technology*, 137–146.

Jones, A. (1960), "A General Theory for Elastically Constrained Ball and Radial Roller Bearings Under Arbitrary Load and Speed Conditions," *Journal of Basic Engineering*, **82** (2), 309–320.

Kirk, R. (1978), "The Influence of Manufacturing Tolerances on Multi-Lobe Bearing Performance in Turbomachinery," in *Topics in Fluid Film Bearing and Rotor Bearings System Design and Optimization*, ASME, New York, pp. 108–129.

Leader, M., Flack, R., and Allaire, P. (1980), "The Experimental Dynamic Response of a Single Mass Flexible Rotor with Three Different Journal Bearings," *ASLE Transactions*, **23** (4), 363.

Li, D., Choy, K., and Allaire, P. (1980), "Stability and Transient Characteristics of Four Multilobe Journal Bearings," *Journal of Lubrication Technology*, **102** (3), 291–299.

Lund, J. (1968), "Calculation of Stiffness and Damping Properties of Gas Bearings," *Journal of Lubrication Technology*, Ser. F, Vol. **90**, 793–803.

Lund, J. (1966), "Self-Excited, Stationary Whirl Orbits of a Journal in a Sleeve Bearing," Ph.D. Thesis, Rensselaer Polytechnic Institute, Troy, NY.

Lund, J. (1965), "The Stability of an Elastic Rotor in Journal Bearings with Flexible, Damped Supports," *Journal of Applied Mechanics*, 911–920.

Lund, J. (1964), "Spring and Damping Coefficients for the Tilting Pad Journal Bearing," *ASLE Transactions*, **7**, 342–352.

Lund, J., and Sternlicht, B. (1962), "Rotor-Bearing Dynamics with Emphasis on Attenuation," *ASME Trans. Journal of Basic Engineering*, Ser. D, **84**, 491–502.

Lund, J., and Thomsen, K. (1978), "A Calculation Method and Data for the Dynamic Coefficients of Oil-Lubricated Journal Bearings," in *Topics in Fluid Film Bearing and Rotor Bearing System Design and Optimization*, ASME, New York, pp. 1–28.

Malanowski, S. (1979), "Case Histories in which Subsynchronous or Synchronous Vibration Amplitudes Have Been Minimized after Employing Custom Designed Damper Bearings," in *Proceedings of the Conference on the Stability and Dynamic Response of Rotors with Squeeze-Film Bearings*, University of Virginia.

Muszynska, A. (1986), "Whirl and Whip-Rotor/Bearing Stability Problems," *Journal of Sound and Vibrations*, **110**, 443–462.

Ng, C. (1964), "Fluid Dynamic Foundation of Turbulent Lubrication Theory," *ASLE Transactions*, **7**, 311–321.

Ng, C., and Pan, C. (1965), "A Linearized Turbulent Lubrication Theory," *Journal of Basic Engineering*, Ser. D, **87**, 675.

Nicholas, J. (1977), "A Finite Element Dynamic Analysis of Pressure Dam and Tilting-Pad Bearings," Ph.D. thesis, University of Virginia, Charlottesville, VA.

Nicholas, J., and Allaire, P. (1980), "Analysis of a Step Journal Bearing-Finite Length, Stability," *ASLE Transactions*, **57** (4), 197–207.

Nicholas, J., and Barrett, L. (1985), "The Effect of Bearing Support Flexibility on Critical Speed Prediction," *ASLE Transactions*, **29** (3), 329–338.

Nicholas, J., Gunter, E., and Barrett, L. (1978), "The Influence of Tilting Pad Bearing Characteristics on the Stability of High-Speed Rotor-Bearing Systems," in *Topics in Fluid Film Bearing and Rotor Bearing System Design and Optimization*, ASME, New York, pp. 55–58.

Nicholas, J., Whalen, J., and Franklin, D. (1986), "Improving Critical Speed Calculations Using Flexible Bearing Support FRF Compliance Data," in *Proceedings, 15th Turbomachinery Symposium*, Texas A&M University, pp. 69–78.

Ocvirk, F. (1952), "Short Bearing Approximation for Full Journal Bearings," NACA TN 20808.

Orcutt, F. (1967), "The Steady-State and Dynamic Characteristics of the Tilting-Pad Journal Bearing in Laminar and Turbulent Regimes," *Journal of Lubrication Technology*, 392–404.

Orcutt, F., and Arwas, E. (1967), "The Steady State and Dynamics of Arc Bearing in the Laminar and Turbulent Flow Regimes," *Journal of Lubrication Technology*, 143–153.

Pinkus, O., and Sternlicht, B. (1961), *Theory of Hydrodynamic Lubrication*, McGraw-Hill, New York.

Reddi, M. (1969), "Finite-Element Solution of the Incompressible Lubrication Problem," *Journal of Lubrication Technology*, Ser. F, **91** (3), 529–533.

Reddi, M., and Chu, T. (1970), "Finite-Element Solution of the Steady-State Compressible Lubrication Problem," *Journal of Lubrication Technology*, Ser. F, **22**, 495.

Robertson, D. (1933), "Whirling of a Journal in a Sleeve Bearing," *Philosophical Magazine*, Vol. **715**, pp. 113–130.

Shemeld, D. (1986), "A History of Development in Rotordynamics—A Manufacturers Viewpoint," in *Rotordynamic Instability Problems in High-Performance Turbomachinery*, NASA Report No. CP 2443, proceedings of a workshop held at Texas A&M University, pp. 1–18.

Smalley, A., Darlow, M., and Mehta, R. (1977), "The Dynamic Characteristics of O-Rings," *Journal of Mechanical Design*.

Szeri, A. (1980), *Tribology—Friction, Lubrication, and Wear*, Hemisphere.

Tecza, J., Darlow, M., Jones, S., Smalley, A., and Cunningham, R. (1979), "Elastometer Mounted Rotors—An Alternative for Smoother Running Turbomachinery," ASME Paper No. 79-G7-149, Gas Turbine Conference, San Diego, CA.

Tecza, J., Smalley, A., Darlow, M., and Cunningham, R. (1979), "Design of Elastometer Dampers for High-Speed Flexible Rotor," ASME Paper No. 79-DET-98, Design Engineering Technical Conference, St. Louis, MO.

Tonnesen, J. (1976), "Experimental Parametric Study of a Squeeze-Film Damper for a Multi-Mass Flexible Rotor," *Journal of Lubrication Technology*, 206–213.

Walton, J., Walowit, E., Zorzi, E., and Schrand, J. (1987), "Experimental Observations of Cavitating Squeeze Film Dampers," *ASME Trans.*, **109**, 290–295.

Zeidan, F., and Vance, J. (1989), "Cavitation Regimes in Squeeze Film Dampers and their Effect on the Pressure Distribution," STLE Paper No. 89-AM-48-1.

Zeidan, F., and Vance, J. (1988), "Cavitation Effects on the Pressure Distribution of a Squeeze-Film Damper Bearing," in *Rotordynamic Instability Problems in High-Performance Turbomachinery*, NASA Report No. 3026, proceedings of a workshop held at Texas A&M University, pp. 111–132.

4

ROTORDYNAMIC MODELS FOR LIQUID ANNULAR SEALS

4.1 INTRODUCTION

Figure 4.1 illustrates a typical multistage centrifugal pump. Observe that the rotor is comparatively slender and, in the absence of fluid, would be expected to have several lightly damped critical speeds in its operating range. However, most pumps have comparatively benign operating characteristics. Moreover, many of the calculated "dry-critical speeds," i.e., critical speeds of the rotor-bearing model (without fluid forces) simply do not appear, or appear at substantially elevated running speeds.

Forces due to the annular seals illustrated in Figure 4.2 are mainly responsible for the large differences between calculated results for a rotor-bearing (only) model and the actual vibration characteristics of pumps. The neckring or wearing-ring seal restricts leakage flow along the front side of the impeller from impeller discharge to impeller inlet. The interstage seal restricts flow along the shaft between stages. The full head rise of the pump is dropped across the balance-piston seal with the resultant leakage flow returned to the pump inlet.

Figure 4.3 from Black (1979) illustrates the calculated vibration response of a double-suction single-stage pump for various wearing-ring seal clearances. The dry-critical speed results for very large seal clearances. As the clearances are reduced and the seal forces increase, the critical speed is elevated and the response amplitudes are reduced. This figure is illustrative of most pumps which rely heavily on the static support and stiffness and damping provided by the seals. *Rotodynamic models for liquid seals and their influence on pump vibrations are the subject of this chapter.*

"Bulk-flow" versions of Navier-Stokes equations are normally used for seal analysis and are the subject of Section 4.2. Seal forces are significantly more linear functions of displacement than bearings and are normally

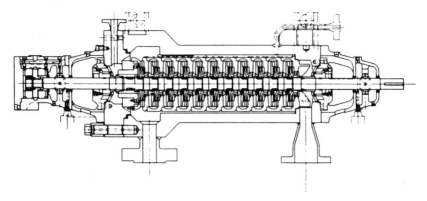

Figure 4.1   Multistage centrifugal pump [Massey (1985)].

represented by a reaction-force/seal-motion model of the form

$$-\left\{\begin{matrix} F_X \\ F_Y \end{matrix}\right\} = \left[\begin{matrix} K & k \\ -k & K \end{matrix}\right]\left\{\begin{matrix} X \\ Y \end{matrix}\right\} + \left[\begin{matrix} C & c \\ -c & C \end{matrix}\right]\left\{\begin{matrix} \dot{X} \\ \dot{Y} \end{matrix}\right\} + M\left\{\begin{matrix} \ddot{X} \\ \ddot{Y} \end{matrix}\right\}. \quad (4.1)$$

This model is valid for small motion about a centered position. The coefficients of Eq. (4.1) are obtained from the first-order perturbation of the bulk-flow governing equations. The development of the perturbation equations is the subject of Section 4.3.

The solution for the zeroth-order solutions consisting of pressure and circumferential-velocity distributions for a centered seal is the subject of

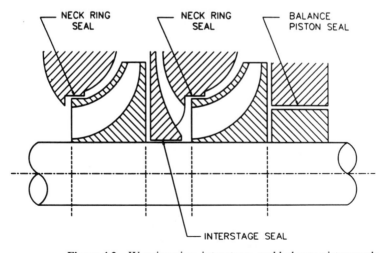

Figure 4.2   Wearing-ring, interstage, and balance piston seals.

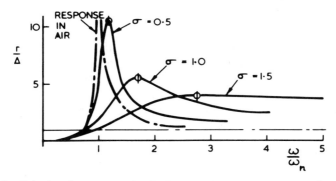

Figure 4.3 Calculated response of a single-stage double-suction pump [Black (1979)].

Section 4.4. A solution procedure is developed for rotordynamic coefficients from Section 4.5, and the influence of various physical parameters on their behavior is the subject of Section 4.6. Only force coefficients due to displacements are considered prior to Section 4.7 where the influence of seal tilt in developing force and moment coefficients is examined. A discussion of alternative computational approaches and comparisons to experimental results are interspersed throughout Sections 4.6 and 4.7. Section 4.8 provides a summary and overview for liquid seals.

4.2 GOVERNING EQUATIONS FOR TURBULENT ANNULAR SEALS, BULK-FLOW MODELS

Introduction

The seal geometry of Figure 4.4 is obviously similar to that of a plain journal bearing; however, there are major differences between bearings and seals.

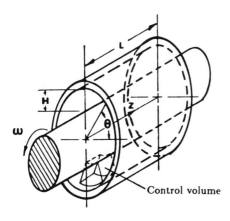

Control volume   **Figure 4.4** Seal geometry [Nelson and Nguyen (1987)].

First, the C_r/R ratio for seals is generally on the order of 0.003 versus 0.001 for bearings. These enlarged clearances, combined with a high pressure drop across the seal, mean that flow is highly turbulent. Hence, the Reynolds equation is completely inappropriate for typical annular pump seals.

Most analyses for pump seals have used "bulk-flow" models for the fluid within the seal. To be brief, in bulk-flow models the variation of fluid velocity components across the clearance is neglected, and average (across the clearance) velocity components are used, hence, the bulk-flow designation. By their nature, bulk-flow models include no definition of shear-stress variation for the fluid within the clearance and only account for shear stress at the boundaries of the bulk-flow model, viz., at the shaft and the seal stator. The model for the shear stress as a function of the local Reynolds number comprises the only turbulence model.

The initial analysis for pump seals was provided by Lomakin (1958), who first explained the large direct stiffness terms which can be developed by pump seals. Most of the subsequent innovations with respect to rotordynamics analysis and phenomena are due to Black (1969), Black and Jenssen (1970), and Black et al. (1981). Black used a series of ad hoc bulk-flow models for his work. Childs (1983) and (1984) has used the Hirs (1973) bulk-flow model with good success as a general basis for seal analysis. The present derivation is based on the work of Nelson (1985).

Governing Equations Derivation

Figure 4.5 illustrates a differential fluid element between a stator and rotor differential areas. The stator at the bottom of this figure is stationary while the rotor has a tangential velocity of $R\omega$. Because of the small C_r/R ratio, curvature is neglected in arriving at Figure 4.5. Note that the differential fluid element extends completely across the fluid film in contrast to the normal fluid element used in obtaining the Navier-Stokes equations and the Reynolds

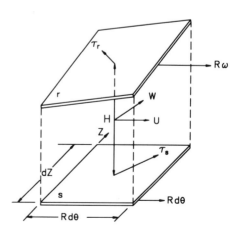

Figure 4.5 Differential-fluid element, after Nelson (1985).

equation in Chapter 3. The shear stresses τ_r, τ_s are acting at the rotor and stator walls. The average (across the film) circumferential- and axial-velocity components are U and W, respectively.

The governing control-volume equations are:

Continuity Equation

$$0 = \frac{\partial}{\partial t} \int_{CV} \rho \, d\mathbf{\forall} + \int_{CS} \rho \mathbf{v} \cdot d\mathbf{A}, \qquad (4.2a)$$

where $\mathbf{\forall}$ is volume.

Axial-Momentum Equation

$$\sum F_z = \frac{\partial}{\partial t} \int_{CV} \rho W \, d\mathbf{\forall} + \int_{CS} \rho W \mathbf{v} \cdot d\mathbf{A}. \qquad (4.2b)$$

Circumferential-Momentum Equation

$$\sum F_\theta = \frac{\partial}{\partial t} \int_{CV} \rho U \, d\mathbf{\forall} + \int_{CS} \rho U \mathbf{v} \cdot d\mathbf{A}. \qquad (4.2c)$$

Direct application of Eq. (4.2a) to the fluid element of Figure 4.5 yields

Continuity Equation

$$\frac{\partial H}{\partial t} + \frac{1}{R}\frac{\partial (HU)}{\partial \theta} + \frac{\partial (HW)}{\partial Z} = 0. \qquad (4.3)$$

Figure 4.6 provides a free-body diagram necessary to account for the forces in the momentum equations yielding

$$\sum F_z = -R \, d\theta \, dZ \left[H \frac{\partial P}{\partial Z} + (\tau_{rz} + \tau_{sz}) \right],$$

$$\sum F_\theta = -R \, d\theta \, dZ \left[\frac{H}{R} \frac{\partial P}{\partial \theta} + (\tau_{r\theta} + \tau_{s\theta}) \right], \qquad (4.4)$$

where $\tau_{rz}, \tau_{r\theta}$ and $\tau_{sz}, \tau_{s\theta}$ are the z and θ components of the rotor and stator shear stresses τ_r, τ_s. Direct application of Eqs. (4.2b) and (4.2c) to the

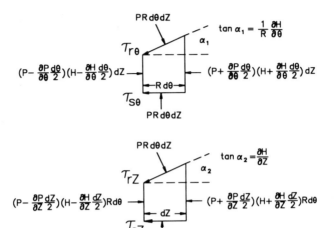

Figure 4.6   Forces on the differential-fluid element of Figure 4.5.

differential element of Figure 4.5 provides (with a lot of algebra)

$$-H\frac{\partial P}{\partial Z} - (\tau_{rz} + \tau_{sz}) = \rho\left[\frac{\partial(HW)}{\partial t} + \frac{\partial(HWU)}{R\,\partial\theta} + \frac{\partial(HW^2)}{\partial Z}\right], \quad (4.5a)$$

$$-\frac{H}{R}\frac{\partial P}{\partial\theta} - (\tau_{r\theta} + \tau_{s\theta}) = \rho\left[\frac{\partial(HU)}{\partial t} + \frac{\partial(HWU)}{R\,\partial\theta} + \frac{\partial(HU^2)}{\partial Z}\right]. \quad (4.5b)$$

These momenta equations can be markedly simplified by the following steps:
(a) multiply the continuity Eq. (4.3) by W and subtract the result from Eq.
(4.5a), and (b) multiply the continuity equation by U and subtract from Eq.
(4.5b). The results are

$$-H\frac{\partial P}{\partial Z} = \tau_{rz} + \tau_{sz} + \rho H\left(\frac{\partial W}{\partial t} + \frac{U}{R}\frac{\partial W}{\partial\theta} + W\frac{\partial W}{\partial Z}\right), \quad (4.6a)$$

$$-\frac{H}{R}\frac{\partial P}{\partial\theta} = \tau_{r\theta} + \tau_{s\theta} + \rho H\left(\frac{\partial U}{\partial t} + \frac{U}{R}\frac{\partial U}{\partial\theta} + W\frac{\partial U}{\partial Z}\right). \quad (4.6b)$$

The equations are now complete except for the shear-stress definitions.
Hirs defined the shear stresses as

$$\tau_s = \rho f_s U_s^2/2, \qquad \tau_r = \rho f_r U_r^2/2, \quad (4.7)$$

where f_s, f_r are the stator and wall friction factors, and U_s and U_r are the bulk-flow velocities relative to the wall and are defined as

$$U_s = (W^2 + U^2)^{1/2},$$
$$U_r = \left[W^2 + (U - R\omega)^2\right]^{1/2}. \tag{4.8}$$

The shear stress components τ_s, τ_r act in a direction opposite to the velocities U_s, U_r and, accordingly, have the components

$$\tau_{rz} = \tau_r(W/U_r), \qquad \tau_{r\theta} = \tau_r(U - R\omega)/U_r,$$
$$\tau_{sz} = \tau_s(W/U_s), \qquad \tau_{s\theta} = \tau_s(U/U_s). \tag{4.9}$$

Substituting these shear-stress components into Eqs. (4.6) yields the final desired form for the momenta equations:

Axial-Momentum Equation

$$-H\frac{\partial P}{\partial Z} = \frac{\rho}{2}WU_sf_s + \frac{\rho}{2}WU_rf_r + \rho H\left(\frac{\partial W}{\partial t} + \frac{U}{R}\frac{\partial W}{\partial \theta} + W\frac{\partial W}{\partial Z}\right). \tag{4.10a}$$

Circumferential-Momentum Equation

$$-\frac{H}{R}\frac{\partial P}{\partial \theta} = \frac{\rho}{2}UU_sf_s + \frac{\rho}{2}(U - R\omega)U_rf_r + \rho H\left(\frac{\partial U}{\partial t} + \frac{U}{R}\frac{\partial U}{\partial \theta} + W\frac{\partial U}{\partial Z}\right). \tag{4.10b}$$

The explicit presence of the friction factors f_s, f_r in these equations is due to Nelson (1984). Szeri (1980) gives an insightful and favorable comparison of the Hirs model to other turbulent lubrication models. Hirs's model is particularly attractive for the higher Reynolds numbers experienced by annular pump seals.

The last term in the momenta equations accounts for temporal and convective acceleration terms. Recall that these terms are entirely neglected in developing the Reynolds equation used for bearing analysis in Chapter 3.

Friction Factor Definitions

While flow in a tight annulus with the inner cylinder rotating has nothing obvious in common with pipe flow, the friction factors f_s, f_r have to date been modeled with friction factors which have been adapted from pipe-flow results. Hirs proposed adopting a Blasius-type pipe friction model of the form

$$f_b = n\mathcal{R}^m, \qquad \mathcal{R} = VD\rho/\mu, \tag{4.11}$$

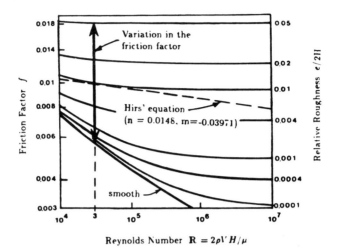

Figure 4.7   Moody diagram [Nelson and Nguyen (1987)].

where n and m are empirical constants, and $\mathcal{R}$ is the Reynolds number defined in terms of the diameter and the average flow velocity.

The Fanning friction-factor definition of Eqs. (4.7) and (4.10) was introduced by Black (1969) and differs from the Darcy-Weisbach friction factors which are commonly used in the Moody diagram for pipe friction. Stating the ΔP due to wall friction for an annulus of length L and clearance C_r in the Fanning format gives

$$\Delta P = 4f_F \left(\frac{L}{2C_r} \right) \frac{\rho V^2}{2},$$

where $2C_r$ is the hydraulic diameter, defined as four times the flow area divided by the wetted perimeter. By comparison, the Darcy-Weisbach format for the Moody diagram is

$$\Delta P = f_{DW} \left(\frac{L}{D} \right) \frac{\rho V^2}{2}.$$

Hence, the friction factors of Eqs. (4.10) must be multiplied by four before they can be plotted directly on a Moody diagram.

Figure 4.7 shows a Blasius-equation model plotted on a modified* Moody diagram for specified values of m and n. Obviously, the Blasius model provides a straight-line approximation to measured values of pipe friction factors. Yamada's (1962) test data for flow in a smooth annulus with the

---

*Fanning friction-factor definition.

inner cylinder rotating yielded a straight-line curve with coefficients $n = 0.079$, $m = -0.25$ which lay on the smooth pipe-friction curve of the modified Moody diagram and provided some comfort for the idea of using pipe-friction models as the basis for annular factors. The Blasius model for Hirs's equations yields the following friction-factor definitions:

$$f_s = ns\left(\frac{2\rho HU_s}{\mu}\right)^{ms}, \qquad f_r = nr\left(\frac{2\rho HU_r}{\mu}\right)^{mr}, \qquad (4.12)$$

where ns, ms and nr, mr characterize the flow resistance characteristics of the stator and rotor surfaces, respectively. The coefficients are empirical and must be determined from experimental data.

Moody produced the following approximate representation for pipe-friction factors:

$$f_m = a_1\left[1 + \left(\frac{b_2 e}{D} + \frac{b_3}{\mathcal{R}}\right)^{1/3}\right], \qquad \mathcal{R} = \frac{VD\rho}{\mu}, \qquad (4.13)$$

where

$$a_1 = 1.375 \times 10^{-3}, \qquad b_2 = 2 \times 10^4, \qquad b_3 = 10^6, \qquad (4.14)$$

and e/D is the relative roughness. This formula gives values which are within 5% of the Moody diagram for $4000 \le \mathcal{R} \le 10^7$ and $e/D \le 0.01$. For $e/D > 0.01$, it significantly underestimates f_m. von Pragenau (1982) initially used Moody's equation for seal analysis, and Nelson and Nguyen (1987) provided a comparison between seal-analysis results using the Blasius and Moody models.

In terms of the annular-seal parameters, the Moody friction factors are

$$f_s = a_1\left[1 + \left(\frac{b_2 e_r}{2H} + \frac{b_3}{\mathcal{R}_s}\right)^{1/3}\right], \qquad \mathcal{R}_s = 2U_s H\rho/\mu,$$

$$f_r = a_1\left[1 + \left(\frac{b_2 e_s}{2H} + \frac{b_3}{\mathcal{R}_r}\right)^{1/3}\right], \qquad \mathcal{R}_r = 2U_r H\rho/\mu, \qquad (4.15)$$

where e_r and e_s are the absolute roughness of the rotor and stator, respectively. For a centered, constant-clearance seal with radial clearance C_r, the relative roughness parameters $e_r/C_r, e_s/C_r$ are the only two parameters which can be varied in the Moody model.

Nondimensionalization

We wish to nondimensionalize the governing equations in terms of the tapered-seal geometry of Figure 4.8. For the centered position, the clearance function is

$$H_0 = \left(\overline{C}_r + \frac{\alpha L}{2}\right) - \alpha Z = \overline{C}_r[1 + q(1 - 2z)] = \overline{C}_r h_0, \quad (4.16)$$

where α is the seal taper angle, and

$$\overline{C}_r = (C_0 + C_1)/2, \quad q = \frac{\alpha L}{2\overline{C}_r} = \frac{C_0 - C_1}{C_0 + C_1}, \quad z = Z/L. \quad (4.17)$$

To complete the nondimensionalization, we introduce the additional variables:

$$W_0 = \dot{Q}/2\pi R\overline{C}_r, \quad T = L/W_0,$$
$$\tau = t/T, \quad h = H/\overline{C}_r,$$
$$u = U/R\omega, \quad w = W/W_0,$$
$$b = R\omega/W_0, \quad p = P/\rho W_0^2, \quad (4.18)$$

where $\dot{Q}$ is the volumetric flow rate, W_0 is the average axial velocity, T is the

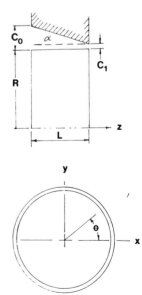

Figure 4.8 Convergent-taper seal geometry.

fluid transit time, and τ is a nondimensional time variable. In terms of these variables, the nondimensional governing equations are

Continuity Equation

$$\frac{\partial h}{\partial t} + b\left(\frac{L}{R}\right)\frac{\partial(hu)}{\partial\theta} + \frac{\partial(hw)}{\partial z} = 0. \tag{4.19a}$$

Axial-Momentum Equation

$$-h\frac{\partial p}{\partial z} = \frac{w}{2}u_s f_s\left(L/\overline{C}_r\right) + \frac{w}{2}u_r f_r\left(L/\overline{C}_r\right)$$

$$+ h\left[\frac{\partial w}{\partial \tau} + b\left(\frac{L}{R}\right)u\frac{\partial w}{\partial\theta} + w\frac{\partial w}{\partial z}\right]. \tag{4.19b}$$

Circumferential-Momentum Equation

$$-\frac{h}{b}\left(\frac{L}{R}\right)\frac{\partial p}{\partial\theta} = \frac{u}{2}u_s f_s\left(L/\overline{C}_r\right) + \frac{(u-1)}{2}u_r f_r\left(L/\overline{C}_r\right)$$

$$+ h\left[\frac{\partial u}{\partial \tau} + b\left(\frac{L}{R}\right)u\frac{\partial u}{\partial\theta} + w\frac{\partial u}{\partial z}\right], \tag{4.19c}$$

where the nondimensional relative velocities are

$$u_s = \left(w^2 + b^2 u^2\right)^{1/2},$$

$$u_r = \left[u_z^2 + b^2(u-1)^2\right]^{1/2}. \tag{4.20}$$

Nondimensional versions of the friction factors are

Blasius

$$f_s = ns\left(\mathcal{R}_0 h u_s\right)^{ms}, \qquad f_r = nr\left(\mathcal{R}_0 h u_r\right)^{mr}. \tag{4.21}$$

Moody

$$f_s = a_1\left[1 + \left(\frac{a_{2s}}{h} + \frac{a_3}{hu_s}\right)^{1/3}\right],$$

$$f_r = a_1\left[1 + \left(\frac{a_{2r}}{h} + \frac{a_3}{hu_r}\right)^{1/3}\right]. \tag{4.22}$$

In these equations,

$$\mathscr{R}_0 = 2\overline{C}_r W_0 \rho / \mu, \qquad a_3 = b_3 / \mathscr{R}_0, \qquad a_{2s} = b_2 \epsilon_s,$$

$$a_{2r} = b_2 \epsilon_r \qquad\qquad \epsilon_s = e_s / 2\overline{C}_r, \qquad \epsilon_r = e_r / 2\overline{C}_r. \qquad (4.23)$$

Note that ϵ_s, ϵ_r are the relative stator and rotor roughnesses.

4.3   PERTURBATION EQUATIONS

The development of zeroth- and first-order governing equations is the objective of this section. The small parameter used for perturbation analysis is ϵ, the perturbation in eccentricity ratio for small motion about a centered position. The zeroth-order equation will be used to define the leakage flow rate and the circumferential velocity distribution for a centered seal. The first-order equation will be used to define rotordynamic coefficients due to small centered motion of the seal rotor. The perturbation variables are introduced by

$$h = h_0 + \epsilon h_1, \qquad w = w_0 + \epsilon w_1,$$

$$u = u_0 + \epsilon u_1, \qquad p = p_0 + \epsilon p_1. \qquad (4.24)$$

Substitution into the governing equations yields:

Zeroth-Order Equations

$$h_0 w_0 = 1,$$

$$\frac{dp_0}{dz} = -\frac{1}{h_0^2}(u_{s0}\sigma_{s0} + u_{r0}\sigma_{r0}) + \frac{1}{h_0^3}\frac{dh_0}{dz},$$

$$\frac{du_0}{dz} = [\sigma_{s0}u_0 u_{s0} + \sigma_{r0}(u_0 - 1)u_{r0}]/2, \qquad (4.25)$$

where

$$\sigma_{s0} = f_{s0}(L/\overline{C}_r), \qquad \sigma_{r0} = f_{r0}(L/\overline{C}_r). \qquad (4.26)$$

First-Order Equations

$$b\left(\frac{\overline{C}_r}{R}\right)h_0\frac{\partial u_1}{\partial \theta} + \frac{\partial}{\partial z}(h_0 w_1) = -\frac{\partial h_1}{\partial t} - b\left(\frac{L}{R}\right)\frac{\partial h_1}{\partial \theta} + \frac{h_1}{h_0^2}\frac{dh_0}{dz}$$

$$\frac{\partial p_1}{\partial z} + A_{2z}u_1 + A_{3z}w_1 + \left[\frac{\partial w_1}{\partial \tau} + b\left(\frac{L}{R}\right)u_0\frac{\partial w_1}{\partial \theta} + w_0\frac{\partial w_1}{\partial z}\right] = A_{1z}h_1,$$

$$\frac{1}{b}\left(\frac{L}{R}\right)\frac{\partial p_1}{\partial \theta} + A_{2\theta}u_1 + A_{3\theta}w_1 + \left[\frac{\partial u_1}{\partial \tau} + b\left(\frac{L}{R}\right)u_0\frac{\partial u_1}{\partial \theta} + w_0\frac{\partial u_1}{\partial z}\right] = A_{1\theta}h_1.$$

$$(4.27)$$

The form of these equations is independent of the friction-factor model. However, formulas for σ_{s0}, σ_{r0}, and the $A_{iz}, A_{i\theta}$ coefficients obviously depend on whether a Blasius or Moody friction law is used, and definitions are provided in Appendix D for these parameters. Note that the coefficients of Eq. (4.27) are functions of the zeroth-order solution. The zeroth-order continuity equation is trivially satisfied and has been used to simplify the zeroth- and first-order axial-momenta equations.

4.4 SOLUTION FOR FLOW AND PRESSURE IN A CENTERED ANNULAR SEAL

The examination of the solution to the zeroth-order equations for the pressure and tangential-velocity distribution is the objective of the present section. From the previous section, the zeroth-order axial and circumferential momenta equations are

$$\frac{dp_0}{dz} = -\frac{1}{2h_0^2}(\sigma_{s0}u_{s0} + \sigma_{r0}u_{r0}) + \frac{1}{h_0^3}\frac{dh_0}{dz},$$

$$\frac{du_0}{dz} = -\frac{1}{2}\left[\sigma_{s0}u_0 u_{s0} + \sigma_{r0}(u_0 - 1)u_{r0}\right]. \qquad (4.28)$$

The continuity equation was used to simplify the momenta equations and is not required.

The momenta equations are coupled and nonlinear and must be solved iteratively to determine $p_0(z)$, $u_0(z)$, and W_0, the average flow velocity. A review of Appendix D will demonstrate that W_0 is a parameter in the friction factors σ_{s0}, σ_{r0}.

The initial condition $u_0(0)$ depends on the seal function and location within the pump. A typical value for an interstage seal is 0.5. However, the radially inward direction of leakage flow approaching wearing-ring and balance-piston seals provides a tangential acceleration of the flow leaving the impeller. Hence, while the exit absolute velocity from the impeller may be around $0.5R_{imp}\omega$, the flow entering the seals can yield near-unity values for $u_0(0)$. As we shall see, the inlet tangential velocity ratio $u_0(0)$ has a critical impact on the destabilizing, cross-coupled coefficient k.

The inlet-pressure boundary condition is defined in terms of the inlet-loss coefficient by

$$P_S - P(0, \theta, t) = \frac{\rho}{2}(1 + \xi)W^2(0, \theta, t). \qquad (4.29)$$

Published values for ξ from measurements vary from 0 to 0.5. The nondimensional, zeroth-order version of Eq. (4.29) is

$$p_S - p_0(0) = \frac{(1 + \xi)}{2}w_0^2(0) = \frac{(1 + \xi)}{2h_0^2(0)}. \qquad (4.30)$$

As will be discussed later, the combined effect of the inlet loss and the axial-pressure gradient accounts for the large direct stiffnesses which can be developed by annular seals, the "Lomakin (1985) effect."

Domm et al. (1967) also noted the possibility of pressure recovery at the seal exit as modeled by the relationship

$$P(1,\theta,t) + \frac{\rho(1 - \xi_e)}{2} W^2(1,\theta,t) = P_e. \tag{4.31}$$

This equation states that the pressure immediately inside the seal exit can be less than the exit pressure downstream of the seal. The zeroth-order nondimensional version of Eq. (4.31) is

$$p_0(1) - p_e = -\frac{(1 - \xi_e)}{2} w_0^2(1) = -\frac{(1 - \xi_e)}{2h_0^2(1)}. \tag{4.32}$$

Domm et al. showed that the exit-recovery phenomenon increases a seal's direct stiffness. Florjancic (1990) demonstrated through measurements of short $(L/D = 0.114)$, smooth, annular seals that the recovery factor significantly increases the accuracy of predictions for direct stiffness. His measured values for ξ_e were 0.85 for new seals $(C_r/R = 0.002)$ and 0.70 for worn seals $(C_r/R = 0.004)$. No pressure recovery is implied by $\xi_e = 1$, and an exit orifice is implied for $\xi_e > 1$.

The zeroth-order solutions provide the possibility for considerable insight with respect to rotordynamic coefficients. To this end, we will be examining variations of a nominal seal defined by the following parameters:

$$\Delta P = 35 \text{ bars}, \qquad\qquad L = 50.8 \text{ mm},$$
$$R = 76.2 \text{ mm}, \qquad\qquad C_0 = C_1 = 0.381 \text{ mm},$$
$$\omega = 3000 \text{ rpm}, \qquad\qquad \rho = 1000 \text{ Kg/m}^3,$$
$$\mu = 1.3 \times 10^{-3} \text{ PaS}, \qquad \xi = 0.10,$$
$$\xi_e = 1.0, \qquad\qquad \epsilon_s = \epsilon_r = 0.0. \tag{4.33}$$

The Moody friction-factor model will be used.

Figure 4.9 illustrates the zeroth-order pressure solution for three clearance ratios. Observe that increasing the clearance increases the leakage velocity and Reynolds number, increases the proportion of ΔP taken at the inlet, and reduces the axial pressure gradient. Conversely, reducing the clearance decreases the proportion of ΔP absorbed by the inlet loss. While not strictly true, the three *centered* solutions of Figure 4.7 can be thought of as an axial-pressure gradient for a centered position $(C_r/R = 0.05)$, the axial pressure gradient at the tight side of an eccentric seal $(C_r/R = 0.025)$, and the axial pressure gradient at the loose side of an eccentric seal $(C_r/R =$

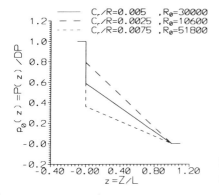

Figure 4.9 $p_0(z)$ for three different clearances.

0.0075). The difference between the pressure gradient on the tight and loose sides provides a restoring force which opposes the shaft displacement, yielding a potentially large direct stiffness. This physical mechanism for the direct stiffness was first explained by Lomakin (1958).

Figure 4.10 illustrates $p_0(z)$ versus z for a range of L/D ratios. Observe that increasing that L/D ratio progressively increases the proportion of ΔP absorbed by wall friction while decreasing the proportion absorbed by the inlet loss. The reduced influence of the inlet loss as L/D is increased eventually causes a reduction in direct stiffness due to a lessening of the Lomakin effect.

Figure 4.11 illustrates that an increase in relative roughness of the stator wall increases the axial pressure gradient and reduces the proportion of ΔP absorbed by the inlet loss.

Figure 4.12 shows the influence of a recovery-factor coefficient on the axial pressure gradient. Reducing the recovery factor from $\xi_e = 1.0$ increases the leakage rate, increases the inlet pressure drop, and increases the axial-pressure gradient.

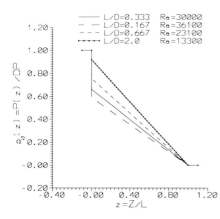

Figure 4.10 $p_0(z)$ for a range of L/D ratios.

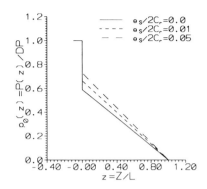

Figure 4.11 $p_0(z)$ for a range of stator surfaces roughnesses.

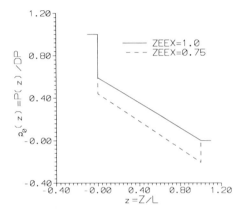

Figure 4.12 $p_0(z)$ with and without pressure recovery at the exit.

Figure 4.13 illustrates $u_0(z)$ for the nominal seal with $L/D = 1.0$, $D = 152.4$ mm, and three different initial conditions. These results are calculated for the same rotor and stator roughnesses. Observe that all of the solutions have the asymptote 0.5. Figure 4.14 shows solutions for the same initial conditions with $\epsilon_s = 0.05$, $\epsilon_r = 0.0$. Observe now that the asymptote is 0.33. Black et al. (1981) first provided analytical solutions which showed the axial

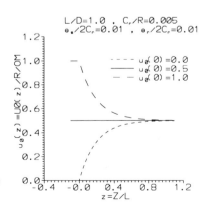

Figure 4.13 $u_0(z)$ for three initial conditions with $\epsilon_s = \epsilon_r = 0.01$.

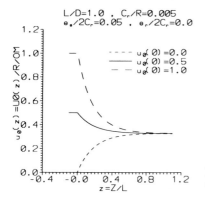

Figure 4.14 $u_0(z)$ for three initial conditions with $\epsilon_s = 0.05$, $\epsilon_r = 0.0$.

development of $u_0(z)$ as the fluid progressed along the seal. von Pragenau (1982) first demonstrated that increasing the stator roughness relative to the rotor roughness would decrease the average tangential velocity of the fluid within the seal.

As shown in the next section, the destabilizing cross-coupled stiffness term k is approximately proportional to the average tangential velocity; hence, any steps which reduce the average tangential velocity, $\bar{u}_0 = \int_0^1 u_0(z)\, dz$, will enhance the stabilizing influence of a seal. Figure 4.13 shows that decreasing the inlet tangential velocity reduces $\bar{u}_0$, and "swirl brakes"* are commonly used upstream of seals to sharply reduce $u_0(0)$, $\bar{u}_0$, and k. Notable examples of the success of this approach in stabilizing otherwise unstable pumps are provided by Massey (1985) and Valantas and Bolleter (1988).

Figure 4.14 shows that a deliberately roughened stator also reduces $\bar{u}_0$, and von Pragenau's "damper seal" approach has been used effectively to improve the stability of both the Space Shuttle main engine (SSME), high-pressure oxygen turbopump (HPOTP) [Childs (1985)], and the high-pressure fuel turbopump (HPFTP) Becht et al. (1991). Iwatsubo and Sheng (1990a) have performed tests which demonstrate the effectiveness of roughened stator surfaces in reducing k.

4.5 SOLUTION OF THE FIRST-ORDER EQUATIONS FOR ROTORDYNAMIC COEFFICIENTS

Our interest here is the solution for the rotordynamic coefficients of Eq. (4.1). To this end, a solution will be developed to the perturbation equations for w_1, u_1, and p_1 due to seal-perturbed motion defined by $h_1(t)$. A separation-of-variables solution approach after Childs (1984) will be used.

---

*See Figure 4.37.

In terms of the seal displacement-vector components, the clearance function is

$$h = h_0 - x(t)\cos\theta - y(t)\sin\theta, \qquad (4.34)$$

where

$$x = X/\overline{C}_r, \qquad y = Y/\overline{C}_r. \qquad (4.35)$$

Hence, by comparison to Eq. (4.24),

$$\epsilon h_1 = -x\cos\theta - y\sin\theta. \qquad (4.36)$$

Guided by Eq. (4.36), the following separation-of-variables solution is assumed for the dependent perturbation variables

$$
\begin{aligned}
w_1(z,\tau,\theta) &= w_{1c}(z,\tau)\cos\theta + w_{1s}(z,\tau)\sin\theta, \\
u_1(z,\tau,\theta) &= u_{1c}(z,\tau)\cos\theta + u_{1s}(z,\tau)\sin\theta, \\
p_1(z,\tau,\theta) &= p_{1c}(z,\tau)\cos\theta + p_{1s}(z,\tau)\sin\theta.
\end{aligned}
\qquad (4.37)
$$

Substituting Eqs. (4.36) and (4.37) into Eq. (4.27) and equating coefficients of $\sin\theta$ and $\cos\theta$ yields six real equations which are reduced to three complex equations by introducing the complex variables

$$
\begin{aligned}
\boldsymbol{w}_1 &= w_{1c} + jw_{1s}, & \boldsymbol{u}_1 &= u_{1c} + ju_{1s}, \\
\boldsymbol{p}_1 &= p_{1c} + jp_{1s}, & \epsilon\boldsymbol{h}_1 &= -(x + jy).
\end{aligned}
\qquad (4.38)
$$

The time dependency in the governing equations is eliminated by assuming a precessional seal motion of the form

$$\epsilon\boldsymbol{h}_1 = -\left(R_0/\overline{C}_r\right)e^{j\Omega t} = -r_0 e^{jf\tau}, \qquad (4.39)$$

where

$$f = \Omega/\omega. \qquad (4.40)$$

The corresponding separation-of-variables solutions for the dependent variables are

$$\boldsymbol{w}_1 = \overline{w}_1 e^{jf\tau}, \qquad \boldsymbol{u}_1 = \overline{u}_1 e^{jf\tau}, \qquad \boldsymbol{p}_1 = \overline{p}_1 e^{jf\tau}. \qquad (4.41)$$

Substitution eliminates τ as a dependent variable and yields the following three complex equations:

$$\frac{d}{dz}\begin{Bmatrix}\overline{w}_1\\\overline{u}_1\\\overline{p}_1\end{Bmatrix} + [A(f,z)]\begin{Bmatrix}\overline{w}_1\\\overline{u}_1\\\overline{p}_1\end{Bmatrix} = \left(\frac{r_0}{e}\right)\begin{Bmatrix}g_1\\g_2\\g_3\end{Bmatrix}, \tag{4.42}$$

where

$$[A] = \begin{bmatrix} \dfrac{1}{h_0}\dfrac{dh_0}{dz} & -j\omega T & 0 \\[2ex] h_0 A_{3\theta} & h_0(A_{2\theta}+j\Gamma T) & -j\dfrac{h_0}{b}\left(\dfrac{L}{R}\right) \\[2ex] \left(A_{3z}-\dfrac{1}{h_0^2}\dfrac{dh_0}{dz}+j\Gamma T\right) & A_{2z}+\dfrac{j\omega T}{h_0} & 0 \end{bmatrix}, \tag{4.43}$$

$$\begin{Bmatrix}g_1\\g_2\\g_3\end{Bmatrix} = \begin{Bmatrix} -\dfrac{1}{h_0^3}\dfrac{dh_0}{dz}+j\dfrac{\Gamma T}{h_0} \\[2ex] -h_0 A_{1\theta} \\[2ex] -\left(A_{1z}-\dfrac{1}{h_0^4}\dfrac{dh_0}{dz}+j\dfrac{\Gamma T}{h_0^2}\right) \end{Bmatrix}, \tag{4.44}$$

and

$$\Gamma T = \omega[f - u_0(z)]T. \tag{4.45}$$

Note that with an appropriate definition for $\overline{C}_r$, these equations apply for seals with a continuous variation in $h_0(z)$, in addition to the convergent-taper geometry of Figure 4.8 and Eq. (4.16). Scharrer (1989) used similar equations in analyzing seals he called "wavy."

Three boundary conditions are required for Eq. (4.42); the first and simplest is

$$\overline{u}_1(0,\theta,\tau) = 0, \tag{4.46a}$$

which states that seal motion does not perturb the fluid tangential velocity immediately upstream of the seal.

The perturbed pressure boundary conditions are obtained from Eqs. (4.30) and (4.31) as

$$p_1(0,\theta,\tau) = -(1+\xi)w_1(0,\theta,\tau)w_0(0)$$
$$= -(1+\xi)w_1(0,\theta,\tau)/h_0(0), \tag{4.46b}$$
$$p_1(1,\theta,\tau) = -(1-\xi_e)w_1(1,\theta,\tau)w_0(1)$$
$$= -(1-\xi_e)w_1(1,\theta,\tau)/h_0(1). \tag{4.46c}$$

In terms of $\bar{u}_1, \bar{w}_1, \bar{p}_1$ these boundary conditions are simply

$$\bar{u}_1(0) = 0, \tag{4.47a}$$

$$\bar{p}_1(0) = -(1 + \xi)\bar{w}_1(0)/h_0(0), \tag{4.47b}$$

$$\bar{p}_1(1) = -(1 - \xi_e)\bar{w}_1(1)/h_0(1). \tag{4.47c}$$

A transition-matrix approach is used to solve Eq. (4.42) while satisfying the boundary conditions of Eq. (4.47) [Meirovitch (1985)]. This approach solves the homogeneous version of Eq. (4.42) successively with the initial conditions $(1, 0, 0), (0, 1, 0), (0, 0, 1)$ to obtain the transition matrix $[\Phi(f, z)]$. The particular solution is then solved for zero initial conditions with $r_0/\epsilon = 1$, yielding the vector $\{v(f, z)\}$. The complete solution is

$$\begin{Bmatrix} \bar{w}_1 \\ \bar{u}_1 \\ \bar{p}_1 \end{Bmatrix} = [\Phi(f, z)] \begin{Bmatrix} \bar{w}_1(0) \\ \bar{u}_1(0) \\ \bar{p}_1(0) \end{Bmatrix} + \left(\frac{r_0}{\epsilon}\right) \begin{Bmatrix} v_1(f, z) \\ v_2(f, z) \\ v_3(f, z) \end{Bmatrix}. \tag{4.48}$$

In fact, given that $\bar{u}_1(0) = 0$, there is no reason to develop the second homogeneous solution. The solution is a function of $f = \Omega/\omega$, the ratio of the precession frequency to the running speed and could be evaluated over the interval $z \in [0, 1]$ if the two unknown initial conditions $\bar{w}_1(0), \bar{p}_1(0)$ were known.

Equation (4.46b) provides one equation for these unknowns. The second is obtained by first evaluating Eq. (4.48) at $z = 1$ to get

$$\bar{w}_1(1) = \Phi_{11}(1)\bar{w}_1(0) + \Phi_{13}(1)\bar{p}_1(0) + \left(\frac{r_0}{\epsilon}\right)v_1(1),$$

$$\bar{p}_1(1) = \Phi_{31}(1)\bar{w}_1(0) + \Phi_{33}(1)\bar{p}_1(0) + \left(\frac{r_0}{\epsilon}\right)v_3(1),$$

and then substituting into Eq. (4.47c) for the final equation. In matrix form the two equations are

$$[B_{ij}] \begin{Bmatrix} \bar{w}_1(0) \\ \bar{p}_1(0) \end{Bmatrix} = \left(\frac{r_0}{\epsilon}\right)(b_i), \tag{4.49}$$

where

$$B_{11} = (1 + \xi)/h_0(0), \qquad B_{12} = 1$$
$$B_{21} = \Phi_{31}(1) + \Phi_{11}(1)(1 - \xi_e)/h_0(1),$$
$$B_{22} = \Phi_{33}(1) + \Phi_{13}(1)(1 - \xi_e)/h_0(1),$$
$$b_1 = 0, \qquad b_2 = -v_3(1) - v_1(1)(1 - \xi_e)/h_0(1).$$

Inversion of Eq. (4.49) gives the initial values $\bar{w}_1(0)$, $\bar{p}_1(0)$ which cause the boundary conditions to be satisfied. The resultant solution can be stated:

$$\begin{Bmatrix} \bar{w}_1 \\ \bar{u}_1 \\ \bar{p}_1 \end{Bmatrix} = \left(\frac{r_0}{\epsilon} \right) \begin{Bmatrix} f_{1c} + jf_{1s} \\ f_{2c} + jf_{2s} \\ f_{3c} + jf_{3s} \end{Bmatrix}. \tag{4.50}$$

The perturbation reaction-force components are defined by

$$F_{X1} = -\epsilon \int_0^L \int_0^{2\pi} P_1 \cos\theta\, R\, d\theta\, dZ,$$

$$F_{Y1} = -\epsilon \int_0^L \int_0^{2\pi} P_1 \sin\theta\, R\, d\theta\, dZ, \tag{4.51}$$

or

$$\frac{F_{X1}}{F_0} = -\frac{\epsilon}{C_d} \int_0^1 \int_0^{2\pi} p_1(\theta, z, t)\cos\theta\, d\theta\, dz = -\frac{\epsilon\pi}{C_d} \int_0^1 p_{1c}(z, t)\, dz,$$

$$\frac{F_{Y1}}{F_0} = -\frac{\epsilon}{C_d} \int_0^1 \int_0^{2\pi} p_1(\theta, z, t)\sin\theta\, d\theta\, dz = -\frac{\epsilon\pi}{C_d} \int_0^1 p_{1s}(z, t)\, dz, \tag{4.52}$$

where

$$F_0 = 2LR\Delta P, \qquad \Delta P = C_d \frac{\rho\bar{V}^2}{2}. \tag{4.53}$$

The $r - \theta$ polar coordinates are precessing with the rotor at the rate $\Omega = f\omega$, and the reaction-force components in this system are

$$\frac{F_{r1} + jF_{\theta1}}{F_0} = \frac{(F_{X1} + jF_{Y1})e^{-jf\tau}}{F_0} = -\frac{\epsilon\pi}{C_d} \int_0^1 (p_{1c} + jp_{1s})e^{-jf\tau}\, dz$$

$$= -\frac{\epsilon\pi}{C_d} \int_0^1 \bar{p}_1\, dz = -\frac{\epsilon\pi}{C_d} \left(\frac{r_0}{\epsilon} \right) \int_0^1 (f_{3c} + jf_{3s})\, dz. \tag{4.54}$$

Hence,

$$f_{r1}(f) = \left(\frac{F_{r1}}{F_0} \right) \cdot \frac{1}{r_0} = -\frac{\pi}{C_d} \int_0^1 f_{3c}(f, z)\, dz,$$

$$f_{\theta1}(f) = \left(\frac{F_{\theta1}}{F_0} \right) \cdot \frac{1}{r_0} = -\frac{\pi}{C_d} \int_0^1 f_{3s}(f, z)\, dz. \tag{4.55}$$

Observe that we have a frequency-response solution for the reaction-force components. The comparable result in terms of the rotordynamic-coefficient definition of Eq. (4.1) is

$$f_{r1}(f) = -\left(\overline{K} + f\overline{c} - f^2\overline{M}\right), \qquad f_{\theta1}(f) = \overline{k} + f\overline{C}, \qquad (4.56)$$

where the nondimensional coefficients are

$$\overline{K} = K\overline{C}_r/F_0, \qquad \overline{k} = k\overline{C}_r/F_0,$$

$$\overline{C} = C\overline{C}_r\omega/F_0, \qquad \overline{c} = c\overline{C}_r\omega/F_0,$$

$$\overline{M} = M\overline{C}_r\omega^2/F_0. \qquad (4.57)$$

To be brief, the rotordynamic coefficients are obtained by calculating $f_{r1}(f)$ and $f_{\theta1}(f)$ for a range of f values and then using a least-squares curve fit of Eq. (4.56). The author generally uses $f \in [0, 2]$ for these calculations.

"Short-seal" solutions provide an alternative to this section's finite-length solutions. The first useful solutions for plain seals were provided by Black's (1969) short-seal solutions which are analogous to Ocvirk's (1952) short-bearing solution in that the pressure-induced perturbation flow u_1 is neglected. With equal surface roughnesses on the rotor and stator this assumption yields an analytical solution for the rotordynamic coefficients. Black's initial solution required that $U_0(z) = R\omega/2$. However, in a result published after his death, a second solution was presented which accounted for the development of circumferential flow due to shear forces along the seal [Black et al. (1981)]. Unfortunately, different physical models were used in the two developments. Childs (1983) developed a short-seal solution based on Hirs's (1973) lubrication equation with a Blasius friction-factor model that incorporated all of Black's various developments. The simplicity of the short-seal solutions are quite attractive. However, their application is limited to truly short seals, $L/D \leq 0.25$. Black and Jenssen (1970) provided "correction" factors to account for long seals; however, comparisons between these results and finite-length calculations are not encouraging [Childs (1984)].

4.6 ROTORDYNAMIC COEFFICIENTS AND THEIR DEPENDENCE ON PHYSICAL PARAMETERS

Introduction

The dependence of rotordynamic coefficients on physical parameters is the subject of this section. We will begin with plain (ungrooved) seals examining the influence of L/D, $\bar{C}_r/R$, ξ, ξ_{ex}, ϵ_s, $u_0(0)$, ΔP, and ω. We will then review the influence of grooving and eccentricity on seals.

Plain (Ungrooved) Annular Seals

The preceding section explained how annular-seal rotordynamic coefficients are derived. This section will examine the functional dependence of the coefficients on various seal parameters.

The Moody wall-friction model will be used for most of these calculations rather than the Blasius. The author has worked primarily with the Blasius model; however, the Blasius m and n parameters have no inherent physical meaning. To determine whether a particular m, n set is reasonable and represents a "smooth" or "rough" seal, one must generally calculate friction factors for a range of Reynolds numbers and plot the result on a Moody diagram. By contrast, the Moody-equation model uses the single relative-roughness parameter which provides *some* immediate physical feel. The reader is advised to think of the Moody relative-roughness parameters as empirical, "effective," relative-roughness values which should be obtained from leakage and axial-pressure-gradient measurements and are not necessarily related to measured wall roughnesses. Nelson and Nguyen (1987) provide a thorough comparison of calculated results for a seal using the Moody and Blasius models with relative stator roughness as a parameter.

The nominal seal to be examined is defined by the parameters

$$\Delta P = 35 \text{ bars,}$$

$$L = 50.8 \text{ mm,} \qquad R = 76.2 \text{ mm,}$$

$$C_0 = C_1 = 0.381 \text{ mm,}$$

$$\omega = 3000 \text{ rpm} = 314.2 \text{ rd/sec,}$$

$$\rho = 1000 \text{ Kg/m}^3, \qquad \mu = 1.3 \times 10^{-3} \text{ PaS,}$$

$$\xi = 0.10, \qquad \xi_e = 1.0,$$

$$\epsilon_r = \epsilon_s = 0.001. \tag{4.58}$$

For $u_0(0) = 0.5$, the calculated rotordynamic coefficients for these data are (in SI units):

$K \times 10^{-7}$	$k \times 10^{-7}$	$C \times 10^{-5}$	$c \times 10^{-4}$	M	$M\omega/c$	$k/C\omega$	
1.567	0.473	0.300	0.210	6.68	0.903	0.500	(4.59)

In contrast to bearings, seals develop significant direct stiffnesses in the centered position, independent of rotation speed. As explained in Section 4.4, the Lomakin (1958) effect, based on the Blasius model, is the physical source for most of this stiffness. Specifically, radial displacement of the seal rotor changes the local Reynolds number (linearly through the clearance change and less directly through the change in leakage velocity) which changes the friction factor and the axial-pressure gradient. The trade-off between the reduced and expanded clearance sides of the seals gives the reaction force and direct stiffnesses.

The Moody friction-factor equation provides an additional mechanism for the direct stiffness. Picture the Moody diagram at a high Reynolds number range where the friction-factor curves for various relative-roughness factors are flat. Changes in the clearance and Reynolds numbers no longer yield a change in the friction factor; hence, the Lomakin effect is largely eliminated. However, changes in the clearance due to a radial seal offset directly change the relative roughness and move one vertically from one nearly horizontal relative-roughness curve to another. On the reduced-clearance side, the relative roughness and friction factor are increased. On the increased-clearance side, the opposite effect holds. Consequent trade-offs between the pressure drop due to inlet loss and the axial-pressure gradient create a Moody effect which is directly analogous to the Lomakin effect. Hence, the Moody-equation model predicts higher direct stiffnesses than the Blasius-equation model. Nelson and Nguyen (1987), in comparing results for the Blasius and Moody models, show that predictions from the two models coincide for smooth seals. However, as ϵ_s and ϵ_r increase, the Moody-equation direct stiffness becomes markedly larger than the Blasius-equation prediction. For $\epsilon_s = \epsilon_r = 0.05$, the Moody-equation stiffness values K and k are about 45 and 25% higher, respectively, than their Blasius-equation counterparts. Predictions for C are about the same for both models, while the Blasius model predicts slightly higher values for c and M.

Returning to the results of (4.59), the cross-coupled stiffness term k arises from fluid rotation in precisely the same fashion as an uncavitated plain journal bearing. For example, moving the seal rotor in the $+X$ direction creates converging and diverging sections in the upper and lower halves of the seal, respectively. The pressure is elevated in the converging section and depressed in the diverging section, yielding a reaction force in the $+Y$ direction.

The damping forces arises from drag terms and can be substantial. The whirl-frequency ratio $\Omega_w = k/C\omega$ is about 0.5 corresponding to the average tangential velocity within the seal of $0.5R\omega$.

The cross-coupled damping and the mass term arise primarily from inertial effects. The M term is simply not obtained from a Reynolds-equation analysis but does arise in seals because of the elevated Reynolds number and is analytically predicted because of the temporal acceleration terms in the governing equations. For long seals, M can be quite large and has a significant impact on pump rotordynamics.

From Eq. (4.1), a seal rotor precessing at Ω with an orbit amplitude A yields radial and circumferential reaction force components

$$\frac{F_\theta}{A} = k - C\Omega = C\omega(\Omega_w - f),$$

$$-\frac{F_r}{A} = K + c\Omega - M\Omega^2 = K + cf\omega\left[1 - \left(\frac{M\omega}{c}\right)f\right],$$

$$f = \Omega/\omega, \qquad \Omega_w = k/C\omega. \tag{4.60}$$

The first of these equations states that the circumferential force will oppose the whirling motion if f/Ω_w is greater than one. If one thinks of Ω as ω_n, the frequency of a rotor precessing at its natural frequency, then the tangential force becomes destabilizing at the speed

$$\omega_s = \omega_n/\Omega_w. \tag{4.61}$$

The result of (4.59) shows Ω_w to be about 0.5, which is the same as a plain journal bearing. Hence, for $\Omega_w \cong 0.5$, a seal becomes destabilizing at running speeds in excess of about twice the rotor's first critical speed. In seals (unlike plain journal bearings) Ω_w can be decreased significantly by reducing the inlet tangential velocity and/or increasing the stator relative roughness.

Note from the second of Eq. (4.60) that the cross-coupled damping force acts as a "gyroscopic stiffening" term in concert with K and opposed by M. For the results of (4.59) $M\omega/c \cong 1$, and c and M tend to cancel for $f = 1$, i.e., synchronous precession at the running speed. This is a typical result for $\bar{u}_0 = \int_0^1 u_0(z)\,dz = 0.5$ and was first demonstrated analytically by Black (1969).

Before we begin a systematic review of the influence of parameter variations on seal rotordynamic coefficients, let us briefly consider some of the main differences between seals and bearings. First, bearings are normally cavitated, an effect which yields their direct stiffness terms and complicates their analysis. Further, bearings are quite nonlinear, yielding rotordynamic coefficients which are strongly dependent on the static eccentricity ratio. Fortunately, seals do not generally cavitate, and they remain linear out to static eccentricity ratios of about 0.5. Hence, the coefficients of Eq. (4.1) for small motion about a centered position are generally satisfactory for the analysis of pump rotordynamics.*

As was shown in the preceding chapter, journal bearings can be characterized in terms of a single parameter, the Sommerfeld number, $S = \mu(\omega/2\pi)(R/C)^2 DL/F_0$. Unfortunately, seal rotordynamic coefficients de-

---

*We consider analytical and test results concerning this point later in this section.

pend on a host of nondimensional parameters, e.g.,

$$\mathscr{R}_0 = \frac{2W_0\overline{C}_r\rho}{\mu}, \qquad \mathscr{R}_\theta = \frac{R\omega\overline{C}_r\rho}{\mu}, \qquad \epsilon_r, \qquad \epsilon_s,$$

$$u_0(0), \qquad L/D, \qquad \overline{C}_r/R, \qquad \Delta P/\rho W_0^2, \qquad \text{etc.}$$

Hence, one cannot simply develop design charts for seals, but must repeatedly calculate rotordynamic coefficients for separate seal operating conditions. In the balance of this section, we will consider the influence of important parameters on the behavior of seal rotordynamic coefficients by examining results of changes in the parameters of the seal defined by (4.58). In particular, we will examine the influence of changes in L/D, $\overline{C}_r/R$, ξ, ξ_e, ϵ_s, $u(0)$, ΔP, and ω.

Starting with L/D, Figures 4.15a–4.15(c) illustrate the rotordynamic coefficients versus L/D for the data of (4.58) with $u_0(0) = 0.5$. These results show that all coefficients increase monotonically with increasing L/D except K, which has a maximum around $L/D = 0.5$ and then steadily declines, becoming negative for $L/D \geq 1.75$. Recall that the zeroth-order solution showed that increasing L/D progressively decreases the influence of the inlet loss, thus reducing the Lomakin effect. Fritz (1972) has previously predicted negative stiffnesses for long annuli (without significant ΔP) around rotors. Some pump manufacturers "break up" their long-L/D balance pistons with deep grooves to achieve the effect of several shorter seals in series instead of a single long seal with reduced stiffness.

Note the "huge" added-mass terms predicted in Figure 4.15(c) for long seals. As we shall see, experimental evidence supports these predictions.

Figure 4.16 illustrates the influence of C_r/R on K and C for the data of (4.58) with $L/D = 0.5$ ($L = 76.2$ mm) and $u_0(0) = 0.5$. The C_r/R range $[0.002, 0.01]$ proceeds from a seal that is quite tight to completely worn out. Observe the rapid drop in the coefficients. As we saw in Figure 4.3, the reduction of K due to wearing out of seals can frequently cause a shift of a pump critical speed downwards into the operating speed range. The parallel loss of damping means much sharper response of the pump rotor motion as well. Because of the loss of stiffness and damping with increasing clearance, prudent practice in pump rotordynamics will include an analysis for new and double-clearance seals. Although not illustrated, the parallel reductions in k, c, and M are 78, 85, and 85%, respectively.

Figure 4.17 illustrates K versus ξ, the inlet-loss coefficient, for data (4.58) with $u_0(0) = 0.5$. The variation in ξ illustrated is consistent with the published range of measurements for this parameter. Despite the strong dependence of the direct stiffness on $(1 + \xi)$ through the Lomakin effect, changing ξ from 0 to 0.25 only increases K by about 4%. Parallel calculated changes to k, C, c, and M are about 6, 6, 8, and 8%. In short, the rotordynamic

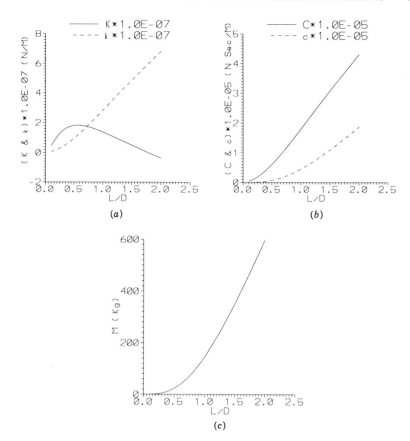

Figure 4.15   (a) K and k versus L/D for data (4.58) with $u_0(0) = 0.5$. (b) C and c versus L/D for data (4.58) with $u_0(0) = 0.5$. (c) M versus L/D for data (4.58) with $u_0(0) = 0.5$.

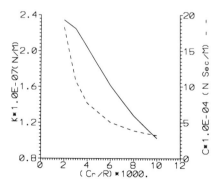

Figure 4.16 K and C versus C_r/R with $u_0(0) = L/D = 0.5$.

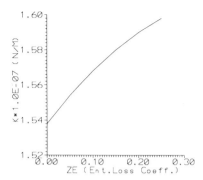

Figure 4.17 K versus ξ for data (4.58) with $u_0(0) = 0.5$.

coefficients simply do not change much with reasonably expected variations in ξ.

The loss coefficient $(1 + \xi)$ accounts for the energy required to accelerate the fluid upstream of the seal from its nominally zero velocity to the fully developed flow at velocity V within the seal. As with pipe flow, a vena contracta is formed very near the seal inlet and has an associated sharp drop in static pressure. The pressure rises as flow proceeds down the seal and becomes fully developed. The $(1 + \xi)$ loss coefficient model concentrates all of the losses at the inlet. In pipes, the flow is assumed to be fully developed after around 20 diameters. The tight clearances and large shearing forces in seals accelerate the convergence of flow toward a fully developed condition. Kundig (1990), in a seal with large clearances, $C_r/R = 0.010$, $L = 62.5$ mm, shows complete pressure recovery after only about three seal hydraulic diameters, i.e., $6C_r$. Nonetheless, concentrating all the losses at the entrance becomes questionable as seals become progressively shorter.

Figure 4.18 shows K versus $(1 - \xi_e)$ for data (4.58) with $u_0(0) = L/D = 0.5$ over the expected-variation range of this parameter. Note that $\xi_e = 1$ $(1 - \xi_e = 0)$ corresponds to no pressure recovery downstream of the seal. K is seen to increase more or less linearly and substantially with increasing

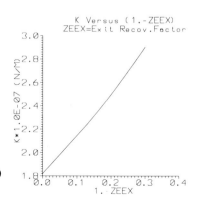

Figure 4.18 K versus $1 - \xi_e$ for data (4.58) with $u_0(0) = L/D = 0.5$.

$(1 - \xi_e)$. Parallel calculated changes for k, C, c, and M are -20, -21, 10, and 10%, respectively. Hence, changes in $(1 - \xi_e)$ have the largest influence on K. The author knows of no attempts to increase K by maximizing $(1 - \xi_e)$ through a Bernoulli-type seal exit. Such a modification would obviously increase leakage flow.

Reliable measured data for the inlet and exit loss coefficients have proven to be surprisingly difficult to obtain. In general, "small" pressure taps are drilled inwards through the stator wall to provide a pressure-measuring port. However, because of the tight clearances within the seal, the diameter of the pressure port will generally be several times larger than the radial clearance. Hence, the pressure which is measured in the port can include a portion of the velocity head in the axial and circumferential directions. Measurement accuracy is also impaired by vortices which are established at the mouth of the port. With these cautions in mind, measured results given by Stampa (1971) are $\xi \cong 0.1$ ($\mathscr{R}_0 > 7000$), $\xi_e \cong 0.75$ ($\mathscr{R}_0 > 5000$). As cited earlier, Florjancic (1990) gives $\xi \cong 0.25$, $\xi_e \cong 0.8$ and showed a marked improvement between measurement and prediction of the direct stiffness coefficient K by incorporating the exit recovery factor in his model. For reasonably expected ratios of V to $R\omega$, Kundig (1990) cites $\xi \cong 0.2$. Iino and Kaneko (1980) give measured results $0 \leq \xi \leq 0.2$, $0.6 \leq \xi_e \leq 0.8$. They show a nominally linear drop in ξ as $\mathscr{R}_0$ is increased from 10^3 to 5×10^3 and a linear drop in ξ_e as $\mathscr{R}_\theta / \mathscr{R}_0$ is increased from 1 to 3. Childs et al. (1988) showed a drop in $(1 + \xi)$ from around 1.2 to 1.0 as $\mathscr{R}_0$ was increased from 100,000 to 300,000.

Kim and Childs (1987) curve-fitted the $(1 + \xi)$ versus $\mathscr{R}_0$ relationship and included it in their perturbation of the inlet-loss boundary condition Eq. (4.29), showing a resultant improvement in their theory-versus-measurement correlation. However, given the impreciseness of available $(1 + \xi)$ and $(1 - \xi_e)$ data, the exercise involved in developing functional relationships for those coefficients and then perturbing them seems questionable.

Figures 4.19 shows the results of changes in the stator relative roughness on the rotordynamic coefficients. The results are for data (4.58) with $u_0(0) = 0.5$ and $L/D = 1.0$ ($L = 152.4$ mm). The range of ϵ_s used is consistent with

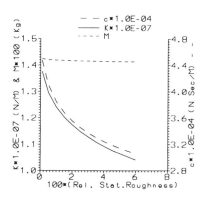

Figure 4.19 (a) K, c, and M versus ϵ_s for data (4.58) with $u_0(0) = 0.5$, $L/D = 1.0$.

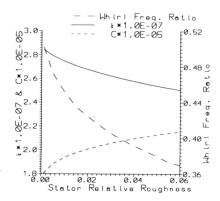

Figure 4.20 k, C, and Ω_w versus ϵ_s for data (4.58) with $u_0(0) = 0.5$, $L/D = 1.0$.

the author's experience on measured *effective* relative roughness in seals whose stators have been roughened by a hole pattern [Childs et al. (1990b)]. The effective roughness is obtained by plotting friction-factor/Reynolds number data on a modified Moody diagram and has nothing to do with physical roughness associated with the hole depths. (In most cases the holes are two or three times deeper than the clearance). Figure 4.19 shows K decreasing monotonically with increasing ϵ_s. The results of this plot are related to those of figure 4.15a of K versus L/D. Increasing ϵ_s and L/D increases the proportion of pressure drop taken by the axial pressure gradient and reduces the proportion taken by the inlet loss; hence, they reduce the Lomakin effect and K. Figure 4.19a also illustrates c dropping steadily as ϵ_s is increased. The cross-coupled damping coefficient c is proportional to $\bar{u}_0 = \int_0^1 u_0(z)\, dz$. The wall roughness decreases* $\bar{u}_0$ and consequently c. As illustrated, M is insensitive to changes in ϵ_s.

Figure 4.20 illustrates k, c, and $\Omega_w = k/C\omega$ versus ϵ_s. The cross-coupled stiffness k drops due to decreasing $\bar{u}_0$, just like c. This result was first predicted by von Pragenau (1982), in what he referred to as a "damper seal." The whirl-frequency ratio drops steadily with increasing ϵ_s. The drop in Ω_w closely follows the drop in k and demonstrates a major improvement in the stabilizing capacity of the seal. Again, the seal becomes destabilizing when the running speed exceeds

$$\omega_s = \omega_{n1}/\Omega_w,$$

where ω_{n1} is the pump's first critical speed. For $u_0(0) = 0.5$ and $\epsilon_s = \epsilon_r$, $\omega_s \cong 2\omega_{n1}$. However, from Figure 4.19(b), $\epsilon_s = 0.06$ yields $\omega_s \cong 2.7\omega_{n1}$. Stator surface roughness treatments become less effective for short seals since the wall shear stress has less axial distance and time to reduce $\bar{u}_0$.

A question which may arise in practice is: Should I use a swirl brake, stator roughness, or a swirl brake in combination with stator roughness to

---

*Review Figure 4.14.

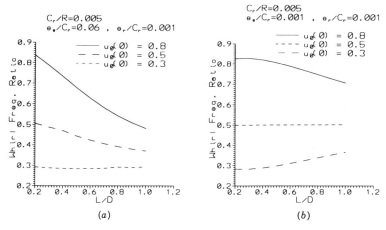

Figure 4.21 (a) Ω_w versus L/D for data (4.58) and $\epsilon_s = 0.006$. (b) Ω_w versus L/D for data (4.58) and $\epsilon_s = \epsilon_r = 0.001$.

enhance pump rotordynamic stability? The frames of Figure 4.21a are for data (4.58) with a rough stator ($\epsilon_s = 0.06$), three inlet tangential velocities $[u_0(0) = 0.8, 0.5, 0.3]$ and a range of L/D ratios. The results show the roughened-stator damper seal to be very effective for long seals with a highly prerotated flow. The stator roughness is progressively less effective in reducing Ω_w as $u_0(0)$ is reduced. A swirl brake which reduces $u_0(0)$ from 0.8 to 0.5 or 0.3 is most effective for short seals but remains substantially effective for all seal lengths.

Figure 4.21(b) provides the same results as Figure 4.21a except the seal stator is now smooth. Again, a swirl brake that reduces $u_0(0)$ from 0.8 to 0.5 or 0.3 is effective for all L/D ratios, but is most effective for short seals.

A comparison of Figures 4.21a and 4.21b shows that a swirl-brake/roughened-stator *combination* would be most effective for long seals with highly prerotated flow. For short seals, a swirl brake alone is adequate without stator roughness.

Figure 4.22 illustrates K, k, and C versus ΔP for $u_0(0) = 0.5$. Although not illustrated, c and M are insensitive to changes in ΔP. Note that k and C rise asymptotically to a linearly increasing function of ΔP, while K is a linear function of ΔP.

Note that ΔP for a pump is generally proportional to the density of the fluid being pumped. Hence, a pump's vibration characteristics can change *markedly* with a change in fluid density.

Given that ΔP is generally a quadratic function of pump running speed, Figure 4.23a illustrates K and k versus running speed in rpm, for $u_0(0) = 0.5$. Both stiffnesses increase quadratically with rpm. However, k increases due to both increasing ΔP and an increase in the inlet tangential velocity $U_0(0) = 0.5R\omega$ which obviously increases with ω. Figure 4.23b demonstrates that C

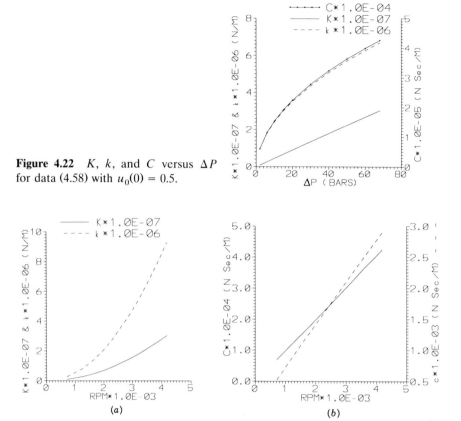

Figure 4.22 K, k, and C versus ΔP for data (4.58) with $u_0(0) = 0.5$.

Figure 4.23   (a) K and k versus running speed in rpm for data (4.58) with $u_0(0) = 0.5$. (b) C and c versus running speed in rpm for data (4.58) with $u_0(0) = 0.5$.

and c are linear functions of rpm. Again, the increase in c mainly follows the increase in $U_0(0)$. The whirl-frequency ratio accompanying the results of Figures 4.23 is relatively constant at 0.5. Hence, the seal would become destabilizing for running speeds slightly above twice the rotor's first critical speed. The mass term M is insensitive to changes in running speed.* Note in reviewing these results that the clearance has remained constant. Circumstances in which the clearance changes with changing speed due to either centrifugal stresses or temperature could yield a contrary result.

Results are presented in Figure 4.24 for K versus the taper parameter q for data (4.58). Analytical results predict the possibility of a substantial increase in direct stiffness if a convergent tapered geometry is used on a seal;

---

*As noted in Chapter 7, this characteristic can be used to good advantage in analyzing pump rotordynamics.

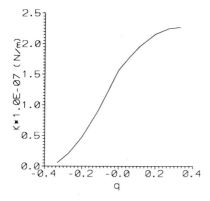

Figure 4.24 K versus q for data (4.58).

viz., $C_0 > C_1$, $q > 0$. However, in the only published tests of convergent tapered *liquid* seals, Childs and Dressman (1985) showed no increase in stiffness accompanying the introduction of a taper. Their results did show an increase in leakage and a loss of direct damping. The reader is cautioned, however, to make sure that operating pressure loads do not cause a seal divergence; i.e., $C_0 < C_1$, $q < 0$. As shown in Figure 2.24, this circumstance yields a prediction of a marked loss in direct stiffness. While no test data are available for divergent seals, the author's experience with liquid-rocket-engine turbopumps supports these predictions.

In summary, liquid annular seals depend on many related parameters and provide a great number of opportunities for favorably modifying pump rotordynamics.

Plain Seals—Experimental Results

Tables 4.1 and 4.2 provide an overview of measured test results for plain smooth seals. Generally speaking, the experimental results in support of the plain, *smooth*, seal models cited above are encouraging. Childs et al. (1988) compare measured and predicted effective direct stiffness and damping coefficients for a seal with $L/D = \frac{1}{2}$ and $C_r/R = 0.005, 0.0075, 0.010$. Using no exit recovery ($\xi_e = 1$), measured inlet-loss coefficients, and a Blasius-friction-factor model with experimentally determined coefficients, effective damping coefficients were calculated to within 10% at elevated Reynolds numbers. Stiffness was closely predicted at larger clearances but overpredicted by around 25% for $C_r/R = 0.005$. Doubling the seal clearances showed a drop in the direct stiffness and damping coefficients of 50 and 40%, respectively, in moving from a C_r/R ratio of 0.005–0.010.

Florjancic (1990) gives theoretical-versus-experimental results for a short seal with $L/D = 0.114$ and $C_r/R = 0.002$. He used measured friction factors to calibrate a Moody-friction-factor model plus measured inlet-loss and exit recovery factors to obtain $K_{th}/K_{ex} \cong 1.20$, $C_{th}/C_{ex} \cong 0.80$. Measured values

TABLE 4.1 Tests for Plain Liquid Seals: Test Conditions, Apparatus, and Results.

Investigators	Apparatus Type	Test Fluid	$u_0(0)$ Control	Data	ΔP_{max} (bars)	ω_{max} (rpm)	Stator[a] Surface	$\mathcal{R}_0$(max)	$\mathcal{R}_\theta$(max)
Childs and Kim (1985)	Synchronous excitation	Halon CBrF$_3$	No	K_{ef}, C_{ef} M_{ef}	23	2,200	Smooth Knurled Post Round Hole	510,000 386,000 384,000 328,000	201,000 201,000 201,000 201,000
Childs and Dressman (1985)	Synchronous excitation	Water	No	K_{ef}, C_{ef} M_{ef}	10	3,500	Smooth Smooth Smooth	26,960 32,700 30,200	10,900 17,100 20,500
Black and Jenssen (1970)	Static loads	Water	No	K, k			Smooth	14,000	3,500
Iino and Kaneko (1980)	External[b] shaker	Water	No	K, C, M	6	1,020	Smooth	3,500	4,800
Childs and Kim (1986a)	Synchronous excitation	Halon CBrF$_3$	No	K_{ef}, C_{ef} M_{ef}	23	7,200	Round Hole	250,000	201,000
Diana et al. (1982)	External[b] shaker	Water	No	K, k	30	4,000	Smooth	6,100[d]	9,300[d]
Childs et al. (1990b)	Synchronous excitation	Halon CBrF$_3$	No	K_{ef}, C_{ef} M_{ef}	38	7,200	Round Hole	550,000	201,000
Falco et al. (1984)	External[c] shaker	Water	No	K, k, C	35	5,000	Smooth	6,100[d]	11,600[d]
Kanki and Kawakami (1984)	Shaker[c]	Water	No	$K_{xx}(\epsilon_0), K_{xy}(\epsilon_0)$ $C_{xx}(\epsilon_0), C_{xy}(\epsilon_0)$ $M_{xx}(\epsilon_0), M_{xy}(\epsilon_0)$ ϵ_0 vs. F_0 K, k, C, c, M	9.8	2,000	Smooth (1) Smooth (2)	10,000 20,000	20,000 30,000
Nordmann and Massmann (1984)	Impulse hammer	Water	No	K, k, C, c, M	2	6,000	Smooth	11,800	5,800
Iwatsubo and Sheng (1990a)	Shaft in a shaft	Water	Yes	K, k, C, c, M	9	4,500	Smooth Damper	8,300 5,300	3,300 3,300
Florjancic (1990)	External[c] shaker	Hot water	Yes	K, k, C, c, M	55	4,000	Smooth	220,000	245,000
Graf and Staubli (1990)	External[e] shaker	Water	$u_0(0) = 0$ $u_0(0) = 1$	K, k, C, c, M	N.A.	2,700	Smooth	61,000	90,000
Kanemori and Iwatsubo (1989)	Shaft in a shaft	Water	No (meas.)	K, k, C, c, M	9	3,300	Smooth	9,060	6,700

[a]See Table 4.2.
[b]Parallel translation giving eccentricity and tilt.
[c]Parallel translation giving only eccentricity.
[d]Estimated by the author.

TABLE 4.2 Tests for Plain Liquid Seals: Test Geometries and Reynolds Number Ranges.

Investigators	Seal Type	D (mm)	L (mm)	C_1 (mm)	L/D	$\bar{C}_r/R$	q	h_d/C_r	γ	$\mathscr{R}_0$(max)	$\mathscr{R}_\theta$(max)
Childs and Kim (1985)	Smooth	101.6	49.9	0.527	0.491	0.0104	0	0	0	510,000	210,000
	Damper[a]							—	—	386,000	
	Damper[b]							—	—	384,000	
	Damper[c]							1.90	0.17	328,000	
Childs and Dressman (1985)	Smooth taper	101.6	50.8	0.508	0.50	0.0138	0.273	0	0	26,960	10,900
				0.508	0.50	0.0156	0.360	0	0	32,700	17,100
				0.508	0.50	0.0160	0.378	0	0	30,200	20,500
Black and Jenssen (1970)	Smooth				0.25						
					0.50						
					1.00						
Iino and Kaneko (1980)	Smooth	260.0	31.8	0.178	0.122	0.0014	0	0	0	3,500	4,800
Childs and Kim (1986a)	Damper[c]	101.6	49.9	0.527	0.491	0.010	0	1.93	0.17	328,000	210,000
								1.93	0.34	369,000	
								1.93	0.50	314,000	
								1.93	0.59	357,000	
								1.93	0.61	354,000	
								0.96	0.34	342,000	
								2.89	0.34	367,000	
								3.86	0.34	362,000	
Diana et al. (1982)	Smooth	160.0	40	0.36	0.25	0.0045	0	0	0	26,000	29,000
			120		0.75					17,000	
			160		1.00					14,000	
			200		1.25					13,000	
Childs et al. (1990b)	Damper[c]	101.6	49.9	0.376	0.491	0.0074	0	1.00	0.34	241,000	149,000
				0.371		0.0073		2.74	0.34	232,000	
				0.376		0.0074		2.92	0.34	235,000	

TABLE 4.2 (*Continued*)

Investigators	Seal Type	D (mm)	L (mm)	C_1 (mm)	L/D	$\bar{C}_r/R$	q	h_d/C_r	γ	$\mathscr{R}_0(\max)$	$\mathscr{R}_\theta(\max)$
Falco et al. (1984)	Smooth	160.0	40.0	0.379	0.25	0.0075	0	4.03	0.34	238,000	29,000
			120.0	0.376	0.75	0.0074		2.92	0.27	258,000	
			160.0	0.381	1.00	0.0075		2.92	0.42	245,000	
			200.0	0.36	1.25	0.0045		0	0	26,000	
										17,000	
										14,000	
										13,000	
Kanki and Kawakami (1984)	Smooth	200.0	200.0	0.50	1.00	0.005	0	0	0	10,000	20,000
			40.0	0.50	1.20	0.005				20,000	30,000
Nordmann and Massmann (1984)	Smooth	42.0	35.0	0.35	0.833	0.017	0	0	0	11,800	5,800
Iwatsubo and Sheng (1990a)	Smooth	70.35	35.17	0.175	0.50	0.005	0	0	0	8,300	3,300
	Damper[d]							1.71	0.80	5,300	
	Damper[d]							2.85	0.80	5,300	
Florjancic (1990)	Smooth	350.0	40	0.37	0.114	0.002	0	0	0	220,000	245,000
Graf and Staubli (1990)	Smooth	280.0	110.0	1.80	0.393	0.013	0	0	0	61,000	90,000
Kanemori and Iwatsubo (1989)	Smooth	240.0	79.3	0.394	3.0	0.010	0	0	0	9,060	6,700

[a]Square post.
[b]Knurled indentation.
[c]Round-hold pattern, see Figure 4.25.
[d]"Isogrid" roughness pattern, see Figure 4.27.

for k and M were small but reasonably well predicted. Graf and Staubli (1990) present theory-versus-experimental results for a long seal, $L/D = 0.785$, $C_r/R = 0.0129$. They show good agreement for all coefficients except for the added-mass term, which is predicted to be around 30 Kg, but measured values were between 33% and 200% greater than predictions, depending on the identification algorithm. Kanemori and Iwatsubo (1989) present results for a long seal, $L/D = 3.0$, $C_r/R = 0.0065$. Most of their test results are in the laminar regime, and no theory-versus-experiment comparison is presented. However, their added-mass coefficient is around 164 Kg versus a theoretical prediction (using a Moody-equation model with $\epsilon_s = \epsilon_r = 0.01$) of approximately 130 Kg. Hence, for long L/D seals, the large predicted added-mass terms clearly exist.

Considerably less conclusive data are available for damper seals. Childs et al. (1985, 1986a, 1990b) have presented extensive results for damper seals; however, their test apparatus does not control the inlet tangential velocity and cannot separately identify the cross-coupled stiffness and direct damping coefficients, yielding instead an effective direct damping coefficient. Childs et al. (1990a) provide friction-factor data versus axial Reynolds number and running speed for the hole-pattern stators of Figure 4.25. As shown in Figure 4.26, these friction-factor results show a significant conflict between theory and experiment. Measured friction factors are more sensitive to changes in running speed than theory. The theory predicts an increase in the friction factor with an increase in running speed; the measurements generally show the reverse.

The effective stiffness for the hole-pattern damper seals of Figure 4.25 are around 75% of comparable smooth seals. When tested at $C_r/R = 0.010$ the best damper seal had an effective damping coefficient about 50% greater than a smooth seal [Childs and Kim (1986a)]. However, for $C_r/R = 0.0075$, the damper seal had about the same effective damping as a smooth seal [Childs et al. (1990a)].

The central virtue of a damper seal is the reduction in tangential velocity and cross-coupled stiffness coefficient, and this virtue can only be conclusively demonstrated by introducing a significant inlet tangential velocity and measuring a reduction in cross-coupled stiffness. Iwatsubo and Sheng (1990a) have presented test results for the damper seal of Figure 4.27* from an apparatus that has precisely this desired capability. Tests are carried out for $C_r/R = 0.005$, $L/D = 0.5$, and hole-depth to clearance ratios $h_d/C_r = 0$, 1.71, 2.86. The proportion of area taken up by holes in Figure 4.27 is about 82% versus 34% for most of the stators in Figure 4.25. Figure 4.28 shows Iwatsubo's results for zero inlet tangential velocity. Observe that introducing holes in the stator reduces $K = K_{XX}$ by about 50%. Direct damping is (surprisingly) about three times greater for the deep-hole damper seal than

---

*To the author's knowledge, von Pragenau (1982) originally proposed this particular damper-seal configuration.

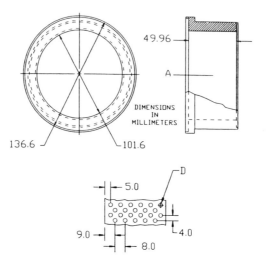

STATOR	D (mm)	Cr. (mm)	h/Cr.	γ
1	3.80	0.376	1.00	0.34
2	3.80	0.371	2.74	0.34
3	3.80	0.376	2.92	0.34
4	3.80	0.379	4.03	0.34
5	3.55	0.376	2.92	0.27
6	4.31	0.381	2.92	0.42

Figure 4.25   Hole-pattern stators [Childs et al. (1990b)].

the smooth seal. The cross-coupled stiffness coefficient is actually slightly higher for the damper seal than the smooth seal. However, Figure 4.29 shows the sharp reduction in cross-coupled stiffness capability of the damper seal *for preswirled flow*. F_θ/e in this figure is related to the rotordynamic coefficients by

$$F_\theta/e = k - C\Omega = k - C\omega\left(\frac{\Omega}{\omega}\right).$$

Hence, the $\Omega/\omega = 0$ intercept of the straight lines is k, and the slope is proportional to damping. *With preswirled flow*, the intercept is smaller (reduced k), and the slope is larger (increased C). This figure clearly confirms von Pragenau's (1982) damper-seal prediction.

Childs et al. (1990a) used a Blasius-friction-factor model based on measured axial-pressure-gradient versus leakage measurements and measured inlet-loss coefficients to predict effective stiffness, damping, and mass coefficients for six damper seals. Damping is generally predicted within 10%, and added mass is underpredicted by a factor of 2–3. Direct stiffness is underpre-

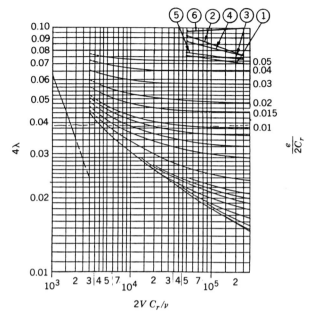

Figure 4.26 Friction factor versus Reynolds number results for stator 1 of Figure 4.21 [Childs et al. (1990b)].

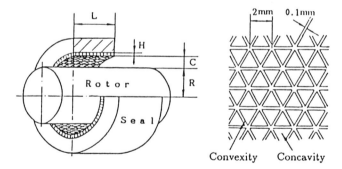

Figure 4.27 Damper seal configuration [Iwatsubo and Sheng (1990a)].

dicted for five of the six stators, but is substantially overpredicted for one of the stators. A Moody-friction-factor model would yield a higher (better) stiffness prediction for most of the stators.

Circumferentially Grooved Seals

Circumferential grooves are frequently produced on either the rotor or stator of seal elements to reduce leakage. There are many possible variations of

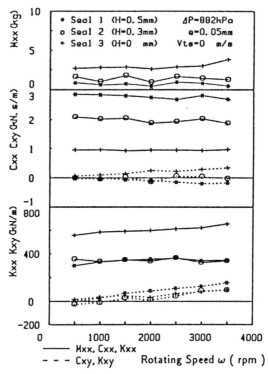

Figure 4.28 Rotordynamic coefficient test results for zero inlet tangential velocity [Iwatsubo and Sheng (1990a)].

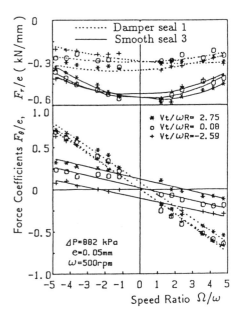

Figure 4.29 F_r/e and F_θ/e versus normalized precession frequency for three inlet tangential velocity ratios [Iwatsubo and Sheng (1990a)].

groove depths, widths, and shapes, and most have been selected by pump companies to reduce leakage without regard for their rotordynamic performance.

Various analyses have been developed for grooved seals based on a variation of the damper-seal model. The most intuitively attractive of these models was developed by Nordmann et al. (1986), who used the minimum clearance when writing the axial-momentum equation and the average clearance when writing the circumferential-momentum equation. The axial-friction-factor parameters can be determined from leakage/axial-pressure-gradient data, and a smooth surface is assumed in the circumferential direction. Unfortunately, despite its intuitive appeal, this model simply does not do a very good job of predicting rotordynamic coefficients of seals.

Kilgore and Childs (1990) made lengthy comparisons between the predictions of this theory for both friction factors and rotordynamic coefficients with discouraging results. The theory (inherently) predicts a much stronger dependency of friction factors on running speed than shown by data. The inadequate static friction-factor model carried over into a generally poor prediction of rotordynamic coefficients. Effective direct stiffnesses were generally underpredicted by about 40%. Effective direct damping coefficients were generally predicted to within 10%.

Nordmann and his co-workers at the University of Kaiserslautern have developed a very general prediction method for seals, including grooved seals, based on a numerical solution of the Navier-Stokes (NS) equations with a $K - \epsilon$ turbulence model [Dietzen and Nordmann (1986, 1987a, 1987b)]. They use finite-difference methods based on the approach of Launder and Spalding (1974) to solve zeroth- and first-order equations in a manner similar to that of the preceding sections. Solutions have been presented for seals with various geometries, including grooved seals. Theory-versus-experiment comparisons for grooved seals have been quite good [Diewald and Nordmann (1988)].

The problem with a NS solution for seals is simple. They are very expensive in terms of computer time and money. Florjancic (1990) developed a bulk-flow model based on an extension of Scharrer's solution (1987) for gas labyrinth seals* to obtain a prediction approach for grooved seals with more reasonable computer costs. He used the "three-control-volume" model illustrated in Figure 4.30. Flow in control volumes 1 and 2 is modeled with a bulk-flow Hirs equation using the Moody-friction-factor model, and a single vortex is assumed to exist in control volume (CV) III, the groove. The jet shear stress between CVs II and III is modeled according to Abramovich (1963) and Wyssman et al. (1984). Entrance losses are introduced between each groove and land. Because these axially distributed inlet-pressure drops account for a substantial portion of the total pressure drop and are not functions of running speed, the resultant friction factors are much less

---

*Chapter 5, Section 5.3.

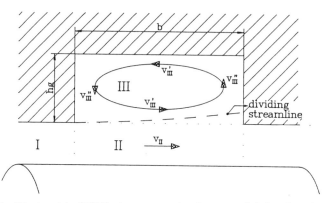

Figure 4.30   Florjancic's (1990) three-control-volume model for flow in a grooved liquid seal.

sensitive to running speed than the model of Nordmann et al. (1986). The single-stable-vortex model of Figure 4.30 is only valid for grooves having approximately equal widths and depths. For deeper grooves, multiple vortices form; for long shallow grooves, the flow can reattach to the rotor. Nonetheless, Florjancic shows generally satisfactory comparison between theory and experiment for two short seals ($L/D = 0.114$, $C_r/R = 0.00211, 0.0042$). The two seal clearances simulated wearout by doubling the minimum clearance. The serration pattern used equal land and groove lengths, and the groove length-to-depth ratios were 4 and 2.86. The groove depth to minimum clearance ratios were $C_g/C_r = 0.95$ and 0.34. Measured rotordynamic-coefficient results for the grooved seals and a smooth seal with comparable dimensions were about the same; i.e., the shallow grooving patterns did not materially change the rotordynamic coefficients.

A different result was obtained by Kilgore and Childs (1990) for a comparison of effective direct stiffness and damping coefficients of a smooth seal and a 12-grooved seal with equal land and groove lengths. Both seals had $C_r/R = 7.6 \times 10^{-3}$, $L/D = \frac{1}{2}$. The smooth seal stiffness and damping coefficients were 2.7 and 2.2 times greater than the grooved seal. Childs and Kim (1986b) present test comparisons for four grooved seals and a smooth seal. Shallow grooving, $C_g/C_r = 0.72$, reduced stiffness and damping by around 40 and 33%, respectively. Deeper grooving $C_g/C_r = 1.42, 3.06$ reduced stiffness and damping by around 80 and 50%, respectively. For all of these seals, grooving accounted for 50% or more of the seal length. Diewald and Nordmann (1988) present results for a smooth and a grooved seal ($C_g/C_r = 2.5$) for which the direct stiffness and damping are reduced by 43 and 28%, respectively. The cross-coupled stiffness was reduced by about 60%.

Kilgore and Childs (1990) examined the influence of seal wear on a grooved seal with equal land and groove lengths beginning with $C_r/R = 0.0076$, $C_g/C_r = 3.0$ and ending with $C_r/R = 0.015$, $C_g/C_r = 1.0$, simulating a seal wearout. Their worn-out seals have about the same stiffness as the original-clearance seals with around a 20% loss in damping. Interestingly, Florjancic (1990) showed a reduction in K and C of around 20% and 40% in going from a C_r/R of 0.002 to 0.004.

Iwatsubo and Sheng (1990b) have also developed a three-control-volume model for grooved seals. They present limited test data for smooth-rotor/grooved-stator and grooved-rotor/smooth-stator seals with $C_r/R =$ 0.005 and a groove-depth to minimum-clearance ratio of 6.9. Their measured values for cross-coupled stiffness are three times greater for the grooved-rotor seals than the grooved-stator seals.

In summary, the best available analysis technique for circumferentially grooved seals is provided by the Navier-Stokes solution approach of Dietzen and Nordmann et al. (1986, 1987a, 1987b). The three-control-volume approaches of Florjancic (1990) and Iwatsubo and Sheng (1990b) appear to provide an economical alternative for reasonable prediction of grooved seals; however, they rely heavily on accurate measurements of inlet- and exit-loss coefficients. Shallow grooving and deep grooving which does not take up a great percentage of the seal length does not markedly reduce the rotordynamic coefficients versus smooth seals. Deep grooving which accounts for a substantial portion of seal length can sharply reduce the direct stiffness and damping of a smooth seal. Grooving reduces cross-coupled stiffness. Grooved-rotor/smooth-stator seals are less stable than grooved-stator/ smooth-rotor seals.

Diewald and Nordmann (1988) showed that grooving would improve the whirl-frequency ratio of a seal since the damping was reduced more than the cross-coupled stiffness. Florjancic (1990) showed a similar result. These changes would enhance stability by reducing $k/C\omega$. (None of these investigators had appreciable inlet tangential velocities; hence their measured values for k are small.) Examples have been cited in the literature of pumps (particularly with tight C_r/R ratios) which have been stabilized by introducing grooves in the wearing-ring seals. Conversely, many contrary examples have been cited in which unstable pumps have been stabilized by removing grooves. Because of the potentially high direct stiffnesses developed by seals and the strong sensitivity of some seal designs to grooving, questions concerning the ultimate impact of seal changes on stability require a full rotordynamic analysis of a particular pump.

Helically Grooved Seals

On occasion, helical grooves have been used on either the rotor or stator or both the rotor and the stator (screw seal). The grooves are arranged to oppose fluid rotation in the direction of shaft rotation. The helical grooving patterns act to pump the fluid "upstream" with respect to the normal leakage flow direction. No satisfactory analysis has been developed for this type of seal.

Kanki and Kawakami (1984, 1987) presented test results for screw seals with a helix angle of 2.2° for $L/D = 0.2, 1.0$; $C_r/R = 2.5 \times 10^{-3}$, groove depth to clearance $C_g/C_r = 2.4$. Their tests consider the influence of running speed and eccentricity and provide direct comparisons between smooth and screw seals. The screw seals leak substantially less than smooth seals and are insensitive to changes in eccentricity. The coefficients depend strongly on running speed, and the direct stiffness becomes (quickly) negative as the speed increases. Direct damping is lower by about a factor of 3 for screw seals than smooth seals. The screw-seal whirl-frequency ratio is 0.5.

Childs et al. (1990b) presented test results for turbulent annular seals with smooth rotors and helically grooved stators. Their seals had helix angles of 0, 15, 30, 40, 50, 60, and 70° with $C_r/R \cong 0.0075$, $L/D = \frac{1}{2}$, $C_g/C_r = 1$. Test results are presented for friction factors, discharge coefficients, and effective stiffnesses, damping, and added-mass coefficients. Their results show leakage increasing as the helix angle increases. Seals with helix angles greater than 30° leak more than smooth seals. Direct stiffnesses are positive; however, seals with helix angles less than 40° have substantially reduced direct stiffnesses. The effective direct damping coefficient is insensitive to helix angles. Childs's results are for small orbits about a centered position.

Iwatsubo et al. (1990) presented test results for a smooth-rotor/helically grooved stator with helix angles 0, 0.83, 3.32, 6.65, 10.0, and 15.1° with $L/D = 0.5$, $C_r/R = 0.0025$, $C_g/C_r = 6.86$. They also presented test results for a smooth-stator/helically grooved rotor seal with a 3.32° helix angle. All results are for small motion about a centered position. Leakage increases with increasing helix angles. However, the leakage rate of helically grooved seals decreases with increasing speed. For a fixed ΔP and running speed, an optimal helix angle exists which yields a minimum (negative) cross-coupled stiffness coefficient. For the data presented, their optimal helix angle is 3.32°. All direct stiffnesses are positive. Direct damping is relatively constant for all seals. The smooth-rotor/helically grooved stator seal is *markedly* more stable than the helically grooved rotor/smooth-stator seal. Grooving the rotor yields a positive (destabilizing) value for k which increases with running speed versus a negative (stabilizing against forward whirl) relative constant value for the grooved stator.

Influence of Eccentricity

The model of Eq. (4.1) assumes that the reaction force is a linear function of displacement, independent of eccentricity ratio. Recall that a quite different situation held for bearings where the rotordynamic coefficients were functions of the static eccentricity ratio ϵ_0. Similarly, if the static eccentricity ratio is important for a seal, the motion/reaction-force model becomes

$$
-\begin{Bmatrix} F_X \\ F_Y \end{Bmatrix} = \begin{bmatrix} K_{XX}(\epsilon_0) & K_{XY}(\epsilon_0) \\ K_{YX}(\epsilon_0) & K_{YY}(\epsilon_0) \end{bmatrix} \begin{Bmatrix} X \\ Y \end{Bmatrix} + \begin{bmatrix} C_{XX}(\epsilon_0) & C_{XY}(\epsilon_0) \\ C_{YX}(\epsilon_0) & C_{YY}(\epsilon_0) \end{bmatrix} \begin{Bmatrix} \dot{X} \\ \dot{Y} \end{Bmatrix}
$$
$$
+ \begin{bmatrix} M_{XX}(\epsilon_0) & 0 \\ 0 & M_{YY}(\epsilon_0) \end{bmatrix} \begin{Bmatrix} \ddot{X} \\ \ddot{Y} \end{Bmatrix}. \tag{4.62}
$$

Obviously, this result is more complicated and less pleasant than the simpler result of Eq. (4.1). The question of interest here is: Under what circumstances is the simpler model valid?

Black and Jenssen (1970) published the first experimental and theoretical reaction-load-versus-eccentricity results for plain smooth seals. Their experimental results are for a long seal $L/D = 1$ with large clearances $C_r/R = 0.0107$. They used a fully developed flow ($u_0 = R\omega/2$) short-seal analytical/numerical solution which overpredicts the reaction forces. At an axial Reynolds number of 14,000 and a running speed of 5000 rpm, measured load-deflection relationships are reasonably linear out to an eccentricity ratio of 0.6. A modest nonlinearity was introduced by reducing the speed which yielded a "softening-spring" effect. Allaire et al. (1978) developed comparable short-seal eccentric solutions for rotordynamic coefficients.

Nelson and Nguyen (1988a, 1988b) developed the first finite-length eccentric solutions for plain incompressible seals. A bulk-flow Moody model is used with a perturbation analysis about a finite static eccentricity ratio. The zeroth order equations are solved by expanding the variables in a separation-of-variables solution format as follows:

$$p_0(Z, \theta) = \text{Real}\left\{ 2 \sum_{n=0}^{N-1} P_n(Z) e^{jn\theta} \right\}.$$

The solution is discretized circumferentially by dividing the circumference into $2N$ evenly spaced sectors. The solution procedure uses a fast-Fourier-transform (FFT) algorithm to calculate the Fourier coefficients at a location Z and a Euler algorithm to march along the seal in the Z direction. A numerical Newton-Raphson algorithm is used to iteratively satisfy the boundary conditions. The (1988b) article presents comparisons to Black and Jenssen's (1970) experimental results, showing a marked improvement in correlation between theory and experiment versus Black and Jenssen's initial short-seal analysis.

Nelson and Nguyen (1988b) compare predictions to the experimental results of Falco et al. and to a finite-element, Reynolds-equation-based solution presented in Falco et al. The Nelson and Nguyen solution does a reasonable job in comparison to the sparse test data and a significantly better job than the Reynolds-equation-based solution. The Falco et al. seal is short $L/D = 0.25$ and fairly loose $C_r/R = 0.009$ with $\Delta P = 10$ bar, $\omega = 4000$ rpm. Water is the test media. For this seal, the stiffness coefficients are predicted to be constant out to an eccentricity ratio of 0.3 and then rise quadratically with increasing eccentricity. For a liquid hydrogen seal, at much higher pressures, speeds, and Reynolds numbers with $L/D = 0.51$, $C_r/R = 3.5 \times 10^{-3}$, the rotordynamic coefficients are predicted to be constant out to eccentricity ratios of 0.5 to 0.6. Black and Jenssen and Nelson and Nguyen both note the possibility of a transition from turbulent to laminar flow in the tight side of an eccentric seal yielding a possibility for a pronounced nonlin-

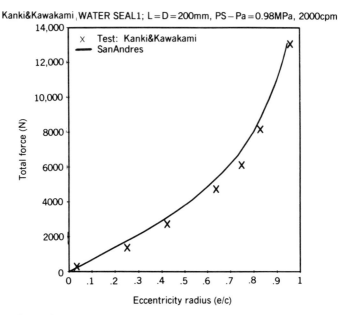

Figure 4.31 Load force versus eccentricity ratio comparisons for San Andres (1991) and Kanki and Kawakami (1984) [Kanki and Kawakami (1984)] and [San Andres (1991)].

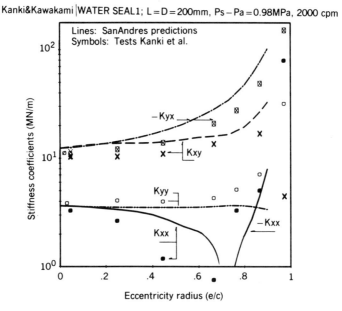

Figure 4.32 Stiffness coefficients versus eccentricity-ratio comparisons for San Andres (1991) and Kanki and Kawakami (1984) [Kanki and Kawakami (1984)] and [San Andres (1991)].

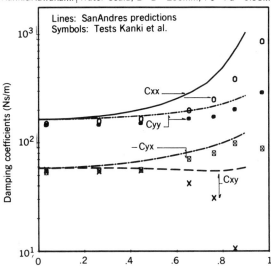

Kanki&Kawakami | Water Seal1; L = D = 200mm, Ps − Pa = 0.98MPa, 2Kcpm

Lines: SanAndres predictions
Symbols: Tests Kanki et al.

Figure 4.33 Damping coefficients versus eccentricity-ratio comparisons for San Andres (1991) and Kanki and Kawakami (1984) [Kanki and Kawakami (1984)] and [San Andres (1991)].

earity. This type of transition is more likely in seals where the nominally centered seal is operating at a low Reynolds number compared to the laminar-turbulent transition regime. Mathematical modeling of this type of transition is possible within a numerical solution but would be poorly supported by available test data.

Nordmann and Dietzen (1990) have presented a CFD solution based on a perturbation of the Navier-Stokes equations for an eccentric plain seal. Their results are similar to those of Nelson and Nguyen. Unfortunately, Nordmann's Kaiserslautern group have not as yet published an eccentric solution for grooved seals.

San Andres* (1991) has developed a CFD solution for plain seals using a bulk-flow model with a Moody friction factor. He uses the finite-difference algorithm of Launder and Leschziner (1978) with the "simple" algorithm of Von Doormal and Raithby (1984). San Andres's solution compares well with predictions by Nelson and Nguyen over their calculated eccentricity range; however, San Andres's procedure readily converges at higher eccentricity ratios which cause problems in the Nelson and Nguyen algorithm. Figures 4.31–4.34 provide load and rotordynamic-coefficient versus eccentricity ratio comparisons between the predictions of San Andres and the measurements

---

*San Andres's analysis considers a compressible barotropic fluid.

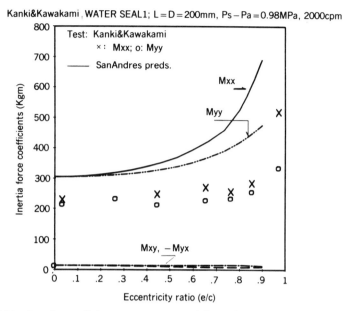

Figure 4.34 Inertia coefficients versus eccentricity-ratio comparisons for San Andres (1991) and Kanki and Kawakami (1984) [Kanki and Kawakami (1984)] and [San Andres (1991)].

of Kanki and Kawakami (1984). The theory does a fairly good job. Kanki and Kawakami's seal has $L/D = 1$, $C_r/R = 0.005$, $\Delta P = 9.8$ bar. The rotordynamic coefficients are reasonably constant out to an eccentricity ratio of 0.3. The numerical solution does not predict the drop in C_{XY} with increasing eccentricity ratio shown by measurement and overpredicts M_{XX}, M_{YY} by around 30%.

Kanki and Kawakami (1987) also provide rotordynamic coefficients versus eccentricity ratio results for grooved seals, showing them to be insensitive to changes in eccentricity ratio.

4.7   TILT AND MOMENT COEFFICIENTS

The analyses presented so far consider the consequences of a parallel displacement of the seal rotor without any "tilting" of the seal. Obviously, when a pump is vibrating the resultant precessional motion of a seal includes both pitching and yawing, and experience has shown that the consequence of this angular motion can be considerable, particularly for long seals. In this section, the reaction forces and moments are considered due to small general motion of a seal rotor about a centered and aligned position.

More specifically, we will be considering a motion/reaction relationship of the form

$$
-\begin{Bmatrix} F_X \\ F_Y \\ M_Y \\ M_X \end{Bmatrix} =
\begin{bmatrix}
K & k & K_{\epsilon\alpha} & -k_{\epsilon\alpha} \\
-k & K & -k_{\epsilon\alpha} & -K_{\epsilon\alpha} \\
K_{\alpha\epsilon} & k_{\alpha\epsilon} & K_{\alpha} & -k_{\alpha} \\
k_{\alpha\epsilon} & -K_{\alpha\epsilon} & k_{\alpha} & K_{\alpha}
\end{bmatrix}
\begin{Bmatrix} X \\ Y \\ \alpha_Y \\ \alpha_X \end{Bmatrix}
$$

$$
+\begin{bmatrix}
C & c & C_{\epsilon\alpha} & -c_{\epsilon\alpha} \\
-c & C & -c_{\epsilon\alpha} & -C_{\epsilon\alpha} \\
C_{\alpha\epsilon} & c_{\alpha\epsilon} & C_{\alpha} & -c_{\alpha} \\
c_{\alpha\epsilon} & -C_{\alpha\epsilon} & c_{\alpha} & C_{\alpha}
\end{bmatrix}
\begin{Bmatrix} \dot{X} \\ \dot{Y} \\ \dot{\alpha}_Y \\ \dot{\alpha}_X \end{Bmatrix}
$$

$$
+\begin{bmatrix}
M & 0 & M_{\epsilon\alpha} & 0 \\
0 & M & 0 & -M_{\epsilon\alpha} \\
M_{\alpha\epsilon} & 0 & M_{\alpha} & 0 \\
0 & -M_{\alpha\epsilon} & 0 & M_{\alpha}
\end{bmatrix}
\begin{Bmatrix} \ddot{X} \\ \ddot{Y} \\ \ddot{\alpha}_Y \\ \ddot{\alpha}_X \end{Bmatrix}.
\qquad (4.63)
$$

Obviously, the rotordynamic-coefficient matrices are now 4×4 instead of 2×2, and one can have forces due to rotations, moments due to displacements, etc. The development used in this section is patterned after Childs (1982).

If rotation is permitted about the X_0 and Y_0 axes of Figure 4.35, the clearance function of Eq. (4.34) becomes

$$
h = h_0 - \left[x + \alpha_Y \left(\frac{L}{\overline{C}_r} \right)(z - z_0) \right] \cos\theta - \left[y - \alpha_X \left(\frac{L}{\overline{C}_r} \right)(z - z_0) \right] \sin\theta.
$$

$$(4.64)$$

Hence, in comparison to Eqs. (4.36) we find

$$
\epsilon h_1 = -\left[x + \alpha_Y \left(\frac{L}{\overline{C}_r} \right)(z - z_0) \right] \cos\theta - \left[y - \alpha_X \left(\frac{L}{\overline{C}_r} \right)(z - z_0) \right] \sin\theta.
$$

$$(4.65)$$

Further, in comparison to the last of Eq. (4.38)

$$
-\epsilon \overline{h}_1 = r + \alpha, \qquad r = x + jy, \qquad \alpha = \alpha_Y - j\alpha_X.
$$

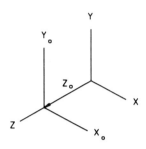

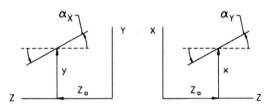

Figure 4.35   Coordinate system for seal pitch and yaw.

Assuming seal precession motion of the form

$$r = r_0 e^{jf\tau}, \qquad \alpha = \alpha_0 e^{jf\tau}, \qquad h_1 = h_{10} e^{jf\tau}$$

and the parallel solution for the dependent variables

$$w_1 = \bar{w}_1 e^{jf\tau}, \qquad u_1 = \bar{u}_1 e^{jf\tau}, \qquad p_1 = \bar{p}_1 e^{jf\tau}$$

yields the following governing equations

$$\frac{d}{dz}\begin{Bmatrix} \bar{w}_1 \\ \bar{u}_1 \\ \bar{p}_1 \end{Bmatrix} + [A]\begin{Bmatrix} \bar{w}_1 \\ \bar{u}_1 \\ \bar{p}_1 \end{Bmatrix} = \left(\frac{r_0}{\epsilon}\right)\begin{Bmatrix} g_1 \\ g_2 \\ g_3 \end{Bmatrix} + \left(\frac{\alpha_0}{\epsilon}\right)\begin{Bmatrix} g_4 \\ g_5 \\ g_6 \end{Bmatrix}, \qquad (4.66)$$

where

$$\begin{Bmatrix} g_4 \\ g_5 \\ g_6 \end{Bmatrix} = \left(\frac{L}{\overline{C}_r}\right)\begin{Bmatrix} 1 + j\Gamma Tz \\ -A_{1\theta} z \\ -[(1 + A_{1z} z) + jz\Gamma T] \end{Bmatrix}. \qquad (4.67)$$

The boundary conditions continue to be provided by Eq. (4.47).

By comparison to Eq. (4.42), pitch and yaw introduce an additional forcing function. Solution of Eq. (4.66) is straightforward, involving successive solutions for displacement and rotation excitation. As stated in Eq. (4.48), a complete solution is the sum of the homogeneous solution and the particular solution due to a right-hand excitation vector. *Complete* solutions are developed separately for the two vectors on the right-hand side of Eq. (4.66). Because the problem is linear, these two complete solutions can be added to obtain the *system* solution or used separately to calculate the rotordynamic coefficients. The solution due to displacement excitation, without rotation ($\alpha_0 = 0$) is given in Eq. (4.48). The solution due to rotation, without displacement ($r_0 = 0$) can be stated

$$\begin{pmatrix} \overline{w}_1 \\ \overline{u}_1 \\ \overline{p}_1 \end{pmatrix} = \left(\frac{\alpha_0}{\epsilon} \right) \begin{pmatrix} f_{4c} + jf_{4s} \\ f_{5c} + jf_{5s} \\ f_{6c} + jf_{6s} \end{pmatrix}. \tag{4.68}$$

The differential reaction moment components about the origin of the X_0, Y_0, Z system are

$$dM_X = -P(Z - Z_0)\sin\theta R\, d\theta\, dZ,$$
$$dM_Y = P(Z - Z_0)\cos\theta R\, d\theta\, dZ, \tag{4.69}$$

and the moment perturbations are

$$M_{X1} = -\epsilon R \int_0^L \int_0^{2\pi} P_1(Z - Z_0)\sin\theta\, d\theta\, dZ,$$

$$M_{Y1} = \epsilon R \int_0^L \int_0^{2\pi} P_1(Z - Z_0)\cos\theta\, d\theta\, dZ. \tag{4.70}$$

In terms of nondimensional Z and P, these equations become

$$M_{X1} = -\epsilon \frac{F_0 L}{C_d} \int_0^1 \int_0^{2\pi} p_1(z - z_0)\sin\theta\, d\theta\, dz,$$

$$M_{Y1} = \epsilon \frac{F_0 L}{C_d} \int_0^1 \int_0^{2\pi} p_1(z - z_0)\cos\theta\, d\theta\, dz.$$

Following the development of Section 4.5, the polar components of the reaction moment can be stated:

$$\frac{M_{\theta 1} - jM_{r1}}{LF_0} = \frac{(M_{Y1} - jM_{X1})e^{-jf\tau}}{LF_0} = -\epsilon \frac{\pi L}{C_d} \int_0^1 \overline{p}_1(z - z_0)\, dz. \tag{4.71a}$$

The polar components of the reaction force was previously derived in

Eq. (4.45) as

$$\frac{F_{r1} + jF_{\theta 1}}{F_0} = -\frac{\epsilon \pi}{C_d} \int_0^1 \bar{p}_1 \, dz.$$ (4.71b)

As outlined in Childs (1981, 1989), the reaction forces and moments due to displacement perturbations are obtained by substituting $\bar{p}_1$ from Eq. (4.50); viz., $\bar{p}_1 = (r_0/\epsilon)(f_{3c} + jf_{3s})$, while the reaction forces and moments due to rotations follow from substituting for $\bar{p}_1$ from Eq. (4.68); $\bar{p}_1 = (\alpha_0/\epsilon)(f_{6c} + jf_{6s})$. The coefficients are calculated in precisely the same manner outlined in Section 4.5; viz., force and moment coefficients are calculated as a function of the normalized precession frequency f. The desired coefficients are obtained by a least-square curve fit of the resulting functions of f.

An obvious question at this point is, *"What difference do moment coefficients make, and when should they be included in a pump rotordynamics analysis?"* A "feel" can be developed for this issue by considering Figure 4.36 which represents an apparatus for measuring seal rotordynamic coefficients [Diana et al. (1982)]. The fluid enters the center of the rig and discharges axially across the two seals. The test seal is on the left. The seal's stator is fixed, and a shaker is applied to the left end of the shaft to provide relative

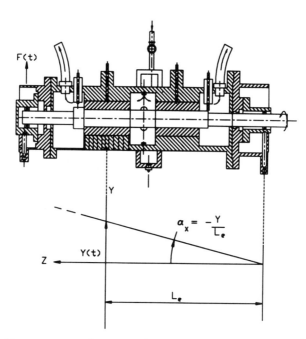

Figure 4.36   Test apparatus for measuring rotordynamic coefficients [Diana et al. (1982)].

motion between the test-seal rotor and its stator. Transient measurements are made of the perturbed motion and the transient pressure within the seal. When the results of Diana et al. (1982) were presented, note was made that measured direct stiffness coefficients were much larger than predicted. The excitation used provides simultaneous displacement and tilt excitation. The rig was subsequently redesigned to eliminate tilting [Falco et al. (1984, 1986)] and generally produced data which were consistent with predictions.

An idea of the change in the predicted reaction force due to tilting can be obtained by examining motion in the plane of Figure 4.36, assuming that the rotor and stator elements are rigid. Restricting our attention to the stiffness-matrix contribution, the Y reaction-force component is

$$-F_Y = KY - K_{\epsilon\alpha}\alpha_X. \tag{4.72}$$

Assuming that the shaft pivots about the right bearing means that a Y displacement in the center of the test seal yields $\alpha_X = -Y/L_e$, where L_e is the distance from the center of the seal to the right bearing. Hence,

$$-F_Y = YK[1 + (K_{\epsilon\alpha}/KL_e)], \tag{4.73}$$

and the "effective" stiffness which would be measured is

$$K_e = K[1 + (K_{\epsilon\alpha}/KL_e)]. \tag{4.74}$$

The parameters defining some of the Diana et al. test cases are

$$D = 0.16 \text{ m}, \qquad C_r = 0.36 \text{ mm}, \qquad P_s = 30 \text{ bar},$$

$$\omega = 5000 \text{ rpm}, \qquad \mu = 1.3 \times 10^{-3} \text{ PaS}, \qquad \rho = 1000 \text{ Kg/m}^3,$$

$$L = 0.04, 0.12, 0.16, 0.20 \text{ m}.$$

These data yield $C_r/R = 4.5 \times 10^{-3}$ and

$$L/D = 0.25, 0.75, 1.0, 1.25,$$
$$L/C_r = 111.0, 333.0, 444.0, 555.0.$$

Calculated results for K, $K_{\epsilon\alpha}$, and K_e are provided in Table 4.3 using a Blasius friction-factor model for smooth surfaces. These calculated results show that the pitching motion radically changes the effective stiffness for seals with appreciable L/D ratios and demonstrate that the full 4×4 rotordynamic-coefficient matrices should be used for seals having appreciable L/D ratios on the order of 0.75 or greater. For most pumps, the full descriptions are never needed for short wearing-ring seals, will occasionally be needed for longer interstage seals, and are almost always needed for long balance-piston seals.

TABLE 4.3   Calculated Results for the Diana et al. (1982) Data, $L_e = 0.66$ m.

L/D	$K \times 10^{-6}$ N/M	$K_{\epsilon\alpha} \times 10^{-6}$ N	$\dfrac{K_{e\alpha}}{KL_e}$	$K_e \times 10^{-6}$ N/M
0.25	13.150	1.364	0.1572	15.20
0.50	17.560	3.684	0.3180	17.88
0.75	14.680	6.288	0.6480	24.20
1.00	8.034	8.914	1.6810	21.50
1.25	− 0.891	11.440	− 19.5000	16.50

Note particularly that $K_{e\alpha}$ does not always add to the stiffness of K. Simply reversing the ΔP direction changes the sign of $K_{e\alpha}$ and would reduce K.

Experimental Measurements for the Influence of Tilt

Iino and Kaneko (1980) stated in their presentation that the introduction of tilt made a significant difference in their measurement of rotordynamic coefficients, similar to the experience of Diana et al. The only published results where reaction moments have been measured are by Kanemori and Iwatsubo (1989). For a seal with $L/D = 3$, $C_r/R = 0.01$ they present results for K, k, C, c, M, $K_{\alpha\epsilon}$, $k_{\alpha\epsilon}$, and $C_{\alpha\epsilon}$. Quoting Kanemori and Iwatsubo, "The dynamic coefficients K, k, C, c, M and $-K_{\alpha\epsilon}$ derived from experimental results coincide fairly well with Childs (1982) theory. The value of $k_{\alpha\epsilon}$ and $C_{\alpha\epsilon}$ coincide qualitatively with the theory" (Kanemori and Iwatsubo (1989)). Results have not been presented for moment or force coefficients due to pitch or yaw.

4.8   SUMMARY, CONCLUSIONS, AND DISCUSSION

As stated at the outset of this chapter, annular seals have a major impact on the rotordynamics of pumps. To try to make sense of the large volume of material in this chapter, we will consider a series of questions that have hopefully been answered.

1   How good is the model of Eq. (4.1), or when should I worry about the influence of eccentricities and use the model of Eq. (4.62)? The model of Eq. (4.1) is generally valid for static eccentricity ratios out to 0.5. The model is better at higher axial Reynolds numbers where the flow is well into the turbulent regime. Eccentricity effects become more important as the Reynolds number approaches the transition regime; e.g., Kaneko et al. (1984) show pronounced eccentricity effects for a tight, smooth seal with an axial Reynolds number around 4000. Experiments have shown that grooved seals are much less sensitive to changes in eccentricity than plain seals. As a point of interest, even if eccentricity effects become important, rotordynamics codes in use today are generally not set up to account for seal eccentricities.

2 When does the influence of tilting motion become important, or equivalently, when should I use the 4 × 4 matrix definitions of Eq. (4.63) instead of the 2 × 2 matrix definitions of Eq. (4.1)? Tilt and moments become important as L/D increases. The 2×2 model can always be used for wearing-ring seals. The 4×4 matrix model should almost always be used for long balance piston seals with $L/D = 0.75$ or greater. Moment coefficients can be important for interstage seals, depending on the slope of a rotor's mode shapes at the seal location. Modern rotordynamic analysis codes for pumps should be able to account for moment coefficients.

3 How good are current analyses for liquid seals? If accurate supporting empirical data are available (inlet-loss coefficient, exit-recovery factor, and wall friction-factor data), the finite-length analyses outlined in this chapter provide reasonable predictions for plain, smooth seals for either parallel seal displacements or combined displacement and tilting. Predictions for the model of Eq. (4.1) are also reasonable for damper seals; however, the Childs et al. (1990b) measurements show poor correlation between damper-seal friction-factor data and either the Moody or Blasius friction-factor models. Moment coefficient data have not been published for damper seals; however, the analysis outlined in Section 4.7 for moment coefficients should give reasonable answers.

Nelson and Nguyen (1988a, 1988b) and San Andres (1991) provide efficient, experimentally validated algorithms for force coefficients of eccentric plain smooth seals. They should give good results for damper seals; however, no data have been published for eccentric damper seals to provide validation. Also, no data have been published for statically misaligned seals.

For small motion about a centered position, circumferentially grooved seals (with rectangular grooves) appear to be adequately modeled by the three-control-volume models of Florjancic (1990) or Iwatsubo and Sheng (1990b). The CFD methods of Dietzen and Nordmann (1987b) and Nordmann and Dietzen (1990) are more generally applicable, but are obviously expensive in terms of computer time and resources. No reliable prediction method has been published for helically grooved seals. Fortunately, available experimental results show that grooved seals are even less sensitive to eccentricity effects than smooth seals; hence, the absence of a validated prediction method for eccentric grooved seals is not a serious deficit.

4 How do smooth seals compare to damper seals? Damper (rough-stator/smooth-rotor) seals leak less than smooth seals. For prerotated flow they have smaller cross-coupled stiffness coefficients. However, they can have substantially lower direct stiffnesses, and a direct choice between a smooth and a damper configuration will normally require a full rotordynamics analysis.

5 How do smooth seals compare to seals with circumferentially grooved stators?
Smooth seals leak more than grooved seals. Deep grooving patterns (groove depth to minimum-clearance ratio $C_g/C_r \cong 1$ or greater) can reduce direct stiffness and damping substantially. Shallow grooves or deep grooves which do not take up a substantial portion of a seal's length do not reduce direct stiffness and damping appreciably. Grooving reduces cross-coupled stiffness.

6 From a rotordynamics viewpoint, should circumferential grooves be on the rotor or the stator? Grooved-stator/smooth-rotor seals have substantially lower cross-coupled stiffnesses than grooved-rotor/smooth-stator seals. Hence, the grooving should be on the stator.

7 What effect does helical grooving have on the rotordynamic characteristics of seals? Helical grooving on seals aims to "pump" against the nominal seal ΔP. Hence, the grooving pattern acts to rotate the leakage flow against shaft rotation. Helical grooving on the stator can be quite effective, actually reversing the sign of the cross-coupled stiffness coefficient. Helical grooving on the rotor is not effective in reducing the cross-coupled stiffness, nor is grooving on both the rotor and the stator (screw seals). The optimum helix angle depends on the seal operating conditions. No satisfactory analysis is available for predicting the optimum helix angle.

8 When should the added-mass terms which are predicted by analysis be included in rotordynamics analysis? Always! A rotordynamics code which does not accept the direct added-mass term is not suitable for pump rotordynamics.

9 How does wearout influence seal rotordynamic coefficients? Wearout, constituting doubling of seal clearances, reduces the direct stiffness and damping of smooth seals by around 50 and 40%, respectively. The data of Childs and Kilgore (1990a) for grooved seals are limited to a single configuration and show a 20% reduction in direct damping with no reduction in direct stiffness. Florjancic (1990) showed a reduction in the direct stiffness and damping coefficients as C_r/R was increased from 0.002 to 0.004 of around 20 and 40%, respectively. Grooved seals are clearly less sensitive to wear than smooth seals. No wearout data are available for damper seals.

10 How can seals be modified to improve pump rotordynamics? Swirl brakes represent the single most effective seal modification for improving pump rotordynamics. Figure 4.37 illustrates a typical balance-piston swirl brake consisting of axial slots* cut into the inlet of the seal. Swirl brakes have historically been introduced into seals to eliminate an existing problem [Massey (1985) and Valantas and Bolleter (1988)]. Swirl brakes have normally

---

*Based on experience with gas annular seals, hydrodynamically designed turning vanes would be more effective.

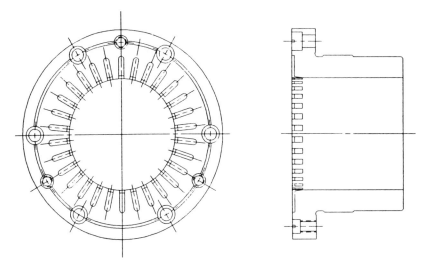

Figure 4.37   Typical balance-piston swirl-brake design [Massey (1985)].

been introduced at the balance piston rather than the wearing-ring or interstage seals because of the dominance of this seal. Figure 4.38 illustrates the dramatic impact of introducing swirl brakes at both the balance piston and wearing-ring seals of a three-stage boiler feed pump. Note the sharp reduction in vibration amplitudes across the running-speed range.

Damper seals using a roughened stator and smooth rotor have been used heavily in the Space Shuttle main engine to improve rotordynamics. Gaffel and Ganter (1992) have recently reported on KSB's success in using conventional honeycomb-seal stators* with smooth rotors in a commercial pump. They have replaced grooved seals with these damper-seal configurations at substantially reduced clearances. They report significant improvements in both efficiency and pump-vibration characteristics.

Helically grooved stators represent a promising mechanism for improving pump rotordynamics. The results of Iwatsubo et al. (1990) show that a properly designed seal with helically grooved stators can produce negative cross-coupled stiffness coefficients. Seals with helically grooved rotors and smooth stators provide no stability advantages over smooth seals. Screw seals have about the same whirl-frequency ratio as smooth seals, and have the disadvantage of developing negative direct stiffnesses.

11   What are the biggest holes in our knowledge concerning liquid seals and rotordynamics? With respect to analysis techniques, a satisfactory analysis is needed for helically grooved seals and plain misaligned seals. More work needs to be done on the exit recovery factor, since no experimentally validated technique exists for predicting this coefficient.

---

*See Section 5.6.

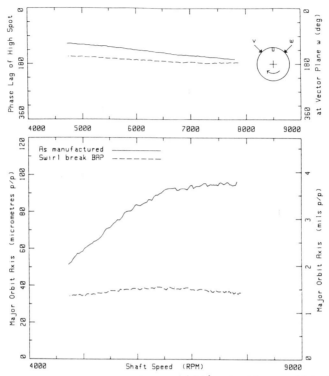

Figure 4.38 Three-stage boiler feed pumps with and without swirl brakes at the balance piston and wearing-ring seals [Frei et al. (1990)].

With respect to test data, more basic friction-factor data are needed for flow between roughened surfaces. More high speed data are needed for all seal configurations.

The results of this chapter will be of considerable value in Chapter 8 when we examine the rotordynamics of a liquid-rocket-engine turbopump. The next chapter considers the related topic of annular seals for compressors and turbines.

REFERENCES: CHAPTER 4

Abramovitch, G. (1963), *The Theory of Turbulent Jets*, MIT Press, Cambridge, MA.

Allaire, P., Lee, C., and Gunter, E. (1978), "Dynamics of Short Eccentric Plain Seals with High Axial Reynolds Numbers," *Journal of Spacecraft and Rockets*, **15**(6), 341–347.

Becht, D., Hawkins, L., Scharrer, J., and Murphy, B. (1991), "Suppression of Subsynchronous Vibration in the SSME HPFTP," *Machinery Dynamics and Element Vibrations*, DE-Vol. 36, ASME, Proceedings 13th Vibration Conference, pp. 11–16.

Black, H. (1979), "Effects of Fluid-Filled Clearance Spaces on Centrifugal Pump and Submerged Motor Vibrations," in *Proceedings of the Eighth Turbomachinery Symposium*, Texas A&M University, pp. 29–38.

Black, H. (1969), "Effects of Hydraulic Forces on Annular Pressure Seals on the Vibrations of Centrifugal Pump Rotors," *Journal of Mechanical Engineering Science*, **11**(2), 206–213.

Black, N., Allaire, P., and Barrett, L. (1981), "The Effect of Inlet Flow Swirl on the Dynamic Coefficients of High-Pressure Annular Clearance Seals," in *Proceedings of the Ninth International Conference in Fluid Sealing*, BHRA Fluid Engineering, Leeuwenborst, The Netherlands.

Black, H., and Jenssen, D. (1971), "Effects of High-Pressure Ring Seals on Pump Rotor Vibrations," ASME Paper No. 71-WA/FE-38.

Black, H., and Jenssen, D. (1970), "Dynamic Hybrid Properties of Annular Pressure Seals," *Journal of Mechanical Engineering*, **184**, 92–100.

Childs, D. (1989), "Fluid-Structure Interaction Forces at Pump-Impeller-Shroud Surfaces for Rotordynamic Calculations," *Journal of Vibration, Acoustics, Stress, and Reliability in Design*, **111**, 216–225.

Childs, D. (1985), "Vibration Characteristics of the High Pressure Oxygen Turbopump (HPOTP) of the Space Shuttle Main Engine (SSME)," *Journal of Engineering for Gas Turbines and Power*, **107**, 152–159.

Childs, D. (1984), "Finite-Length Solutions for Rotordynamic Coefficients of Constant-Clearance and Convergent-Tapered Annular Seals," in *Vibrations in Rotating Machinery*, Proceedings, IMechE, Third International Conference on Vibrations in Rotating Machinery, York, England, pp. 223–231.

Childs, D. (1983), "Dynamic Analysis of Turbulent Annular Seals Based on Hirs' Lubrication Equation," *Journal of Lubrication Technology*, **105**, 437–444.

Childs, D. (1982), "Rotordynamic Moment Coefficients for Finite-Length Turbulent Seals," in *Proceedings of the IFToMM Conference on Rotordynamic Problems in Power Plants*, Rome, Italy, pp. 371–378.

Childs, D., and Dressman, J. (1985), "Convergent-Tapered Annular Seals: Analysis and Testing for Rotordynamic Coefficients," *Journal of Tribology*, **107**, 307–317.

Childs, D., and Kim, C.-H. (1986a), "Test Results for Round-Hole-Pattern Damper Seals: Optimum Configurations and Dimensions for Maximum Net Damping," *Journal of Tribology*, **108**, 605–611.

Childs, D., and Kim, C.-H. (1986b), "Testing for Rotordynamic Coefficients and Leakage: Circumferentially Grooved Seals," in *Proceedings of the Second IFToMM International Conference on Rotordynamics*, Tokyo, Japan, pp. 609–618.

Childs, D., and Kim, C.-H. (1985), "Analysis and Testing for Rotordynamic Coefficients of Turbulent Annular Seals with Different, Directionally-Homogeneous Surface-Roughness Treatment for Rotor and Stator Elements," *Journal of Tribology*, **107**, 296–306.

Childs, D., Nolan, S., and Kilgore, J. (1990a), "Test Results for Turbulent Annular Seals Using Smooth Rotors and Helically-Grooved Stators," *Journal of Tribology*, **112**, 254–258.

Childs, D., Nolan, S., and Kilgore, J. (1990b), "Additional Test Results for Round-Hole-Pattern Damper Seals: Leakage, Friction Factors, and Rotordynamic Force Coefficients," *Journal of Tribology*, **112**, 365–371.

Childs, D., Nolan, S., and Nunez, D. (1988), "Clearance Effects on Leakage and Rotordynamic Coefficients of Smooth, Liquid Annular Seals," in *Vibrations in Rotating Machinery*, International Conference Proceedings, IMechE, Edinburgh, Scotland, pp. 371–378.

Diana, G., Marenco, G., Mimmi, G., and Saccenti, P. (1982), "Experimental Research on the Behavior of Hydrodynamic Plain Seals by Means of a Specific Testing Device," in *Proceedings of the IFToMM International Conference on Rotordynamic Problems in Power Plants*, Rome, Italy, pp. 355–360.

Dietzen, F., and Nordmann, R. (1987a), "Finite Difference Analysis for the Rotordynamic Coefficients of Turbulent Seals in Turbopumps," ASME FED-Vol. 48, Symposium on Thin Fluid Films, pp. 31–42.

Dietzen, F., and Nordmann, R. (1987b), "Calculating Rotordynamic Coefficients by Finite Difference Techniques," *Journal of Tribology*.

Dietzen, F., and Nordmann, R. (1986), "Calculating Rotordynamic Coefficients of Seals by Finite Difference Techniques," Rotordynamic Instability Problems in High Performance Turbomachinery, NASA CP No. 2443, proceedings of a workshop held at Texas A&M University, pp. 77–96.

Diewald, W., and Nordmann, R. (1988), "Influence of Different Types of Seals on the Stability Behavior of Turbopumps," Rotordynamic Instability Problems in High-Performance Turbomachinery, NASA CP No. 3026, proceedings of a workshop held at Texas A&M University, pp. 197–210.

Domm, V., Dernedde, R., and Handwerker, Th. (1967), "Der Einfluss der Stufenabdicktung auf die Kritische Drezahl von Kessel Speisepumpen," *VDI-Berichte*, No. 113, pp. 25–28.

Falco, M., Mimmi, G., and Marenco, G. (1986), "Effects of Seals on Rotor Dynamics," in *Proceedings of the IFToMM International Conference on Rotordynamics*, Tokyo, Japan, pp. 655–662.

Falco, M., Mimmi, G. Pizzigoni, B., Marenco, G., and Negri, G. (1984), *Vibrations in Rotating Machinery*, Proceedings of the Third IMechE International Conference, Paper No. C30384, York, England, pp. 151–158.

Florjancic, S. (1990), "Annular Seals of High Energy Centrifugal Pumps: A New Theory and Full Scale Measurement of Rotordynamic Coefficients and Hydraulic Friction Factors," dissertation, Swiss Federal Institute of Technology, Zürich, Switzerland.

Frei, A., Gulich, J., Eichhorn, G., Ebeiel, J., and McCloskey, T. (1990), "Rotordynamic and Dry Running Behavior of a Full Scale Test Boiler Feed Pump," in *Proceedings of the Ninth International Pump Users Symposium*, Texas A&M University, pp. 81–92.

Fritz, R. (1972), "The Effect of Liquids on the Dynamic Motion of Immersed Solids," *Journal of Engineering for Industry*, **92**, 923–929.

Gaffel, K., and Ganter, M. (1992), "Zellenprofil, Anwendung einer High-tech-Entwichlung aus der Roumfahrt in Industriellen Pumpenbau," *Brenrenstoft, Wärme, Kraft*, Band 44.

Graf, K., and Staubli, T. (1990), "Comparison of Two Identification Models for Rotordynamic Coefficients of Labyrinth Seals," in *Interfluid*, First International Congress on Fluid Handling Systems, Essen BRD.

Hirs, G. (1973), "A Bulk-Flow Theory for Turbulence in Lubricant Films," *Journal of Lubrication Technology*, 137–146.

Iino, I., and Kaneko, H. (1980), "Hydraulic Forces Caused by Annular Pressure Seals in Centrifugal Pumps," Rotordynamic Instability Problems in High Performance Turbomachinery, NASA CP No. 2133, proceedings of a workshop held at Texas A&M University, pp. 213–225.

Iwatsubo, T., and Sheng, B. (1990a), "An Experimental Study on the Static and Dynamic Characteristics of Damper Seals," in *Proceedings of the Third IFToMM International Conference on Rotordynamics*, Lyon, France, pp. 307–312.

Iwatsubo, T., and Sheng, B. (1990b), "Evaluation of Dynamic Characteristics of Parallel Grooved Seals by Theory and Experiment," in *Proceedings of the Third IFToMM International Conference on Rotordynamics*, Lyon, France, pp. 313–318.

Iwatsubo, T., Sheng, B., and Ono, M. (1990), "Experiment of Static and Dynamic Characteristics of Spiral Grooved Seals," Rotordynamic Instability Problems in High-Performance Turbomachinery, NASA CP No. 3122, proceedings of a workshop held at Texas A&M University, pp. 223–234.

Kaneko, S., Hori, Y., and Tanaka, M. (1984), "Static and Dynamic Characteristics of Annular Seals," in *Vibrations in Rotating Machinery*, Proceedings of the Third IMechE. Conference on Vibrations in Rotating Machinery, pp. 205–214.

Kanemori, Y., and Iwatsubo, T. (1989), "Experimental Study of Dynamic Fluid Forces and Moments for a Long Angular Seal," *Machinery Dynamics—Applications and Vibration Control Problems*, DE-Vol. 18-2, ASME, pp. 141–148.

Kanki, H., and Kawakami, T. (1987), "Experimental Study on the Dynamic Characteristics of Screw Grooved Seals," *Rotating Machinery Volume 1*, ASME DE-Vol. 2, pp. 273–278.

Kanki, H., and Kawakami, T. (1984), "Experimental Study on the Dynamic Characteristics of Pump Annular Seals," in *Vibrations in Rotating Machinery*, proceedings of the third IMechE International Conference on Vibrations in Rotating Machinery, York, England, pp. 159–166.

Kilgore, J., and Childs, D. (1990), "Rotordynamic Coefficients and Leakage Flow of Circumferentially-Grooved Liquid Seals," *Journal of Fluids Engineering*, **112**, 250–256.

Kim, C.-H., and Childs, D. (1987), "Analysis for Rotordynamic Coefficients of Helically-Grooved Turbulent Annular Seals," *Journal of Tribology*, **109**(1), 136–143.

Kundig, P. (1990), "Labyrinthdichtungen Hydraulischer Machinen," Ph.D. dissertation, Federal Technical Institute of Switzerland, Zürich, Switzerland.

Launder, B., and Leschziner, M. (1978), "Flow in Finite Width Thrust Bearings Including Inertial Effects, I—Laminar Flow, II—Turbulent Flow," *Journal of Lubrication Technology*, **100**, 330–345.

Launder, B., and Spalding, D. (1974), "The Numerical Computation of Turbulent Flows," *Computer Methods in Applied Mechanics and Engineering*, **3**, 269–289.

Lomakin, A. (1958), "Calculation of Critical Number of Revolutions and the Conditions Necessary for Dynamic Stability of Rotors in High-Pressure Hydraulic Machines when Taking into Account Forces Originating in Sealings," *Power and Mechanical Engineering*, April 1958 (in Russian).

Massey, I. (1985), "Subsynchronous Vibration Problems in High-Speed Multistage Centrifugal Pumps," in *Proceedings of the 14th Turbomachinery Symposium*, Texas A&M University, pp. 11–16.

Meirovitch, L. (1985), *Introduction to Dynamics and Control*, Wiley, New York.

Nelson, C., and Nguyen, D. (1988a), "Analysis of Eccentric Annular Incompressible Seals: Part 1—A New Solution Using Fast Fourier Transforms for Determining Hydrodynamic Forces," *Journal of Tribology*, **110**, 361–366.

Nelson, C., and Nguyen, D. (1988b), "Analysis of Eccentric Annular Incompressible Seals: Part 2—Effects of Eccentricity on Rotordynamic Coefficients," *Journal of Tribology*, **110**, 361–366.

Nelson, C. (1984), "Analysis for Leakage and Rotordynamic Coefficients of Surface-Roughened Tapered Annular Gas Seals," *Journal of Engineering for Gas Turbines and Power*, **106**, 927–934.

Nelson, C. (1985), "Rotordynamic Coefficients for Compressible Flow in Tapered Annular Seals," *Journal of Tribology*, **107**, 318–325.

Nelson, C., and Nguyen, D. (1987), "Comparison of Hirs' Equation with Moody's Equation for Determining Rotordynamic Coefficients of Annular Pressure Seals," *Journal of Tribology*, **109**, 144–148.

Nordmann, R., and Dietzen, F. (1990), "A Three-Dimensional Finite-Difference Method for Calculating the Dynamic Coefficients of Seals," in J. Kim and W. Yang (Eds.), *Dynamics of Rotating Machinery*, Hemisphere, pp. 133–151.

Nordmann, R., Dietzen, F., Janson, W., Frei, A., and Florjancic, S. (1986), "Rotordynamic Coefficients and Leakage Flow for Smooth and Grooved Seals in Turbopumps," in *Proceedings of the Second IFToMM International Conference on Rotordynamics*, Tokyo, Japan, pp. 619–627.

Nordmann, R., and Massman, H. (1984), "Identification of Dynamic Coefficients of Annular Turbulent Seals," Rotordynamic Instability Problems in High-Performance Turbomachinery-1984, NASA CP No. 2338, Proceedings of a Workshop held at Texas A&M University, pp. 295–311.

Ocvirk, F. (1952), "Short Bearing Approximation for Full Journal Bearings," NACA TN 20808.

San Andres, L. (1991), "Analysis of Variable Fluid Properties, Turbulent Annular Seals," *Journal of Tribology*, 684–702.

Scharrer, J., and Nunez, D. (1989), "The SSME HPFTP Wavy Interstage Seal: Part I—Seal Analysis," *Machinery Dynamics—Applications and Vibration Control Problems*, DE-Vol. 18-2, ASME, pp. 95–100.

Scharrer, J. (1987), "A Comparison of Experimental and Theoretical Results for Labyrinth Gas Seals," Ph.D. dissertation, Texas A&M University.

Stampa, B. (1971), Untersuchungen an axial durchströmten Ringspalten, Dissertation, Braunschweig University.

Szeri, A. (1980), *Tribology—Friction, Lubrication, and Wear*, McGraw-Hill, New York.

Valantas, R., and Bolleter, U. (1988), "Solutions to Abrasive Wear-Related Rotordynamic Instability Problems on Prudhoe Bay Injection Pumps," in *Proceedings of the Fifth International Pump Users Symposium*, Texas A&M University, pp. 3–10.

Von Doormal, J., and Raithby, G. (1984), "Enhancements of the Simple Method for Predicting Incompressible Fluid Flows," *Numerical Heat Transfer*, **7**, 147–163.

von Pragenau, G. (1982), "Damping Seals for Turbomachinery," NASA Technical Paper No. 1987.

Wyssman, H., Pham, T., and Jenny, J. (1984), "Prediction of Stiffness and Damping Coefficients for Centrifugal Compressor Labyrinth Seals," *Journal of Engineering for Gas Turbines and Power*, **106**, 920–926.

Yamada, Y. (1962), "Resistance of Flow Through an Annulus with an Inner Rotating Cylinder," *Bull J.S.M.E.*, **5**(18), 302–310.

5

ROTORDYNAMIC MODELS FOR ANNULAR GAS SEALS

5.1 INTRODUCTION

Annular gas seals meet the same leakage-control requirements for compressors and turbines as the liquid annular seals of Chapter 4 did for pumps. Like liquid seals, the forces developed by gas seals are roughly proportional to the ΔP across the seals and the fluid density within the seal. Because of the density dependency, gas seals have had a greater impact on steam turbines and high-pressure compressors where densities are higher than on gas turbines.

High-pressure compressors can use either the "series" or "back-to-back" designs of Figure 5.1. In the throughflow design, flow enters from the left and proceeds directly from impeller to impeller, discharging on the right. For the back-to-back design, flow enters at the left and proceeds from left to right through the first four stages, then follows a crossover duct to the right-hand side of the machine, and proceeds from right to left through the last four stages, discharging at the center. Back-to-back machines obviously react a smaller axial thrust than series machines.

Figure 5.2 illustrates a typical sealing arrangement for a last-stage compressor impeller and can be profitably compared to Figure 4.2 for pumps. The eye-packing seal limits return-flow leakage down the front of the impeller, and the shaft seal restricts leakage along the shaft to the preceding stage. In a series or throughflow compressor, leakage flow through the balance drum is returned to the inlet; hence, the balance drum absorbs the full ΔP of the compressor, and the fluid within the seal has an average density that is approximately proportional to the average of inlet and discharge pressures. For a back-to-back machine, the balance drum absorbs the ΔP between the last stage of the compressor and the last stage of the initial

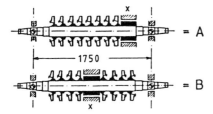

Figure 5.1 (a) Flowthrough or series and (b) back-to-back or parallel compressor designs [Wachter and Benckert (1980)].

series of impellers, i.e., about one half of compressor ΔP. For the same inlet and discharge pressures, the *average* density is higher in the center labyrinth of a back-to-back machine than in the balance-drum labyrinth of a series machine. Historically, back-to-back compressors are more sensitive to the forces from the central labyrinth than are series machines to forces from the balance-drum labyrinth. This result would be expected because of the first-critical-speed mode shape of the rotors, with a much larger amplitude at a center labyrinth than a balance piston.

As noted, there is a strong correlation between the location and function of pump and centrifugal-compressor seals. The notation, as compared below, is different but the functions are identical.

Pump	Compressor
Neck or wearing ring	Eye packing
Interstage	Shaft
Balance piston	Balance drum

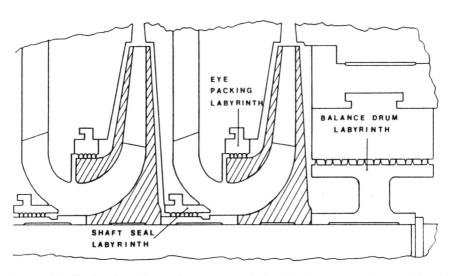

Figure 5.2 Typical impeller seal arrangements for the last stage of a centrifugal compressor [Kirk (1987)].

The forces developed by compressor-seal labyrinths are at least one order of magnitude lower than their liquid-seal counterparts. They have negligible added-mass terms and are typically modeled by the following reaction-force/motion model:

$$-\left\{\begin{array}{c} F_X \\ F_Y \end{array}\right\} = \left[\begin{array}{cc} K & k \\ -k & K \end{array}\right] \left\{\begin{array}{c} X \\ Y \end{array}\right\} + \left[\begin{array}{cc} C & c \\ -c & C \end{array}\right] \left\{\begin{array}{c} \dot{X} \\ \dot{Y} \end{array}\right\}. \tag{5.1}$$

Unlike the pump seal model of Eq. (4.1), the direct stiffness term is typically negligible and is negative in many cases. *Most of this chapter concerns analysis and tests to define the coefficients of Eq. (5.1).*

Figure 5.3 is taken from Greathead and Bostow (1976) and illustrates the seal leakage flow from a high-pressure steam turbine. Greathead and Bostow were concerned with a "load-dependent" instability. Specifically, the unit could be operated at full speed up to about 90% of power. However, at higher power levels, the rotor began to whirl subsynchronously at its lowest natural frequency. Pollman and Termeuhlen (1975) provide a parallel and illuminating discussion of "steam whirl." This result is also typical of instabilities in centrifugal compressors and leads to an "onset *power level* of instability" versus the onset *speed* of instability for rotors due to hydrodynamic bearings. The density dependency of labyrinth-seal coefficients provides one explanation for power-level-dependent instability. As discussed in Chapter 6, turbines also provide a destabilizing force which is proportional to power levels. The eye-packing and shaft seal configurations illustrated in Figure 5.2 are called "see-through" or half-labyrinth seals. The balance-drum seal configuration is called an interlocking or full labyrinth. Sections 5.2 and 5.3 consider the available analysis and test results for these types of *labyrinth* seals.

Figure 5.4 illustrates a "honeycomb" seal. This seal has an obvious similarity to the hole-pattern damper seal of Figure 4.21. The roughened stator is provided to reduce leakage and has the major additional benefit of reducing the average circumferential velocity within the seal and the destabilizing cross-coupled stiffness coefficient k of Eq. (5.1). This seal type has been used profitably for balance-drum applications in compressors and as a

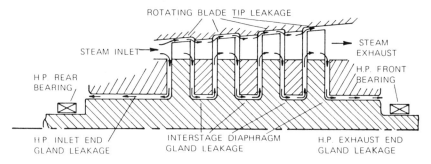

Figure 5.3 Steam turbine sealing path [Greathead and Bostow (1976)].

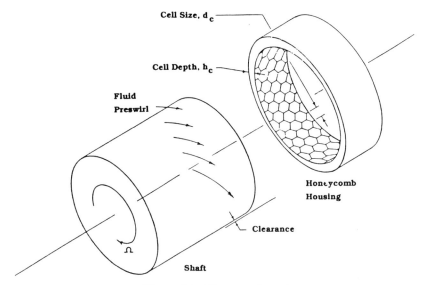

Figure 5.4 Honeycomb seal.

turbine-interstage seal for the high-pressure oxygen turbopump (HPOTP) of the Space Shuttle main engine (SSME).

Figure 5.5 illustrates the HPOTP rotating assembly. Flow is from right to left through the two turbine stages. Leakage flow from the second-stage turbine proceeds along the shaft through the two "floating-ring" seals illustrated in Figure 5.6. These seals use stators which are supposed to center themselves with respect to the shaft at comparatively low speeds and ΔPs. At

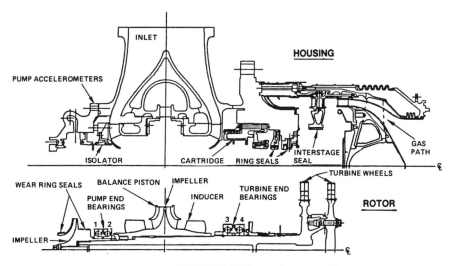

Figure 5.5 SSME-HPOTP rotating assembly.

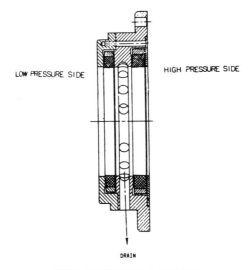

DOUBLE CIRCUMFERENTIAL FLOATING RING
TURBINE SEAL ASSEMBLY (SSME HPOTP)

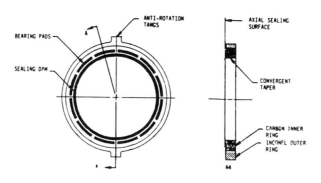

TAPERED BORE FLOATING RING CARBON SEAL (SSME HPOTP)

Figure 5.6 Floating-ring gas seal.

higher ΔPs Coulomb friction forces associated with the axial seal forces "lock" them radially. Floating-ring seals typically use smooth stators with constant, convergent-tapered, or stepped clearances to maximize the seal's direct stiffness and centering capability. Sections 5.6 and 5.7 examine the available analyses and test results for smooth and honeycomb seals.

Figure 5.7 illustrates a "brush" seal. This seal uses a biased pattern of wires in contact with a ceramic coating on the shaft and has a sharply reduced leakage flow as compared to a labyrinth or honeycomb seal. Based on initial tests, the brush seal has favorable rotordynamic characteristics as compared to other gas seal configurations. Section 5.8 provides comparisons between measured test results for brush and see-through labyrinth-seal configurations.

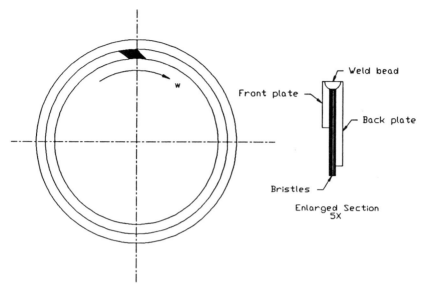

Figure 5.7 Brush-seal configuration.

Section 5.9 provides a summary and overview of known results for annular gas seals.

5.2 INITIAL TESTS AND ANALYSIS FOR ANNULAR LABYRINTH SEALS

As discussed in Chapter 4 for liquid seals, the cross-coupled stiffness coefficient k of Eq. (5.1) arises primarily because of the circumferential velocity within the seal. From a rotordynamics viewpoint, this is the central, crucial fact related to labyrinth seals, and it was first demonstrated conclusively by Benckert and Wachter (1978, 1979, 1980a, 1980b). Benckert and Wachter *statically* displaced the labyrinth rotor relative to its stator, measured the pressure distribution to define the reaction force, and then resolved the reaction force into components which were parallel and perpendicular to the displacement vector. From Eq. (5.1), these components yield the direct and cross-coupled stiffness coefficients. Benckert and Wachter's measurements were made for the following conditions:

(a) nonrotating seal with prerotated flow, and
(b) rotating seal with nonprerotated flow.

They tested see-through, interlocking, and comb-groove seals and showed the rotordynamic coefficients to be generally constant and independent of eccentricity ratio out to 0.5. This result supports the "eccentricity-independent" model of Eq. (5.1). Benckert and Wachter demonstrated the effectiveness of

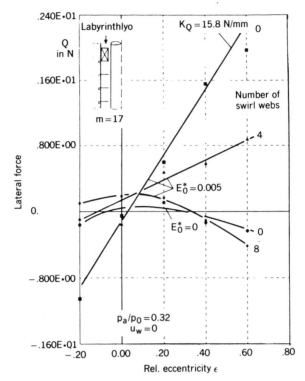

Figure 5.8 Reduction of seal cross-coupled stiffness force by placing swirl webs immediately upstream of a labyrinth, [Wachter and Benckert (1980)]; $m = 17 = $ number of labyrinth chambers.

"swirl webs" placed immediately upstream of their labyrinths to reduce or destroy the inlet tangential velocity. These webs, consisting of axially directed fins, are illustrated in Figure 5.8 and are the first published demonstration of swirl brakes. The destabilizing force $Q = F_Y = kX$ due to a radial displacement in the X direction is shown for zero, four, and eight swirl webs. Observe that four swirl webs reduce the destabilizing force, and eight actually reverse the destabilizing force coefficient at higher eccentricities. Reversing the sign of k in Eq. (5.1) means that it would oppose forward whirl and drive a reverse whirl.

Benckert and Wachter measured negative direct stiffness coefficients for their (comparatively long) seals. Wright (1983) measured a positive direct stiffness coefficient for a single-cavity labyrinth seal. Leong and Brown (1984) and Brown and Leong (1984) developed a test rig that is quite similar to that of Benckert and Wachter to produce test data for seals whose dimensions are representative of steam turbines. They present a rich body of test data combining both preswirl and shaft rotation, yielding results which are basi-

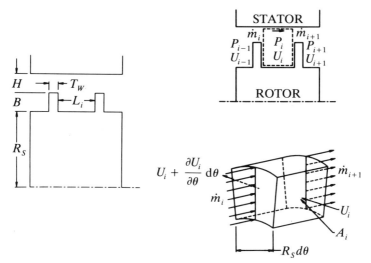

Figure 5.9 One-control-volume model for a see-through labyrinth cavity [Childs and Scharrer (1986)].

cally similar to Benckert and Wachter. Most of their seals had negative direct stiffnesses; however, a shorter seal with five cavities had a positive stiffness.

Governing Equations: Iwatsubo (1980), Iwatsubo et al. (1982), Childs and Scharrer (1986a)

The first model containing the essential physical elements which are necessary to explain the Benckert and Wachter results was by Iwatsubo (1980) and Iwatsubo et al. (1982). Iwatsubo wrote the circumferential-momentum and continuity equations for the control volume of Figure 5.9 to define the average (bulk-flow) circumferential velocity within a labyrinth cavity. He used a leakage equation to define the axial velocity.

The continuity equation for the control volume of Figure 5.9 is obtained from

$$0 = \frac{\partial}{\partial t} \int_{CV} \rho \, d\forall + \int_{CS} \rho \boldsymbol{v} \cdot d\boldsymbol{A}, \tag{5.2a}$$

where $\forall$ is the volume, to be

$$\frac{\partial}{\partial t}(\rho_i A_i) + \frac{\partial}{\partial \theta}\left(\frac{\rho_i U_i A_i}{R_s}\right) + \dot{m}_{i+1} - \dot{m}_{i-1} = 0, \tag{5.2b}$$

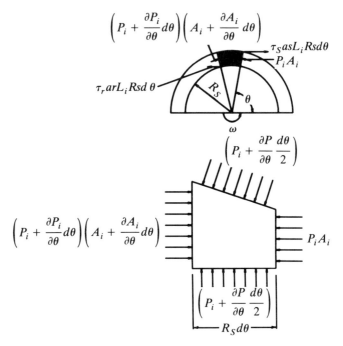

Figure 5.10   Forces on control volume of Figure 5.9.

where U_i is the average circumferential velocity, and $\dot{m}_i$ is the (axial) mass flow rate per unit circumference in cavity i. The transverse area A_i is defined by

$$A_i = (B + H)L_i. \tag{5.3}$$

The circumferential-momentum equation is obtained from

$$\sum F_\theta = \frac{\partial}{\partial t} \int_{CV} U_i \rho \, d\forall + \int_{CS} \rho U_i \boldsymbol{v} \cdot d\boldsymbol{A} \tag{5.4a}$$

and Figure 5.10 to be

$$\frac{\partial(\rho_i U_i A_i)}{\partial t} + \frac{\partial(\rho_i A_i U_i^2)}{\partial \theta} = -\frac{A_i}{Rs}\frac{\partial P_i}{\partial \theta} + (\tau_{ri} ar_i - \tau_{si} as_i)L_i. \tag{5.4b}$$

Equations (5.2b) and (5.4b) are from Childs and Scharrer (1986a) and include area derivatives with respect to θ which were not accounted for by Iwatsubo et al. Shear stresses on the right side of Eq. (5.4b) act on the nondimensional lengths ar_i and as_i, respectively. For tooth-on-rotor and tooth-on-stator labyrinths they are defined, respectively, by

$$as_i = 1, \qquad ar_i = (2B + L_i)/L_i, \tag{5.5a}$$

and

$$as_i = (2B + L_i)/L_i, \qquad ar_i = 1. \tag{5.5b}$$

Observe from Eqs. (5.4) and (5.5) that the shear stress is assumed to be constant at the base and sides of the labyrinth channel.

Following the procedure of Section 4.2 for liquid seals, the Blasius shear-stress model is used, yielding*

$$\tau_{ri} = \frac{\rho_i}{2} nr \left(\frac{|Rs\omega - U_i|Dh_i}{\nu_i} \right)^{mr} |Rs\omega - U_i|(Rs\omega - U_i),$$

$$\tau_{si} = \frac{\rho_i}{2} ns \left(\frac{|U_i|Dh_i}{\nu_i} \right)^{ms} |U_i|U_i, \tag{5.6}$$

where Dh_i is the hydraulic diameter defined by

$$Dh_i = \frac{2(H + B)L_i}{(H + B + L_i)}.$$

Following Childs and Scharrer, and subtracting Eq. (5.2b) times U_i from Eq. (5.4b) yields the following reduced form of the momentum equation:

$$\rho_i A_i \frac{\partial U_i}{\partial t} + \rho_i U_i A_i \frac{\partial U_i}{Rs\, \partial\theta} + \dot{m}_i(U_i - U_{i-1}) = \frac{-A_i}{Rs} \frac{\partial P_i}{\partial\theta} - \tau_{si} as_i L_i + \tau_{ri} ar_i L_i. \tag{5.7}$$

The density terms are replaced via the ideal gas law; i.e.,

$$\rho_i = P_i/ZRT \tag{5.8}$$

where R is the gas constant, Z is the compressibility factor, and T is temperature, which is assumed to be constant throughout the seal. Subsequent perturbation analysis is simplified by the substitution

$$\dot{m}_{i+1} - \dot{m}_i = \frac{\dot{m}_{i+1}^2 - \dot{m}_i^2}{2\dot{m}_0}, \tag{5.9}$$

where $\dot{m}_0$ is the steady-state mass flow rate.

---

*Note in comparison to Eq. (4.9) that the axial velocity is neglected in this definition.

The leakage rate is defined by a modified form of Neumann's (1964) *empirical* leakage equation

$$\dot{m}_i = \mu_0 \mu_i H \sqrt{\frac{P_{i-1}^2 - P_i^2}{RT}} = \rho_i H W_i, \tag{5.10}$$

where μ_i is the flow coefficient and μ_0 is the "kinetic-energy carryover factor." For see-through labyrinth seals, μ_0 is defined by

$$\mu_0 = \left(\frac{NT}{(1 - \alpha)NT + \alpha} \right)^{1/2},$$

$$\alpha = 1 - (1 + 16.6C_r/L)^{-2}, \tag{5.11}$$

where NT is the number of teeth. For the first tooth of see-through seals and all teeth of interlocking seals, μ_0 is unity. The flow coefficient is defined by Chaplygin's formula, Gurevich (1966) as

$$\mu_i = \frac{\pi}{\pi + 2 - 5\beta_i + 2\beta_i^3}, \qquad \beta_i = \left(\frac{P_{i-1}}{P_i} \right)^{(\gamma - 1)/\gamma} - 1, \tag{5.12}$$

where γ is the ratio of specific heats. For choked flow, Fliegner's formula is used for the last seal strip

$$\dot{m}_{NT} = 0.51 \mu_0 P_{NC} H / \sqrt{RT}, \tag{5.13}$$

where $NC = NT - 1$ is the number of labyrinth cavities [John (1979)]. These empirical leakage equations were used by Childs and Scharrer. Iwatsubo et al. combined $\mu_0 \mu_i$ into a single empirical parameter. Note that the leakage equation replaces the axial-momentum equation of the liquid-seal models in defining the axial-velocity and pressure distributions. Hence, leakage does not depend on the stator roughness, and the same leakage is predicted for a smooth or honeycomb stator. Further, the leakage rate does not depend on the circumferential velocity.

In summary, the governing equations for cavity i are the continuity equation (5.2b), the circumferential-momentum equation (5.4b), the equation of state (5.8), and the leakage equations (5.10) or (5.13). The following assumptions were used to obtain these equations:

(a) The fluid is modeled by the ideal-gas law.
(b) The pressure is constant within a chamber.
(c) The lowest circumferential acoustic-resonance frequency within the cavity is much higher than the rotor speed.

(d) The axial velocity component is neglected. Only the circumferential velocity is used in calculating wall shear stress.

(e) The temperature is constant throughout the seal.

Using Eqs. (5.8)–(5.10) to eliminate ρ_i and $\dot{m}_i$, $\dot{m}_{i+1}$ from the continuity equation, Eq. (5.2b), yields

$$\frac{\partial(P_i A_i)}{\partial t} + \frac{\partial(P_i U_i A_i)}{Rs\,\partial\theta} - \frac{\mu_0^2 H^2}{2\dot{m}_0}\left[\mu_i^2\left(P_{i-1}^2 - P_i^2\right) - \mu_{i+1}^2\left(P_i^2 - P_{i+1}^2\right)\right] = 0.$$

$$(5.14a)$$

For unchoked flow, this equation applies for all chambers; however, for choked flow the exit chamber pressure is defined by Eq. (5.13), and the governing equation is

$$\frac{\partial(P_i A_i)}{\partial t} + \frac{\partial(P_i U_i A_i)}{Rs\,\partial\theta} - \frac{\mu_0^2 H^2}{2\dot{m}_0}\left[\mu_i^2\left(P_{i-1}^2 - P_i^2\right) - 0.260 P_{NC}^2\right] = 0,$$

$$i = NC. \quad (5.14b)$$

Similarly, eliminating ρ_i and $\dot{m}_i$ from the circumferential-momentum equation (5.7) gives

$$P_i A_i \frac{\partial U_i}{\partial t} + \frac{P_i A_i U_i}{Rs}\frac{\partial U_i}{\partial\theta} + (U_i - U_{i-1})\mu_0\mu_i H\left[RT\left(P_{i-1}^2 - P_i^2\right)\right]^{1/2}$$

$$= RT\left(\frac{-A_i}{Rs}\frac{\partial P_i}{\partial\theta} - \tau_{si} as_i L_i + \tau_{ri} ar_i L_i\right), \qquad i = 1, 2, \ldots, NC. \quad (5.15)$$

Perturbation Analysis

The rotordynamic coefficients of Eq. (5.1) apply for small motion of the seal rotor about a centered position. The coefficients are obtained by expanding the governing equations (5.14) and (5.15) in terms of the perturbation variables

$$P_i = P_{0i} + \epsilon P_{1i}, \qquad H = Cr_i + \epsilon H_1$$

$$U_i = U_{0i} + \epsilon U_{1i}, \qquad A_i = A_{0i} + \epsilon L_i H_1, \qquad (5.16)$$

where $\epsilon = e/C_r$ is the eccentricity ratio. Perturbation for ρ_i and $\dot{m}_i$ are not required since they have been eliminated from the equations. The zeroth-order equations define $\dot{m}_0$ and the cavity-to-cavity distribution of pressure

P_{0i} and circumferential velocity U_{0i}. The first-order equations define perturbations in P_i and U_i due to a radial perturbation of the seal rotor. Strictly
speaking, the results are only valid for small motion about a centered
position.

Zeroth-Order Solution

The zeroth-order solution for leakage yields $\dot{m}_0 = \dot{m}_i$, $i = 1, 2, \ldots$, NT for a
given supply pressure P_s and back pressure P_b. $\dot{m}_0$ must be obtained
iteratively, beginning with a determination of whether or not the flow is
choked. As a first step, choked flow is assumed at the exit strip. Then,
knowing the pressure ratio for choked flow, P_{NC} is found, and an estimate
for $\dot{m}_0$ is obtained from Eq. (5.13). Equation (5.10) is used in the form

$$P_{0i-1}^2 = P_{0i}^2 + RT \left(\frac{\dot{m}_0}{\mu_0 \mu_i C_r} \right)^2 \tag{5.17a}$$

to work "upstream" from chamber to chamber until a calculated supply
pressure P_s' is obtained. If $P_s \geq P_s'$, the flow is choked, and the steady-state
solution is obtained by guessing P_{NC}, using Eq. (5.13) to estimate $\dot{m}_0$, and
using Eq. (5.17a) to move upstream and recalculate P_{0i}. These steps are
repeated until P_s' converges to P_s.

For unchoked flow, P_{01} is guessed, Eq. (5.10) is used to calculate an initial
estimate for $\dot{m}_0$, and then Eq. (5.10) is used in the form

$$P_{0i}^2 = P_{0i-1}^2 - RT \left(\frac{\dot{m}_0}{\mu_0 \mu_i C_r} \right)^2 \tag{5.17b}$$

to march downstream and provide a calculated estimate P_b' for the back
pressure P_b. These steps are repeated until P_b' converges to P_b.

First-Order Equations and Solutions

The first-order continuity and circumferential-momentum equations corresponding to Eqs. (5.14) and (5.15) are

$$G_{1i} \frac{\partial P_{1i}}{\partial t} + G_{1i} \frac{U_{0i}}{Rs} \frac{\partial P_{1i}}{\partial \theta} + G_{1i} \frac{P_{0i}}{Rs} \frac{\partial U_{1i}}{\partial \theta} + G_{3i} P_{1i} + G_{4i} P_{1i-1} + G_{5i} P_{1i+1}$$

$$= -G_{6i} H_1 - G_{2i} \frac{\partial H_1}{\partial t} - G_{2i} \frac{U_{0i}}{Rs} \frac{\partial H_1}{\partial \theta} \tag{5.18a}$$

and

$$X_{1i}\frac{\partial U_{1i}}{\partial t} + \frac{X_{1i}U_{0i}}{Rs}\frac{\partial U_{1i}}{\partial \theta} + \frac{A_{0i}}{Rs}\frac{\partial P_{1i}}{\partial \theta} + X_{2i}U_{1i}$$

$$- \dot{m}_0 U_{1i-1} + X_{3i}P_{1i} + X_{4i}P_{1i-1} = X_{5i}H_1, \qquad (5.18b)$$

where the G_{1i}'s and X_{1i}'s are provided by Childs and Scharrer.

Recall that our objective is the solution for the force coefficients of Eq. (5.1); K, k, C, and c. We could follow the approach of Chapter 4* and assume a circular precessional orbit for the seal rotor at a frequency Ω. However, this approach yields only two equations per frequency Ω,[†] and at least two values for Ω would be needed. Alternatively, we can assume the following elliptical orbit at the rotor running speed ω

$$X = a \cos \omega t, \qquad Y = b \sin \omega t,$$

$$\dot{X} = -a\omega \sin \omega t, \qquad \dot{Y} = b\omega \cos \omega t, \qquad (5.19)$$

with its associated clearance function

$$\epsilon H_1 = -a \cos \omega t \cos \theta - b \sin \omega t \sin \theta$$

$$= -\frac{a}{2}[\cos(\theta - \omega t) + \cos(\theta + \omega t)] - \frac{b}{2}[\cos(\theta - \omega t) - \cos(\theta + \omega t)]. \qquad (5.20a)$$

Subsequent results will show that this approach yields four independent equations for the four unknowns. The perturbed pressure and velocity can be stated in the comparable solution format

$$P_{1i} = P_{ci}^+ \cos(\theta + \omega t) + P_{si}^+ \sin(\theta + \omega t) + P_{ci}^- \cos(\theta - \omega t)$$

$$+ P_{si}^- \sin(\theta - \omega t),$$

$$U_{1i} = U_{ci}^+ \cos(\theta + \omega t) + U_{si}^+ \sin(\theta + \omega t) + U_{ci}^- \cos(\theta - \omega t)$$

$$+ U_{si}^- \sin(\theta - \omega t). \qquad (5.20b)$$

Substituting Eqs. (5.20) into (5.18) and equating coefficients of $\cos(\theta + \omega t)$, $\cos(\theta - \omega t)$, $\sin(\theta + \omega t)$, $\sin(\theta - \omega t)$ eliminates the θ and t dependency and yields eight linear independent equations per cavity. The equations for the ith cavity can be stated:

$$[A_{i-1}](X_{i-1}) + [A_i](X_i) + [A_{i+1}](X_{i+1}) = \frac{a}{\epsilon}(B_i) + \frac{b}{\epsilon}(C_i), \qquad (5.21a)$$

---

*Section 4.5, Eq. (4.39).
[†]Equation (4.56).

where

$$(X_{i-1}) = \left(P_{si-1}^+, P_{ci-1}^+, P_{si-1}^-, P_{ci-1}^-, U_{si-1}^+, U_{ci-1}^+, U_{si-1}^-, U_{ci-1}^- \right)^T,$$

$$(X_i) = \left(P_{si}^+, P_{ci}^+, P_{si}^-, P_{ci}^-, U_{si}^+, U_{ci}^+, U_{si}^-, U_{ci}^- \right)^T,$$

$$(X_{i+1}) = \left(P_{si+1}^+, P_{ci+1}^+, P_{si+1}^-, P_{ci+1}^-, U_{si+1}^+, U_{ci+1}^+, U_{si+1}^-, U_{ci+1}^- \right)^T. \quad (5.21b)$$

The A matrices and column vectors B and C are defined by Childs and Scharrer. To solve Eq. (5.21) for the entire seal, a system matrix must be formed which is block tridiagonal in the $[A]$ matrices. The size of this resultant matrix is (8NC × 8NC), since pressure and velocity perturbations at the inlet and the exit are assumed to be zero. This system is easily solved by various linear equation algorithms and yields a pressure solution of the form

$$P_{si}^+ = \frac{a}{\epsilon} F_{asi}^+ + \frac{b}{\epsilon} F_{bsi}^+,$$

$$P_{si}^- = \frac{a}{\epsilon} F_{asi}^- + \frac{b}{\epsilon} F_{bsi}^-,$$

$$P_{ci}^+ = \frac{a}{\epsilon} F_{aci}^+ + \frac{b}{\epsilon} F_{bci}^+,$$

$$P_{ci}^- = \frac{a}{\epsilon} F_{aci}^- + \frac{b}{\epsilon} F_{bci}^-. \quad (5.22)$$

Determination of Rotordynamic Coefficients

Returning to Eq. (5.1) and substituting the assumed elliptical orbit solution of Eq. (5.19) yields

$$F_X = -(Ka + Cb\omega)\cos \omega t + (-kb + Ca\omega)\sin \omega t$$

$$= F_{XC} \cos \omega t + F_{XS} \sin \omega t,$$

$$F_Y = -(ka + Cb\omega)\cos \omega t - (Kb + ca\omega)\sin \omega t$$

$$= F_{YC} \cos \omega t + F_{YS} \sin \omega t. \quad (5.23)$$

In terms of the perturbed cavity pressures, the reaction-force definition is

$$F_X = -Rs\epsilon \sum_{i=1}^{NC} \int_0^{2\pi} P_{1i} L_i \cos \theta \, d\theta,$$

$$F_Y = Rs\epsilon \sum_{i=1}^{NC} \int_0^{2\pi} P_{1i} L_i \sin \theta \, d\theta. \quad (5.24)$$

Only one of these components is needed to obtain the rotordynamic coefficients, and we choose (arbitrarily) the X component. Substituting the first of Eq. (5.20b) into the first of Eq. (5.24) and integrating gives

$$F_X = -\epsilon\pi Rs \sum_{i=1}^{NC} L_i\big[(P_{si}^+ - P_{si}^-)\sin \omega t + (P_{ci}^+ + P_{ci}^-)\cos \omega t\big]. \quad (5.25)$$

Substituting the solution of Eq. (5.22) gives

$$F_X = F_{XC}\cos \omega t + F_{XS}\sin \omega t,$$

where

$$F_{XS} = -\pi Rs \sum_{i=1}^{NC} L_i\big[a(F_{asi}^+ - F_{asi}^-) + b(F_{bsi}^+ - F_{bsi}^-)\big],$$

$$F_{XC} = -\pi Rs \sum_{i=1}^{NC} L_i\big[a(F_{aci}^+ - F_{aci}^-) + b(F_{bci}^+ - F_{bci}^-)\big]. \quad (5.26)$$

Equating the definitions for F_{XS}, F_{XC} in Eqs. (5.26) and (5.25), and setting the coefficients of the linearly independent parameters a and b equal to zero yields

$$K = \pi R \sum_{i=1}^{NC} (F_{aci}^+ + F_{aci}^-) L_i,$$

$$k = \pi R \sum_{i=1}^{NC} (F_{bsi}^+ - F_{bsi}^-) L_i,$$

$$C = \frac{-\pi Rs}{\omega} \sum_{i=1}^{NC} (F_{asi}^+ - F_{asi}^-) L_i,$$

$$c = \frac{\pi Rs}{\omega} \sum_{i=1}^{NC} (F_{bci}^+ + F_{bci}^-) L_i. \quad (5.27)$$

Childs and Scharrer (1986a) showed good agreement between their predictions for k and the Benckert and Wachter (1980a) measurements for a see-through teeth-on-stator labyrinth.

Additional One-Control-Volume Analyses

Scharrer (1988) extended the preceding analysis to "stepped" labyrinth seals in which the radius of the labyrinth teeth can increase or decrease while the clearance remains constant. Scharrer presents predictions for convergent, straight (constant-radius), and divergent seals with 5, 10, and 15 teeth.

Tooth-on-rotor and tooth-on-stator seals are considered. He predicts an increase in direct stiffness as one moves from diverging, to straight, to converging designs, but only a modest sensitivity of the whirl-frequency ratio. No experimental results are available for converging or diverging seal designs.

Rajakumar and Sisto (1988) solve a variation of the model in this section for small motion about an arbitrary eccentricity position. They predict a divergence from the symmetric model of Eq. (5.1) as static eccentricities are introduced. The stiffness coefficients are predicted to be generally insensitive to changes in the static eccentricity ratio in agreement with Benckert and Wachter (1980b). A sharp drop in direct damping is predicted at higher eccentricity ratios. As yet, no data are available for comparison to those predictions.

Kurohashi et al. used the basic model of Iwatsubo (1980) but made the flow coefficient μ_i a function of the local clearance $H(\theta, t)$ based on a coefficient-of-contraction model of Kamotori (1973). Unfortunately, no case has been made that this modification actually improves the prediction accuracy. Thieleke and Stetter (1990) also make μ_i variable with θ but provide no information concerning its form. Interestingly, they use the assumed complex, circular-orbit-solution approach of section 4.5* with multiple precession frequency rather than the elliptical-orbit approach of Iwatsubo. They present extensive additional test results for labyrinth seals using the University of Stuttgart apparatus of Benckert and Wachter (1978).

5.3   ADDITIONAL MEASUREMENTS AND TWO-CONTROL-VOLUME MODELS FOR LABYRINTH SEALS

The one-control-volume model of the preceding section fails to account for the known nature of the flow field in labyrinth seals. Iwatsubo et al. (1982) presented flow visualization results and Stoff (1980) presented computational-fluid-dynamic results showing the flow pattern illustrated in Figure 5.11. Specifically, the main leakage flow consists of a jet of high-axial-velocity fluid which emerges from a labyrinth strip and expands slightly as it traverses the labyrinth cavity. A single vortex exists within the cavity and is driven by shear stresses between the jet and the vortex. S in Figure 5.11 denotes the stagnation point of a separating streamline between the expanding jet flow and the vortex. The vortex pattern of Figure 5.11 holds for $B \cong L$. For deep cavities, multiple "stacked" vortices exist in the vertical direction. For shallow cavities the jet flow can reattach.

Wyssman et al. (1984) presented the first "two-control-volume" model to account for these known flow features within a labyrinth. Figure 5.12 is taken from their work and shows a "box-in-a-box" model to account for the *leakage*

---

*Childs (1984), Chapter 4.

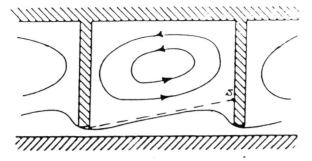

Figure 5.11   Flow pattern in a labyrinth-seal cavity [Scharrer (1987)].

throughflow and the vortex regions. Scharrer (1987) used the superposed two-control-volume model of Figure 5.13 to extend the Wyssman et al. analysis, and the following discussion is based on his analysis. The separate contributions of the two works will be addressed as the development progresses.

From Eq. (5.2a) and Figure 5.13, the following continuity equations hold

Continuity: Control Volume I

$$L_i \frac{\partial(\rho_i H)}{\partial t} + L_i \frac{\partial(\rho_i H U_{1i})}{Rs_1 \partial\theta} + \dot{m}_{i+1} - \dot{m}_i + \dot{m}_{ri} = 0. \qquad (5.28)$$

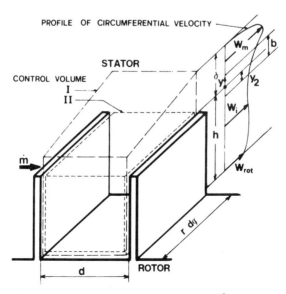

Figure 5.12   "Box-in-a-box," two-control-volume model [Wyssman et al. (1984)].

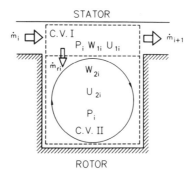

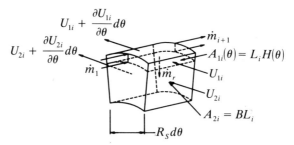

Figure 5.13 Scharrer's (1987) two-control-volume model.

Continuity: Control Volume II

$$L_i B \left[\frac{\partial \rho_i}{\partial t} + \frac{\partial (\rho U_{2i})}{Rs_2 \, \partial \theta} \right] - \dot{m}_{ri} = 0, \tag{5.29}$$

where $Rs_2 = Rs$, $Rs_1 = R_s + B$. Note from Figure 5.13 that $\dot{m}_i = \rho_i H W_{1i_s}$, $\dot{m}_{i+1} = \rho_i H_i W_{1i+1}$ are mass leakage rates per unit circumference entering and exiting control volume I of cavity i. In Eq. (5.29), $\dot{m}_{ri}$ is the flow from control volume I into control volume II. If a streamline were used as a boundary between the throughflow and vortex-flow regions, $\dot{m}_{ri}$ would be (by definition) zero. For the geometric boundary between control volumes I and II of Figure 5.13, $\dot{m}_{ri}$ is zero for a centered position, but not zero for a perturbed position.

From Eq. (5.4a) and Figures 5.14, the following momenta equations hold

Circumferential Momentum: Control Volume I

$$L_i \frac{\partial (\rho_i H U_{1i})}{\partial t} + \frac{L_i}{Rs_1} \frac{\partial (\rho_i H U_{1i}^2)}{\partial \theta} + \dot{m}_{ri} U_{0i} + \dot{m}_{i+1} U_{1i} - \dot{m}_i U_{1i-1}$$

$$= -\frac{L_i H}{Rs_1} \frac{\partial P_i}{\partial \theta} + L_i (\tau_{j\theta i} - as_i \tau_{s\theta i}). \tag{5.30a}$$

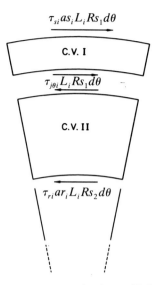

Figure 5.14 Forces on control volumes [Scharrer (1987)].

Circumferential Momentum: Control Volume II

$$
L_i B \frac{\partial(\rho_i H U_{2i})}{\partial t} + \frac{L_i B}{Rs_2} \frac{\partial(\rho_i H U_{2i}^2)}{\partial \theta} - \dot{m}_{ri} U_{0i}
$$

$$
= -\frac{L_i B}{Rs_2} \frac{\partial P_i}{\partial \theta} + L_i(\tau_{j\theta i} - ar_i \tau_{r\theta i}). \tag{5.30b}
$$

Here, ar_i and as_i are defined by Eqs. (5.5), and U_{0i} *is the circumferential velocity between the control volumes.* These governing equations can be simplified considerably. First, Eq. (5.29) is used to eliminate $\dot{m}_{ri}$ from the other equations. For Eq. (5.28) this yields

$$
L_i \frac{\partial(\rho_i H)}{\partial t} + L_i \frac{\partial(\rho_i H U_{1i})}{Rs_1 \partial \theta} + \dot{m}_{i+1} - \dot{m}_i + L_i B \left[\frac{\partial \rho_i}{\partial t} + \frac{\partial(\rho_i U_{2i})}{Rs_2 \partial \theta} \right] = 0. \tag{5.31}
$$

Next, Eq. (5.28) times U_{1i} is subtracted from the control volume I momentum equation (5.30a), giving

$$
\rho_i L_i H \left(\frac{\partial U_{1i}}{\partial t} + \frac{U_{1i}}{Rs_1} \frac{\partial U_{1i}}{\partial \theta} \right) + L_i B \left[\frac{\partial \rho_i}{\partial t} + \frac{\partial(\rho_i U_{2i})}{Rs_2 \partial \theta} \right] (U_{0i} - U_{1i})
$$

$$
+ \dot{m}_i(U_{1i} - U_{1i-1}) = -\frac{L_i H}{Rs_1} \frac{\partial P_i}{\partial \theta} + L_i(\tau_{j\theta i} - as_i \tau_{s\theta i}). \tag{5.32}
$$

Similarly, Eq. (5.29) times U_{2i} is subtracted from the control volume II momentum equation (5.30b) to obtain

$$
\rho_i L_i B \left(\frac{\partial U_{2i}}{\partial t} + \frac{U_{2i}}{Rs_s} \frac{\partial U_{2i}}{\partial \theta} \right) + L_i B \left[\frac{\partial \rho_i}{\partial t} + \frac{\partial (\rho_i U_{2i})}{Rs_2 \, \partial \theta} \right] (U_{2i} - U_{0i})
$$

$$
= \frac{-L_i B}{Rs_2} \frac{\partial P_i}{\partial \theta} + L_i (\tau_{j\theta i} - ar_i \tau_{r\theta i}). \tag{5.33}
$$

The governing equations can be further reduced by using the ideal gas law to eliminate density

$$
\rho_i = P_i / ZRT. \tag{5.34}
$$

Scharrer used the leakage equation from the previous section,

$$
\dot{m}_i = \mu_0 \mu_i H \sqrt{\frac{P_{i-1}^2 - P_i^2}{RT}} = P_i H W_{1i}, \tag{5.10}
$$

for all of the strips, providing that flow is unchoked, and

$$
\dot{m}_{NT} = 0.51 \mu_0 P_{NC} H / \sqrt{RT} \tag{5.13}
$$

for the last strip if the flow is choked. However, μ_0 is no longer constant but is now defined according to Vermes (1961) by

$$
\mu_0 = (1 - \alpha)^{-1/2}, \tag{5.35}
$$

where

$$
\alpha = 8.52 H / [(L_i - T_{wi}) + 7.23H],
$$

where T_{wi} is the ith tooth width. Obviously, with this definition μ_0 now becomes a function of the local clearance H, similar to the results of Kurohashi et al. (1980) and Thieleke and Stetter (1990). μ_0 is unity for the first strip of each seal and all teeth of interlocking seals. As with the one-control-volume developments, these equations replace axial-momentum equations.

Substitution from Eqs. (5.34), (5.10), and (5.13) into Eqs. (5.31) yields

$$
L_i \frac{\partial (P_i H)}{\partial t} + L_i \frac{\partial (P_i H U_{1i})}{Rs_1 \, \partial \theta} + \mu_0 \mu_{i+1} H \left[RT (P_i^2 - P_{i-1}^2) \right]^{1/2}
$$

$$
- \mu_0 \mu_i H \left[RT (P_{i-1}^2 - P_i^2) \right]^{1/2} + L_i B \left[\frac{\partial P_i}{\partial t} + \frac{\partial (P_i U_{2i})}{Rs_2 \, \partial \theta} \right] = 0. \tag{5.36a}
$$

This equation holds for all cavities if flow is unchoked; however, for choked flow, the last chamber equation is

$$L_i\frac{\partial(P_iH)}{\partial t} + L_i\frac{\partial(P_iHU_{1i})}{Rs_1\,\partial\theta} + \mu_0\mu_{i+1}H\left[RT\left(P_i^2 - P_{i-1}^2\right)\right]^{1/2}$$

$$- 0.51\mu_0H\sqrt{RT}\,P_i + L_iB\left[\frac{\partial P_i}{\partial t} + \frac{\partial(P_iU_{2i})}{Rs_2\,\partial\theta}\right] = 0, \qquad i = NC. \quad (5.36b)$$

Similar elimination of ρ_i and $\dot{m}_i$ from the control volume I momentum equation gives

$$L_iHP_i\left(\frac{\partial U_{1i}}{\partial t} + \frac{U_{1i}}{Rs_1}\frac{\partial U_{1i}}{\partial\theta}\right) + L_iB\left[\frac{\partial P_i}{\partial t} + \frac{\partial(P_iU_{2i})}{Rs_2\,\partial\theta}\right](U_{0i} - U_{2i})$$

$$+ \mu_0\mu_iH\left[RT\left(P_{i-1}^2 - P_i^2\right)\right]^{1/2}(U_{1i} - U_{1i-1})$$

$$= RTL_i\left[\frac{-H}{Rs_1}\frac{\partial P_i}{\partial\theta} + \tau_{j\theta i} - as_i\tau_{s\theta i}\right] \qquad (5.37a)$$

for all cavities if the flow is unchoked, and

$$L_iHP_i\left(\frac{\partial U_{1i}}{\partial t} + \frac{U_{1i}}{Rs_1}\frac{\partial U_{1i}}{\partial\theta}\right) + L_iB\left[\frac{\partial P_i}{\partial t} + \frac{\partial(P_iU_{2i})}{Rs_2\,\partial\theta}\right](U_{0i} - U_{2i})$$

$$+ 0.51\mu_0H\sqrt{RT}\,P_i(U_{1i} - U_{1i-1}) = RTL_i\left[\frac{-H}{Rs_1}\frac{\partial P_i}{\partial\theta} + \tau_{j\theta i} - as_i\tau_{s\theta i}\right],$$

$$i = NC \quad (5.37b)$$

for the last cavity if the flow is choked.

The comparable reduced form of the control volume II momentum equation (5.33) is

$$L_iBP_i\left(\frac{\partial U_{2i}}{\partial t} + \frac{U_{2i}}{Rs_2}\frac{\partial U_{2i}}{\partial\theta}\right) + L_iB\left[\frac{\partial P_i}{\partial t} + \frac{\partial(P_iU_{2i})}{Rs_2\,\partial\theta}\right](U_{2i} - U_{0i})$$

$$= RTL_i\left(\frac{-B}{Rs_2}\frac{\partial P_i}{\partial\theta} + \tau_{j\theta i} - ar_i\tau_{r\theta i}\right). \qquad (5.38)$$

In contrast to Scharrer, Wyssman et al. assumed (a) the flow is incompressible ($\partial\rho_i/\partial t = \partial\rho_i/\partial\theta = 0$), and (b) $\dot{m}_i = \dot{m}_{i+1}$.

The stator $\tau_{s\theta i}$ and rotor $\tau_{r\theta i}$ shear stresses can be defined in terms of either the Moody or Blasius models of Section 4.2. Scharrer used the following Blasius model:

$$\tau_{r\theta i} = \frac{\rho_i}{2} nr \left(\frac{\rho_i U_{r2i} Dh_{2i}}{\mu_i} \right)^{mr} U_{r2i} (Rs_2\omega - U_{r2i}),$$

$$\tau_{s\theta i} = \frac{\rho_i}{2} ns \left(\frac{\rho_i U_{s2i} Dh_{1i}}{\mu_i} \right)^{ms} U_{s2i} (Rs_2\omega - U_{s2i}), \qquad (5.39)$$

where

$$U_{r2i} = \left[W_{2i}^2 + (Rs_2\omega - U_{2i})^2 \right]^{1/2},$$

$$U_{s2i} = \left(W_{2i}^2 + U_{2i}^2 \right)^{1/2}, \qquad (5.40)$$

and

$$Dh_{2i} = 2BL_i/(B + L_i),$$

$$Dh_{1i} = 2HL_i/(H + L_i). \qquad (5.41)$$

Note that U_{r2i} is a function of W_{2i}, the axial component of the average vortex velocity, which was neglected by Wyssman et al., who considered only the circumferential-velocity components.

At this point, note that [after substituting for the shear stresses of Eq. (5.39) into the momenta equations] each cavity has six* unknowns (U_{2i}, U_{1i}, P_i, W_{2i}, U_{0i}, and $\tau_{j\theta i}$) but only three equations, viz., one continuity equation and two momenta equations. Three additional equations are obviously required to obtain closure.

Wyssman et al. noted the similarity between flow exiting a labyrinth tooth and jet flow to begin a definition for $\tau_{j\theta i}$. They introduced the following empirical equation from Abramovich (1963) for the velocity profile of a semicontained, one-dimensional, turbulent jet with a co-flowing stream

$$W_i(y) = W_{1i} + (W_{2i} - W_{1i}) \left[1 - \left(\frac{y + y_2}{b} \right)^{1.5} \right]^2. \qquad (5.42a)$$

The coordinate y, the mixing thickness b, and the boundary-layer thickness y_2 are illustrated in Figure 5.15. The axial flow picture of Figure 5.15 is intuitively congruent with flow exiting a one-dimensional jet; however, the applicability of the same model for flow in the circumferential direction is not immediately obvious. Nonetheless, absent a jet-flow model with co-flow and cross-flow, Eq. (5.42a) is assumed to hold for the circumferential direction as

---

*W_{1i} is defined by Eq. (5.10).

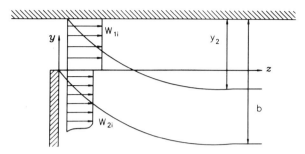

Figure 5.15 Model of semicontained turbulent jet with a co-flowing stream.

well, yielding

$$U_i(y) = U_{1i} + (U_{2i} - U_{1i})\left[1 - \left(\frac{y + y_2}{b}\right)^{1.5}\right]^2. \tag{5.42b}$$

Hence, the *vector* velocity distribution is

$$V_i(y) = V_{1i} + (V_{2i} - V_{1i})\left[1 - \left(\frac{y + y_2}{b}\right)^{1.5}\right]^2, \tag{5.42c}$$

where

$$V_i = KW_i + u_\theta U_i, \text{ etc.} \tag{5.43}$$

Note that the assumed velocity distribution equation (5.42b) defines the unknown U_{0i}; i.e.,

$$U_{0i} = U_i(0) = U_{1i} + (U_{2i} - U_{1i})\left[1 - \left(\frac{y_2}{b}\right)^{1.5}\right]^2.$$

Wyssman et al. assumes $y_2/b = \frac{1}{2}$. Scharrer obtained the same result with an approximate analytical analysis; hence

$$U_{0i} = 0.582U_{2i} - 0.418U_{1i}. \tag{5.44}$$

Wyssman et al. introduced Prandtl's mixing-length hypothesis

$$\tau_{ji} = \rho_i l^2 \left|\frac{\partial V_i}{\partial y}\right| \frac{\partial V_i}{\partial y} = \rho_i \beta^2 b^2 \left|\frac{\partial V_i}{\partial y}\right| \frac{\partial V_i}{\partial y} \tag{5.45}$$

to calculate the free shear stress at the jet boundary. For a free jet, l is normally chosen to be b; hence, $\beta = 1$. For an obstructed jet, β must be calculated. Wyssman et al. did not include W_{2i} in their velocity development and used a two-dimensional CFD solution to arrive at an empirical correlation of β versus C_r/L_i and Rs/L_i. Scharrer included W_{2i} and used an axisymmetric CFD solution to arrive at the simple result

$$\beta = 0.275, \qquad (5.46)$$

independent of C_r/L_i and Rs/L_i.

Further, Scharrer's CFD solutions yielded

$$W_{2i} = 0.206 W_{1i}. \qquad (5.47)$$

Specifically, the *spatial-average* axial velocity in the vortex adjacent to the streamline interface is 20.6% of the average throughflow axial velocity W_{1i}. Equation (5.47) eliminates an additional unknown.

The final unknown $\tau_{j\theta i}$ is obtained by first finding

$$\left| \frac{\partial V_i}{\partial y} \right|_0 = -\frac{3}{b}(V_{1i} - V_{2i})\left[1 - \left(\frac{y_2}{b} \right)^{1.5} \right]\left(\frac{y_2}{b} \right)^{0.5}$$

$$= 1.37(V_{2i} - V_{1i})/b \qquad (5.48)$$

from Eq. (5.42c) with $y_2/b = 0.5$. Next, substituting from Eqs. (5.46) and (5.48) into Eq. (5.45) gives

$$\tau_{ji} = 0.142\rho_i|V_{2i} - V_{1i}|(V_{2i} - V_{1i}). \qquad (5.49)$$

This result is intuitively appealing in that the shear stress between the two control volumes is found to depend on the difference between the bulk-flow velocities adjacent to their interface. Strictly speaking, the shear-stress definition only holds on the stream surface between the throughflow region and the vortex region, not the geometric boundary of Figure 5.13. However, Scharrer cites experimental results which show only about a 6° divergence between the horizontal geometric boundary and the stream surface. The θ and Z components[*] of τ_j are

$$\tau_{jzi} = 0.142\rho_i|V_{2i} - V_{1i}|(W_{2i} - W_{1i}), \qquad (5.50a)$$

$$\tau_{j\theta i} = 0.142\rho_i|V_{2i} - V_{1i}|(U_{2i} - U_{1i}), \qquad (5.50b)$$

where

$$|V_{2i} - V_{1i}|^2 = (W_{2i} - W_{1i})^2 + (U_{2i} - U_{1i})^2. \qquad (5.50c)$$

[*]Florjancic (1990), Chapter IV used both those components for the analysis of circumferentially grooved liquid seals.

In review, cavity i has six unknowns: U_{1i}, U_{2i}, U_{0i}, W_{1i}, W_{2i}, and P_i. The governing equations for these unknowns are:

(a) Combined continuity equations (5.36a) or (5.36b) for control volumes I and II.

(b) Circumferential-momentum equations (5.37a) or (5.37b) for control volume I.

(c) Circumferential-momentum equation (5.38) for control volume II.

(d) Leakage equations (5.10) or (5.13) which define W_{1i} in terms of P_i, P_{i-1}.

(e) Interface circumferential-velocity definition of Eq. (5.44) which defines U_{0i} in terms of U_{1i} and U_{2i}.

(f) The average axial velocity of the vortex which defines W_{2i} as a function of W_{1i}, Eq. (5.47).

The shear stresses $\tau_{r\theta i}, \tau_{s\theta i}, \tau_{j\theta i}$ are defined by Eqs. (5.39) and (5.49). Substitution for $W_{1i}, U_{0i}, \tau_{r\theta i}, \tau_{s\theta i}, \tau_{j\theta i}$ into the continuity and momenta equations yields three equations per cavity in the fundamental variables: U_{1i}, U_{2i}, P_i.

Solution Procedure

Scharrer solves the governing equation via the perturbation-analysis procedure of the preceding section, starting with the perturbation-variable definitions:

$$H = C_r + \epsilon H_1, \qquad U_{1i} = U_{10i} + \epsilon U_{11i},$$
$$P_i = P_{0i} + \epsilon P_{1i}, \qquad U_{2i} = U_{20i} + \epsilon U_{21i},$$

where $\epsilon = e/C_r$ is the eccentricity ratio. Substitution of these definitions into the continuity and momenta equations yields three zeroth-order and first-order equations *for each cavity*. The *system* of zeroth-order equations defines the leakage, pressure, and velocity distributions for a centered seal. The *system* of first-order equations defines P_{1i}, U_{11i}, U_{21i} caused by the clearance perturbation H_1. The solution procedure of the preceding section for both the zeroth- and first-order equations continues to apply for the present two-control-volume model.

Test Results

Tables 5.1 and 5.2 summarize all of the tests which have been performed for labyrinth seals which include the direct damping coefficient C as well as the direct damping and cross-coupled stiffness coefficients K and k. Table 5.1 contains the geometrical data for the seals tested, while Table 5.2 presents the operating conditions. A summary of the results of these tests, an

TABLE 5.1 Labyrinth Seal Tests, Seal Geometry, Dimensions in Millimeters. All Surfaces Are Smooth Except Hawkins (1988), Who Tested a TOR[a] See-Through with a Honeycomb Stator; dc = 1.57 mm, hc = 1.91 mm.

Investigators	NC	Geometry	Pitch L_i	Height B	Tooth Width	Radius	Clearance C_r
Wright (1983)	1	TOR, straight	12.92	5.03	Sharp	101.6	0.1585
		Converging	12.92	4.998, 5.062	Sharp	101.6	0.1915, 0.1272
		Diverging	12.92	5.058, 4.993	Sharp	101.6	0.1311, 0.1963
Kanemitsu and Osawa (1986)	21	TOS[b]—see through	6.0	4.0		80.0	0.50
	21	Interlocking	6.0	4.0		80.0	0.50
Childs and Scharrer (1986)	16	TOS—see through	3.175	3.175	0.152	75.6	0.406
Childs and Scharrer (1986)	16	TOR—see through	3.175	3.175	0.152	72.5	0.406
Childs and Scharrer (1988)	16	TOS—see through	3.175	3.175	0.152	75.6	0.33
							0.40
							0.50
	16	TOR—see through	3.175	3.175	0.152	72.5	0.30
							0.40
							0.55
Hawkins et al. (1988)	16	TOR—see through Honeycomb stator	3.175	3.175	0.152	75.7	0.203
							0.304
							0.406
		TOR—see through Smooth stator	3.175	3.175	0.152	75.7	0.203
							0.304
							0.406
Childs et al. (1988)	11	TOS—see through Interlocking	4.0	4.5	0.5	76.0 73.5	0.25 0.25
Millsaps and Martinez-Sanchez (1990)	1	TOR—Straight	10.0	5.0	N.A.	150	0.813
Pelletti and Childs (1991)	7	TOS	3.175	3.175	0.152	75.58 75.68 75.58	0.229 0.305 0.419
	7	TOR	3.175	3.175	0.152	75.679	0.241 0.318 0.406

[a]Tooth on rotor. [b]Tooth on stator

TABLE 5.2 Labyrinth Seal Tests: Apparatus and Conditions. All Investigators ControlInlet Circumferential Velocity Except Wright (1983). All Tests Are with Air.

Investigators	Apparatus Type	Dynamic Data	ω RPM	$R\omega$ (max) m/sec	P_b bar	P_b/P_s	$P_s - P_b$ bars
Wright (1983)	Unique	$K_\theta(\Omega), K_r(\Omega)$ & K_θ, K_r	1800	19.2	1.02–4.08	0.93–0.357	0.068–3.1
Kanemitsu and Osawa (1986)	Shaft in a shaft	$K_\theta(\Omega), K_r(\Omega)$	1600, 2400	20.1	1.0	0.33–0.20	2–4
Childs and Scharrer (1986)	External shaker	K, k, C	500–8000	63.3	1.6	0.245–0.11	3.1–8.3
Childs and Scharrer (1988)	External shaker	K, k, C	3000–16,000	127.0	1.6	0.248–0.11	3.1–8.3
Hawkins et al. (1988)	External shaker	K, k, C	3000–16,000	127.0	1.6	0.288–0.11	3.1–8.3
Childs et al. (1988)	External shaker	K, k, C	3000–16,000	127.0	1.0	0.288–0.11	3.1–8.3
Millsaps and Martinez-Sanchez (1990)	Shaft in a shaft	k, C	−2500–2500	39.3	1.0	0.676–0.592	0.48–0.69
Pelletti and Childs (1991)	External shaker	K, k, C	3000–16,000	127.0	3.2–12.3	0.67–0.40	6.9, 12.1, 18.3

317

indication of their agreement with predictions, and physical lessons learned from the tests are the subjects of this subsection.

First some points concerning the test approaches. Wright's (1983) test rig is unique in that an active feedback system was used to maintain a constant-radius circular orbit of the test shaft at a frequency Ω near the test rig's natural frequency. Forward- and backward-precessing orbits are generated yielding precession frequencies Ω^+, $-\Omega^-$. Data are presented in the form of the force coefficients

$$K_\theta = F_\theta/A = k - C\Omega,$$

$$K_r = -F_r/A = K + c\Omega, \tag{5.51}$$

where A is the orbit radius.

For most of the cases K_θ^+, K_θ^- and K_r^+, K_r^- are given for forward and backward precession without their accompanying forward- and backward-precessional frequencies. For a limited number of cases, the associated frequencies are also provided, and cross-coupled stiffness and direct damping coefficients can be independently calculated. Complete data sets are only presented for quite low ΔP ranges. For $P_b = 3.1$ bar, $\Delta P = 0.34$ bar, a representative result is

	Converging	Straight	Diverging
$C(NS/\text{mm})$	-0.0094	0.053	0.107
$k(N/\text{mm})$	-12.10	-2.73	0.713
$K(N/\text{mm})$	1175	787	356

The negative damping value cited for the converging seal is unique in the published literature. No comparisons between theory and experiments have been published for Wright's data.

Kanemitsu and Ohasawa (1986)[*] provide complete data sets of K_θ^+, K_θ^-, $K_r^+, K_r^-, \Omega^+, -\Omega^-$ for all their data points. Hence, direct damping and cross-coupled coefficients can be obtained from their results. No comparisons between theory and experiments have been published for their data.

The shaft-in-a-shaft[†] test rigs used by Kanemitsu and Ohasawa (1986), and Millsaps and Martinez-Sanchez (1990) have an inner shaft spinning at the running speed ω and an outer shaft with a controllable running speed Ω. An eccentric of magnitude A between the two shafts causes the inner shaft to spin at ω while precessing in a circular orbit of radius A and precession frequency Ω. *Centered* circular orbits are produced. Both research groups measured the dynamic pressure caused by the circular motion and integrated

---

[*]Unfortunately, this interesting work has not been subsequently published in a more accessible venue.

[†]This apparatus design was first developed by researchers at the California Institute of Technology for impeller testing [Section 6.3].

to get force components which are parallel and perpendicular to the rotating vector $A\boldsymbol{u}_r$.

The apparatus used by Childs and his co-workers at Texas A&M University supports the test-seal housing in a pendulum arrangement and then uses a hydraulic shaker to excite the housing in the horizontal direction. As a result, while the test seal is spinning at the running speed ω, it is oscillated horizontally about the center of the stator. A swept-sine-wave excitation is used, and transient measurements are made of $F_X(t), F_Y(t), X(t)$ in Eq. (5.1). An FFT analysis produces the transfer functions

$$Z = K + jC\omega,$$
$$z = -k - jc\omega,$$

and curve fitting yields the desired coefficients K, C, k.

Note in reviewing Table 5.2 that only Kanemitsu and Ohasawa (1986) and Pelletti and Childs (1991) have independently varied the back pressure P_b and ΔP.

We will not review the data in detail but will instead use it as evidence in examining the following two questions:

(a) How do the rotordynamic coefficients depend on physical parameters?
(b) How good are the theories developed in this section in predicting rotordynamic coefficients?

The next section examines point (a).

5.4   DEPENDENCE OF K, k, C, AND $k/C\omega$ ON PHYSICAL PARAMETERS

As with liquid seals, the rotordynamic coefficients of labyrinth seals depend on many physical parameters. Unfortunately, data are not available to examine the influence of all parameters; e.g., systematic variations in tooth heights or pitches or seal radii have not been considered. Based on the available data, the influence of the following parameters will be considered:

ω: running speed,

$u_0(0) = U(0)/R\omega$: inlet tangential velocity ratio,

Ps: supply pressure,

P_b/Ps: pressure ratio,

$\Delta P = Ps - P_b$: pressure differential,

NC: number of cavities.

Comparisons will also be made between alternative seal configurations, e.g., teeth-on-rotor (TOR) and teeth-on-stator (TOS) seals. Only K, k, and C will be discussed. The cross-coupled damping coefficient c tends to be both poorly defined by test data and (fortunately) of little influence in rotordynamics.

K: Direct Stiffness for See-Through Seals

As noted in the introduction to this chapter, the direct stiffness for labyrinth seals is generally quite low and tends to have a negligible influence on rotordynamics of compressors or turbines. Sometimes, rotordynamic calculations for back-to-back compressors predict that negative stiffness values for the center labyrinth will have a perceptibly bad influence on rotordynamics of the first bending mode. For throughflow compressors, predicted negative magnitudes for K have a minor influence. The mode-shape locations for the two seals accounts for this difference.

Direct stiffness depends strongly on the number of chambers. Single-cavity labyrinths normally have modest positive stiffness values. For NC greater than 5, K is almost always negative.

K generally decreases (becomes more negative) as running speed increases.

Childs and Scharrer (1988) reported K increasing (smaller negative values) with increasing C_r; however, Pelletti and Childs (1991) showed no clear trend with respect to changes in C_r.

Concerning ΔP, for longer seals all cited results show K decreasing as ΔP increases (or pressure-ratio decreases).

No clear trends exist concerning the dependency of K on $u_0(0)$, the inlet circumferential-velocity ratio. Childs and Scharrer (1988) show K generally having a minimum around $u_0(0) = 0$ and increasing as $u_0(0)$ is either decreased or increased. Pelletti and Childs show no consistent trend with changes of $u_0(0)$.

k: Cross-Coupled Stiffness for See-Through Seals

Because of its destabilizing influence, k tends to be the most important rotordynamic coefficient. Except for Kanemitsu and Osawa, all of the test results cited show k varying more or less linearly with $u_0(0)$. However, Childs and Scharrer (1988) and Pelletti and Childs show a reduction of this effect as running speed increases. Figure 5.16 illustrates this effect. Note that k is generally negative for $u_0(0) = 0$.

Figure 5.16 also shows an increasing dependence of k on $P_{ra} = P_b/P_s$ as the supply pressure increases. At elevated supply pressures, k increases with increasing P_{ra} (decreasing ΔP). P_{ra} is increased by holding the supply pressure constant and elevating the back pressure. Hence increasing P_{ra} implies an increase of the pressure (and density) within the seal. This result

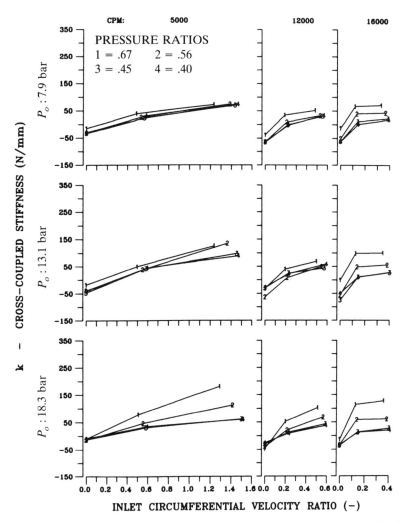

Figure 5.16 k versus $u_0(0)$ with P_b/Ps as a parameter for a TOS seal, $C_r/Rs =$ 3.0×10^{-3} [Pelletti (1990)].

supports the suggestion of Kirk and Donald (1983) that the stability of high-pressure compressors depends directly on both ΔP and discharge pressure. More specifically, compressors tend to become unstable as their discharge density increases.

Pelletti and Childs showed k increasing with increases in P_s for preswirled flow and higher pressure ratios (reduced ΔP).

Childs and Scharrer (1988) showed that k increased with increasing running speed for TOR seals and decreased with increasing running speed for TOS seals; $L/D = \frac{1}{3}$. Childs et al. (1986) obtained the same results for

TOS seals also with $L/D = \frac{1}{3}$. Hisa et al. (1986) reported a similar result for a short, three-cavity, TOS steam turbine seal. For a shorter seal ($L/D = \frac{1}{6}$), Pelletti and Childs (1991) showed an increase in k for both TOR and TOS seals with increasing ω; however, the result is most pronounced at higher pressure ratios (lower ΔP's) and lower supply pressures.

Pelletti and Childs show a decrease in k with decreasing clearance for a TOS seal at 16,000 rpm. For lower speeds the seal was insensitive to changes in clearance. Their TOR seals showed k decreasing as the clearance was increased. Childs and Scharrer (1988) showed no clear trends in k with changes in C_r.

Benckert and Wachter (1980a) show an increase in k with increasing NC.

C: Direct Damping for See-Through Seals

Damping values for labyrinth seals are typically low, but are of significant importance to the extent that they counteract k.

Pelletti and Childs show C increasing with decreasing pressure ratio (increasing ΔP, decreasing average density) for both TOR and TOS designs. They also show a reduction in C with increasing clearances. No clear relationship is found between C and ω or between C and $u_0(0)$.

No data are available for direct comparison between labyrinth seals looking at the influence of seal length (or NC) on direct damping. Unfortunately, the data of Childs and Scharrer (1988) and Pelletti and Childs (1990) are taken at markedly different back pressures.

$\Omega_w = k/C\omega$: Whirl-Frequency Ratios for See-Through Seals

The whirl-frequency ratio is remarkably useful in characterizing the stability of a labyrinth seal; for a synchronous circular orbit it is the ratio of the destabilizing force due to k and the stabilizing force due to C. Figure 5.17 illustrates Ω_w versus $u_0(0)$ with pressure ratio as a parameter for a TOS seal with three clearances. Figure 5.18 shows the same results for a TOR seal with three clearances. In all cases, Ω_w increases with increasing $u_0(0)$; however, the effect is moderated for increasing clearances. At higher speeds and tighter clearances Ω_w tend to approach 0.5. The whirl-frequency ratio increases as the pressure ratio increases (decreasing ΔP, increasing density). Increasing the seal clearance increases Ω_w and degrades stability for both TOS and TOR seals. These findings for the TOR seal are in conflict with Childs and Scharrer (1988), who found that an increase in clearance improved the stability of a longer ($L/D = \frac{1}{3}$) TOR seal with atmospheric back pressure.

Labyrinth-Seal Configuration Comparisons

Test results to be used in comparing *labyrinth* configurations are not in complete agreement. There is agreement that TOS labyrinths are modestly

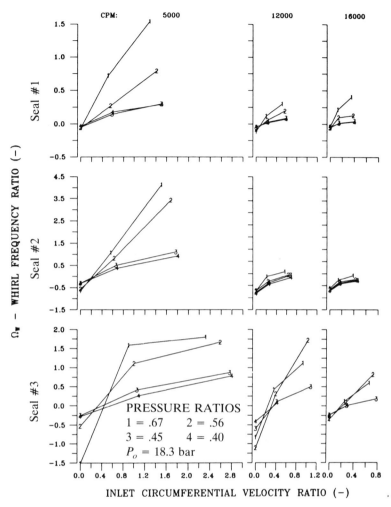

Figure 5.17 $k/C\omega$ versus $u_0(0)$ for a TOS seal with three clearances and three values for $P_{ra} = P_b/P_s$ [Pelletti (1990)].

more stable than TOR labyrinths. However, there is disagreement between the relative rotordynamic virtues of see-through and interlocking seals. Kanemitsu and Osawa report direct damping values with $\Delta P = 0.4$ MPa of C (interlocking) = 480 NS/mm and C (see-through) = 170 NS/mm. Conversely, Childs et al. (1988) cite higher direct damping values (by a factor of 2) for a see-through versus an interlocking labyrinth.

In contrast to the results of Section 5.7 for plain seals, Hawkins et al. (1988) found that labyrinths with honeycomb stators are no more stable than labyrinths with smooth stators.

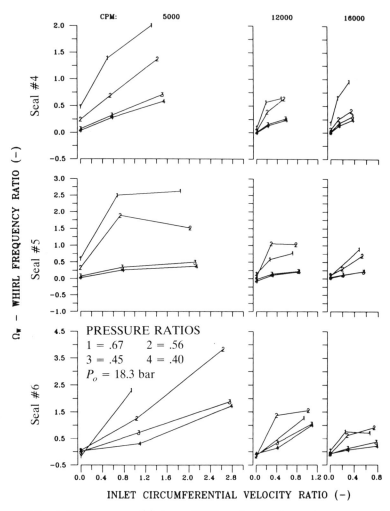

Figure 5.18 $k/C\omega$ versus $u_0(0)$ for a TOR seal with three clearances and three values for $P_{ra} = P_b/Ps$ [Pelletti (1990)].

Summary and Overview of Measured Results

The *important* and agreed-upon facts concerning the rotordynamic coefficients of labyrinth seals are as follows:

(a) For more than five cavities, the direct stiffness will be negative and increasingly negative as the number of cavities increases. For a back-to-back compressor, the predicted negative stiffness for the center labyrinth can adversely impact rotordynamics. For a series (or through-flow) compressor it does not generally make any difference.

(b) The cross-coupled stiffness coefficient increases with the number of cavities and is mainly a function of the inlet tangential velocity. Decreasing the inlet velocity with a swirl brake is the best way to eliminate k. The cross-coupled stiffness coefficient increases with increasing density of the gas within the seal.

(c) The damping coefficient is small but *quite* important for rotordynamic calculations. Absent the damping, rotordynamics analysis predicts instability for many stably operating compressors.

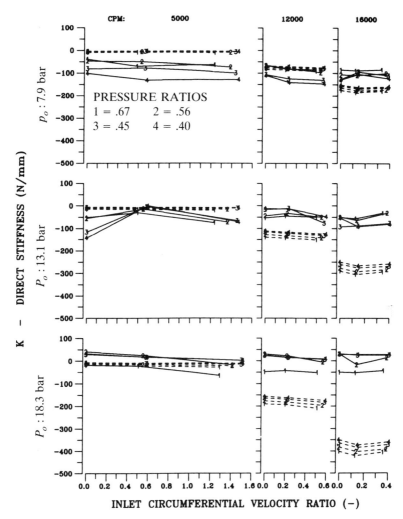

Figure 5.19 K versus $u_0(0)$ for a TOS seal with three speeds, three supply pressures, and four pressure ratios; $C_r/Rs = 3.0 \times 10^{-3}$ [Pelletti (1990)]; theory (dashed), measurement (solid).

5.5 BULK-FLOW MODEL PREDICTIONS VERSUS EXPERIMENT FOR LABYRINTH SEALS AND ALTERNATIVE ANALYSIS PROCEDURES

Reversing the normal narrative order, the answer to the question, "How do the predictions from the theory of Section 5.2 compare to measured results?", is fairly well at larger clearances ($C_r/R \cong 5.5 \times 10^{-3}$), lower supply pressures, and lower running speeds. Reasonably favorable comparisons between theory and experiment are given by Childs and Scharrer (1988), Millsaps and

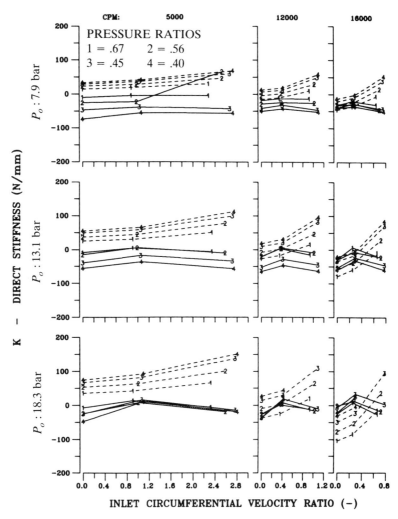

Figure 5.20 K versus $u_0(0)$ for a TOS seal with three speeds, three supply pressures, and four pressure ratios; $C_r/Rs = 5.5 \times 10^{-3}$ [Pelletti (1990)]; theory (dashed), measurement (solid).

Martinez-Sanchez (1990), Hawkins et al. (1988), and Thieleke and Stetter (1990). However, poorer correlations and systematic errors are reported by Pelletti and Childs (1991) and Pelletti (1990) at reduced clearances as speeds and supply pressures are increased. Most of the subsequent results considered here are from Pelletti's work.

Concerning static characteristics, the theory does a good job of predicting the pressure distribution for Pelletti's seals, ($L = 25.4$ mm, $D = 152.4$ mm, $L/D = \frac{1}{6}$) but underpredicts leakage rates by 5 to 25%. Childs and Scharrer

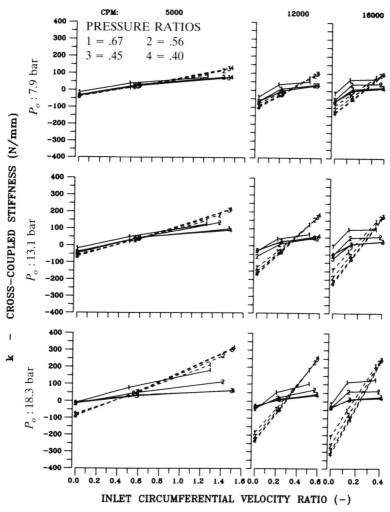

Figure 5.21 k versus $u_0(0)$ for a TOS seal with three speeds, three supply pressures, and four pressure ratios; $C_r/Rs = 3.0 \times 10^{-3}$ [Pelletti (1990)]; theory (dashed), measurement (solid).

(1988) found the leakage underpredicted by about 25%. A substantial portion of this error is caused by Scharrer's (1988) "clearance-dependent" kinematic carryover coefficient of Eq. (5.35). Using the constant carryover coefficient definition of Eq. (5.11), the Childs and Scharrer (1986a) predictions of leakage were within 5%. Scharrer is knowingly sacrificing leakage-prediction accuracy to introduce a clearance-dependent carryover factor.

Figures 5.19 and 5.20 illustrate K versus $u_0(0)$ for $C_r/Rs = 0.003$ and 0.0055, respectively, for a range of running speeds, supply pressures, and

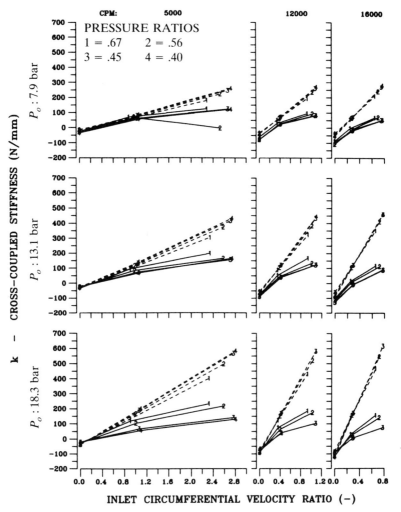

Figure 5.22 k versus $u_0(0)$ for a TOS seal with three speeds, three supply pressures, and four pressure ratios; $C_r/Rs = 5.0 \times 10^{-3}$ [Pelletti (1990)]; theory (dashed), measurement (solid).

pressure ratios. For $C_r/Rs = 0.0055$, the results are reasonable for all cases. However, for $C_r/Rs = 0.003$ there is a systematic deviation between theory and experiment for increasing supply pressure and running speed. Parallel results for a TOR seal are similar but not as bad.

Figures 5.21 and 5.22 show k versus $u_0(0)$ for $C_r/Rs = 0.003$ and 0.0055, respectively, for a range of running speeds, supply pressures, and pressure ratios. As with K, note that the theoretical predictions are reasonable for the larger clearances but poor for the tight-clearance seals. Again, there is a

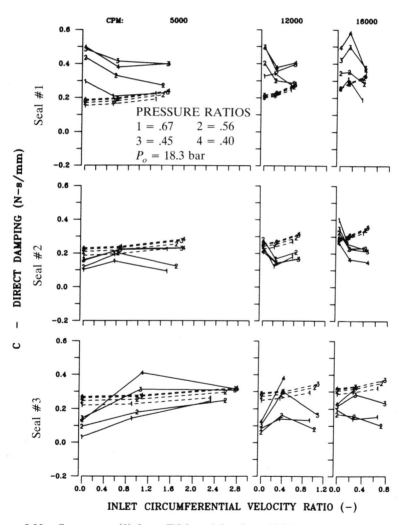

Figure 5.23 C versus $u_0(0)$ for a TOS seal for $Ps = 18.3$ bars, three speeds, three clearances, and four pressure ratios [Pelletti (1990)]; theory (dashed), measurement (solid). Clearances for seals 1, 2, and 3 are 0.229, 0.305, and 0.419 mm, respectively.

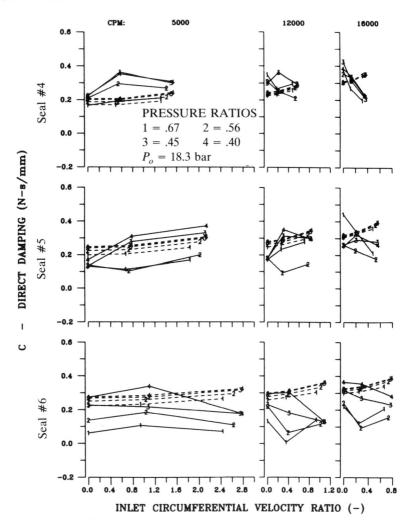

Figure 5.24 *C* versus $u_0(0)$ for a TOR seal for $Ps = 18.3$ bars, three speeds, three clearances, and four pressure ratios [Pelletti (1990)]; theory (dashed), measurement (solid). Clearances for seals 1, 2, and 3 are 0.241, 0.318, and 0.406 mm, respectively.

systematic divergence between theory and experiment as speed and supply pressure increase. Comparable results are obtained for TOR seals.

Figures 5.23 and 5.24 show *C* versus $u_0(0)$ for $P_s = 18.3$ bars, three clearances, three speeds, and four pressure ratios. The theory underpredicts damping at the large clearance but does a "reasonable" job overall. From these results, the theory provides reasonable predictions for C_r/Rs on the order of 0.005. For tight clearances, high speeds, and high pressures, the model seriously underestimates *C*.

Alternative Calculation Approaches

The comparisons considered in this section are not encouraging at reduced clearances, higher speeds, and supply pressures. Alternative models which have been proposed for labyrinths include the work of Nordmann et al. (1987) and Nordmann and Weiser (1988), who use a perturbation of the Navier-Stokes (NS) equations, and Weiser and Nordmann (1989), who perform a direct solution to the NS equations for a centered circular orbit. Finally, Nordmann and Weiser (1990) have presented a three-control-volume model for gas labyrinths that is quite similar to Florjancic's (1990) model for liquid labyrinth seals. Unfortunately, no systematic comparisons between theory and experiment have been published for any of these approaches.

5.6 BULK-FLOW MODELS FOR PLAIN ANNULAR GAS SEALS WITH SMOOTH OR HONEYCOMB STATORS

As noted in the introduction of this chapter, smooth-bore seals are sometimes used as floating-ring seals and smooth-rotor honeycomb stators are used as direct replacements for TOS labyrinth seals. This section focuses on models, analysis, and test results for these types of seals.

Bulk-Flow Governing Equations

The initial analysis for annular gas seals were given by Fleming (1979), (1980) who used a one-dimensional duct-flow analysis to separately calculate K and C. Nelson (1984), (1985) provided the first complete analysis for the coefficients, and this section is based on his work.

For the control volume of Figure 4.4, the continuity equation is

$$\frac{\partial(\rho H)}{\partial t} + \frac{\partial(\rho W H)}{\partial Z} + \frac{\partial(\rho U H)}{R\,\partial\theta} = 0. \tag{5.52}$$

After simplification using the continuity equation, the momenta equations are

$$-H\frac{\partial P}{\partial Z} = \frac{\rho}{2}WU_s f_s + \frac{\rho}{2}WU_r f_r + \rho H\frac{DW}{Dt}, \tag{5.53}$$

$$-\frac{H}{R}\frac{\partial P}{\partial\theta} = \frac{\rho}{2}UU_s f_s + \frac{\rho}{2}(U - R\omega)U_r f_r - \rho H\frac{DU}{Dt}, \tag{5.54}$$

where

$$\frac{D}{Dt} = \frac{\partial}{\partial t} + \frac{U}{R}\frac{\partial}{\partial\theta} + W\frac{\partial}{\partial Z}. \tag{5.55}$$

These momenta equations basically coincide with Eqs. (4.2b) and (4.2c) for incompressible fluids except that $\rho(Z, \theta, t)$ is now a dependent variable instead of a constant.

For adiabatic flow the energy equation is

$$0 = \frac{\partial(e\rho H)}{\partial t} + \frac{1}{R}\frac{\partial(e\rho HU)}{\partial \theta} + \frac{\partial(e\rho HW)}{\partial Z} + \frac{R\omega}{2}(U - R\omega)\rho f_r, \quad (5.56)$$

where

$$e = c_v T + \frac{U^2}{2} + \frac{W^2}{2}$$

is the internal energy per unit mass, c_v is the specific heat at constant volume, and T is the temperature. The friction factors are defined using either the Blasius- or Moody-equation definitions of Section 4.2.

The gas within the seal is assumed to be governed by the perfect gas law

$$c_v T = \frac{P}{\rho(\gamma - 1)}, \quad (5.57)$$

where γ is the ratio of specific heats. Equation (5.57) can be used to eliminate temperature as a variable. From Eq. (5.57) and the continuity equation (5.52), the reduced energy equation is

$$\frac{P}{H}\frac{\partial H}{\partial t} = U\left(\frac{1}{R}\frac{\partial P}{\partial \theta} + \rho\frac{DU}{Dt}\right) + W\left(\frac{\partial P}{\partial Z} + \rho\frac{DW}{Dt}\right)$$

$$+ \frac{1}{\gamma - 1}\left(\frac{DP}{Dt} - \frac{\gamma P}{\rho}\frac{D\rho}{Dt}\right) + \frac{R\omega}{2}(U - R\omega)f_r. \quad (5.58)$$

Reviewing, the dependent variables are U, W, P, and ρ, and the governing equations are Eqs. (5.52)–(5.54) and (5.58).

Boundary Conditions

Fleming (1979) introduced the upstream pressure-loss boundary conditions as

$$P_s - P(0) = k\rho(0)W^2(0)/2, \quad (5.59)$$

with the entrance loss coefficient to be defined by Deissler's (1953) empirical relationship:

$$k = [5.3/\log_{10} \mathscr{R}_e(0)]^{1/2} - 1,$$

$$\mathscr{R}_e(0) = \frac{2\rho(0)W(0)H(0)}{\mu}. \tag{5.60}$$

From the Mach number definition

$$M = W\sqrt{\frac{\rho}{\gamma P}} \tag{5.61}$$

and the energy equation between upstream supply pressure Ps (reservoir or stagnation pressure), the entrance boundary conditions are

$$p_0(0) = \frac{P_0(0)}{Ps} = \frac{1}{\left[1 + (\gamma - 1)(1 + k_1)M_0^2(0)/2\right]^{\gamma/(\gamma-1)}}, \tag{5.62}$$

$$\tilde{\rho}_0(0) = \frac{\rho_0(0)}{\rho_s} = \frac{\left[1 + (\gamma - 1)M_0^2(0)/2\right]}{\left[1 + (\gamma - 1)(1 + k_1)M_0^2(0)/2\right]^{\gamma/(\gamma-1)}}. \tag{5.63}$$

The inlet circumferential velocity $U(0) = U_0$ is specified, which completes the inlet boundary conditions.

At the exit, if flow is choked the boundary condition

$$M(L) = 1 \tag{5.64}$$

holds. For unchoked flow, one has

$$P(L) = P_b. \tag{5.65}$$

Analysis

Nelson nondimensionalizes the problem by introducing the variables

$$u = U/R\omega, \qquad w = W/R\omega,$$
$$p = P/P_s, \qquad \tilde{\rho} = \rho/\rho_s,$$
$$h = H/\overline{C}, \qquad \overline{C}_r = (C_0 + C_1)/2,$$
$$\tau = t\omega, \qquad z = Z/L, \tag{5.67}$$

and then introduces a perturbation about the centered position defined by the variables

$$h = h_0(z) + \epsilon h_1,$$

$$w = w_0 + \epsilon w_1, \qquad u = u_0 + \epsilon u_1,$$

$$p = p_0 + \epsilon p_1, \qquad \tilde{\rho} = \tilde{\rho}_0 + \epsilon \tilde{\rho}_1. \tag{5.68}$$

Substitution into Eqs. (5.52)–(5.54) and (5.58) gives zeroth- and first-order equations which are quite similar to the equations of Section 4.3.

The zeroth-order equations define $w_0(z)$, $u_0(z)$, $p_0(z)$, and $\tilde{\rho}_0(z)$ and are solved iteratively, given P_s, P_b, and U_0. The first-order equations define $w_1(z,\theta,\tau)$, $u_1(z,\theta,\tau)$, $p_1(z,\theta,\tau)$, and $\tilde{\rho}_1(z,\theta,\tau)$ resulting from the clearance perturbation $h_1(\tau)$. Nelson eliminates the θ and time dependency by assuming an elliptical orbit for the seal rotor at running speed ω, similar to the approach of Eqs. (5.19) and (5.20). This approach yields 16 real equations in z. First-order boundary conditions are obtained by specifying $u_1(0) = 0$ and perturbing Eqs. (5.61), (5.63), and either (5.64) or (5.65). The transition-matrix approach described in Section 4.5 is used to obtain a first-order solution which satisfies the boundary conditions. Finally, the dynamic force coefficients are obtained by integrating the pressure perturbation solution.

Additional and Alternate Models and Analyses

Nelson's model applies to a constant-clearance or convergent-tapered seal with arbitrary rotor and stator relative roughness. The analysis applies for small motion about a centered position. The convergent-taper geometry is of interest because Fleming predicted that a tapered seal would be substantially stiffer than a constant-clearance seal. The ability to specify different friction factors for the rotor and stator surfaces is included to account for the presumed higher stator roughness of a honeycomb-stator seal.

Nelson et al. (1990) and Dunn (1990) extended Nelson's analysis to the stepped seal of Figure 5.25. This seal design was initially proposed by Fleming (1977), whose analysis demonstrated that it would be stiffer than a constant-clearance seal.

The CFD analyses of Nordmann et al. (1987) and Weiser and Nordmann (1989) clearly apply for plain seals as well as labyrinth seals.

Scharrer and Nelson (1991) extended Nelson's original analysis for a tapered seal to a "partially tapered" seal. The clearance for this seal begins with a convergent taper and merges into a constant cross-section at some point (generally near the exit) of the seal's axis. This configuration was of interest because it was predicted to have a direct stiffness comparable to a conventional tapered seal with enhanced wearout resistance.

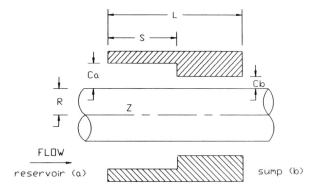

Figure 5.25   Stepped seal [Dunn (1990)].

5.7   THEORY VERSUS EXPERIMENTS AND COMPARATIVE PERFORMANCES OF PLAIN ANNULAR GAS SEALS

As noted previously, smooth annular gas seals are normally used as floating ring seals. Their direct stiffnesses should be large enough to cause the seal to center itself. The convergent-taper and stepped-seal geometries were proposed specifically by Fleming (1977) to provide larger values for K and enhanced centering capability. Nelson et al. (1986) initially presented K, k, and C data for constant-clearance and convergent-tapered seals with the geometry and operating conditions of Table 5.3. The correlation between theory and experiment for the direct stiffness coefficient K is only fair. The theory predicts a weak dependency of K on the inlet circumferential velocity ratio

$$u_0(0) = U_0(0)/R\omega,$$

whereas the tests showed a strong sensitivity. Test results confirmed that the tapered seal had a higher direct stiffness than the constant-clearance seal; however, forward prerotation of the flow reduced this effect. Cross-coupled stiffness was reasonably predicted. Direct damping was very well predicted for the constant-clearance seal and underpredicted for the convergent-tapered

TABLE 5.3   Test Data for Nelson et al. (1986).

R = 76.2 mm,	L = 50.8 mm,	C_1 = 0.737 mm
C_0 = 0.737 mm, 1.114 mm		
P_s = 1.7–7.2 bar,	P_b = 1.0 bar	
ω = 2000–8000 rpm		
Test fluid: Air at 305°K (37°C)		

seal as the supply pressure was increased ($C_{th}/C_{meas} = 0.67$ for $P_b/P_s = 0.188$).

Dunn (1990) showed reasonable qualitative agreement between theory and experiment for stepped seals. Introducing a step at the center of the seal increased leakage. Inlet to exit clearance ratios of 1.21 and 3.5 yielded direct stiffness values which were, respectively, higher and much lower than a constant-clearance seal. Figures 5.26, 5.27, and 5.28 provide theory-versus-experiment comparisons for a smooth constant-clearance seal with $R =$

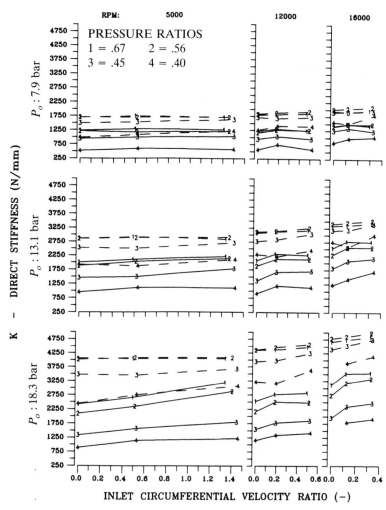

Figure 5.26 Theory (dashed) versus measurement (solid) for a smooth constant-clearance seal. K versus $u_0(0)$ for three speeds, three supply pressures, and four pressure ratios [Dunn (1990)].

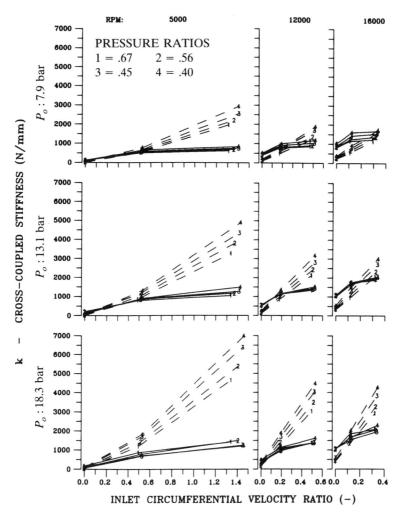

Figure 5.27 Theory (dashed) versus measurement (solid) for a smooth, constant-clearance seal. k versus $u_0(0)$ for three speeds, three supply pressures, and four pressure ratios [Dunn (1990)].

76.2 mm, $L = 50.8$ mm, and $C_0 = C_1 = 0.2786$ mm. K is generally overpredicted by the theory. The cross-coupled stiffness coefficient is reasonably predicted by theory; however, k is predicted to increase linearly with increasing $u_0(0)$, which is contrary to measured results at higher values of $u_0(0)$. C is reasonably predicted except at the highest supply pressure, where it is substantially overpredicted. Predictions for the rotordynamic coefficients of stepped seals were generally poorer than for the constant-clearance seal.

To be direct, theory-versus-experiment experiences for honeycomb seals have been generally poor and disappointing. Elrod et al. (1989, 1990) altered

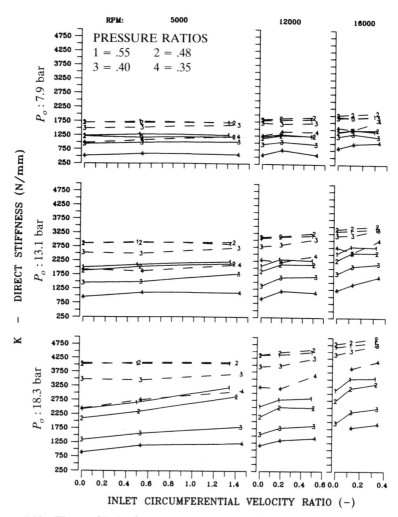

Figure 5.28 Theory (dashed) versus measurement (solid) for a smooth, constant-clearance seal. C versus $u_0(0)$ for three speeds, three supply pressures, and four pressure ratios [Dunn (1990)].

Nelson's basic model by first (1989) using the vector definition for the Mach number,

$$M^2 = \frac{(U^2 + W^2)\rho}{\gamma P},$$

instead of the unidirectional definition of Eq. (5.61). Next (1990), an empirical definition was introduced for the friction factor variation along the seal based on measurements for a nonrotating seal. These measurements showed a high friction factor at the inlet which dropped rapidly through the first quarter, remained stationary through the midsection, and then dropped rapidly through the exit quarter of the seal. While these modifications improved the k predictions for specific seals (1990), they did not improve (the generally poor) predictions for K, and gave about the same (adequate) predictions for C. For other honeycomb-seal configurations, Childs et al. (1990), the predictions were generally poor for all rotordynamic coefficients.

Measurement of rotordynamic coefficients for honeycomb annular seals has yielded erratic and puzzling results. Childs et al. (1989) presented test results for seven honeycomb stators while varying cell sizes and depths and provided comparisons to results for a smooth and a 15-cavity-TOS labyrinth seal with comparable dimensions ($L = 50.8$ mm, $R = 151.36$ mm, $C_r = 0.41$ mm). The honeycomb seals had comparable direct damping and direct stiffness values (for forward prerotated flow) to a smooth seal. The labyrinth seal had negative values for K and much lower C values than either the smooth or honeycomb seals. The honeycomb seals typically had larger k values than the labyrinth in agreement with Benckert and Wachter (1980a). *The honeycomb seal's whirl frequency ratio* $\Omega_w = k/C\omega$ *was much better (lower) than the smooth or labyrinth seal for prerotated flow.*

Hence (as predicted) the honeycomb stators reduce the average circumferential velocity and decrease k. However, there was a large variation in results for the honeycomb seals. Small changes in cell dimensions yielded large changes in rotordynamic coefficients. Also, the honeycomb dimensions which yielded the minimum leakage (and presumably a minimum $u_0(z)$) did not give the lowest whirl-frequency ratio.

Kleynhans (1991) tested shorter smooth and honeycomb-stator seals ($L = 25.4$ mm, $R = 151.36$ mm). The smooth seal was tested at $C_r = 0.23$ and 0.30 mm. The three honeycomb stators used a cell depth* of 2.29 mm and three cell widths,* $d_c = 0.40, 0.79$, and 1.59 mm. Comparisons are made to Pelletti's (1990) measurements for a seven-cavity tooth-on-stator labyrinth. For these shorter seals with $C_r = 0.30$ mm Kleynhans found that the best ($d_c = 0.40$ mm) honeycomb seals had lower k values than either the smooth or the labyrinth seals, had damping which was comparable or slightly less than the smooth seal but much higher than the labyrinth, and had a lower whirl-frequency ratio. At the tighter clearance of 0.23 mm, the results were mixed. At the lowest speed of 5000 rpm, the honeycomb seal had lower whirl-frequency-ratio values; however, at 12,000- and 16,000-rpm values there was no clear "winner" between the three seal configurations.

The fact that Childs et al. (1989) found honeycomb-stator seals to be much superior in rotordynamic performance to smooth and labyrinth seals at $L/D = \frac{1}{3}$, while Kleynhans found a mixed comparison at $L/D = \frac{1}{6}$ has a message similar to the predictions of Figure 4.21 for liquid damper seals.

---

*See Figure 5.4.

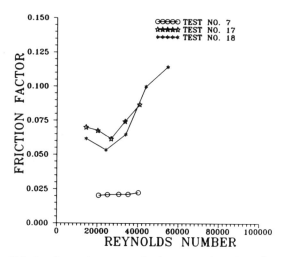

Figure 5.29   Friction-factor-jump case for honeycomb surfaces [Ha et al. (1991)].

Specifically, the roughened stator surface can be very effective for long seals, but ineffective for shorter seals.

To try to explain the erratic behavior of honeycomb-stator seals, a flat-plate tester was developed to measure the friction factors for flow between apposed parallel honeycomb surfaces [Ha and Childs (1991a, 1991b)]. Friction-factor test results for honeycomb test surfaces showed the same erratic dependency on cell sizes, depths, and clearances that rotordynamic results had shown for honeycomb seals. Figure 5.29 illustrates one of the puzzling results for these test series, showing an increase (or jump) in the friction factor for increasing Reynolds* number. This phenomenon was present in 15 out of 39 test cases.

An explanation for the phenomenon was obtained by placing a miniature pressure transducer at the base of a honeycomb cell and examining the associated frequency spectra of the transient pressure signal [Ha et al. (1991)]. The spectra showed that the friction-factor-jump-phenomenon cases were always accompanied by frequency peaks at the fundamental (one end open, one end closed organ pipe) frequency of the individual cavity tubes *and harmonics of this frequency*. The fundamental frequency itself was present in friction-factor cases which did not include the jump. The presence of harmonics of the fundamental frequency suggests that the resonance excitation becomes nonlinear and sufficiently large to retard the main flow. While Ha et al. are able to explain the friction-factor jump cases, they are not able to predict if or for what conditions they will occur. *Given the uncertainty of the friction-factor models and the strong dependency of the rotordynamic coefficients upon them, rotordynamic predictions for honeycomb seals remain problematical and should be anchored by test data.*

---

*Twice the clearance is the characteristic length.

Nonetheless, honeycomb-stator/smooth-rotor seals provide a remarkable opportunity for stabilizing rotors by replacing labyrinth seals. They are rugged, withstanding impacts that would destroy a labyrinth. Moreover, they leak less than a tooth-on-stator labyrinth.

5.8 BRUSH SEALS

As noted in the introduction of Section 5.1, the brush seal of Figure 5.7 uses a biased pattern of wire bristles in contact with a ceramic coated shaft. This type of seal is remarkably effective in reducing leakage. Ferguson (1988) reported that the leakage of a single-stage brush seal was only 5% of a four-cavity, tooth-on-rotor labyrinth. An additional advantage of the brush seal is its compliant nature, which minimizes damage due to transient impact between the housing and rotor.

The side view of Figure 5.7 shows that the bristles are sandwiched between a front plate (high-pressure side) and a back plate (low-pressure side). They are variously cited as tolerating ΔP's between 3.4 and 6.8 bars. Higher ΔP's can permanently deform bristles.

Conner and Childs (1990) reported rotordynamic-coefficient measurements for a four-stage brush seal with a radius of 64.7 mm. Comparisons are made to an eight-cavity tooth-on-stator (TOS) labyrinth with radius 75.68 mm, radial clearance 0.24 mm, and equal cavity lengths and tooth heights of 3.8 mm at a supply pressure of 18.3 bar and pressure ratio of 0.55.

Figure 5.30 compares K for the two seals. Note that K is quite small but positive for the brush seal, while it is negative for the labyrinth at higher running speeds.

Figure 5.31 compares k for the two seals. Note particularly that while k increases monotonically with $u_0(0)$ for the labyrinth, it is low and indepen-

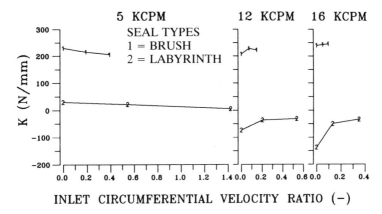

Figure 5.30 Direct stiffness K versus $u_0(0)$ for a four-stage brush seal and an eight-cavity labyrinth [Conner and Childs (1990)].

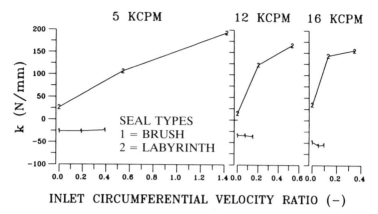

Figure 5.31 Cross-coupled stiffness k versus $u_0(0)$ for a four-stage brush seal and an eight-cavity labyrinth [Conner and Childs (1990)].

dent of $u_0(0)$ for the brush seal. The low value and absence of dependence on $u_0(0)$ is an obvious and immediate advantage for the brush seal.

Figure 5.32 illustrates C for the two seals. The brush seal damping is generally lower. Figure 5.33 illustrates the whirl-frequency ratio for the two seals, and (predictably) the whirl-frequency ratio of the brush seal is negative or nearly zero, independent of running speed or inlet tangential velocity ratio.

The Conner and Childs data are the only published results for brush seals and are preliminary to the extent that they do not address the questions, "What are the rotordynamic coefficients of a single-stage brush seal, or how much will the coefficients change by adding or deleting a stage?" However,

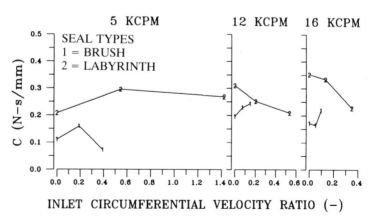

Figure 5.32 Direct-damping C versus $u_0(0)$ for a four-stage brush seal and an eight-cavity labyrinth [Conner and Childs (1990)].

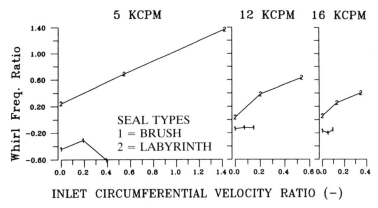

Figure 5.33 Whirl-frequency ratio versus $u_0(0)$ for a four-stage brush seal and an eight-cavity labyrinth [Conner and Childs (1990)].

their results suggest that replacing a labyrinth with a brush seal will not degrade rotordynamics. To the extent that a labyrinth is developing a large cross-coupled stiffness due to $u_0(0)$, a brush seal may materially enhance rotordynamics.

At present, no analysis for the rotordynamic coefficients of brush seals has been proposed.

5.9 SUMMARY, CONCLUSIONS, AND DISCUSSION

Introduction

This chapter has presented a considerable volume of material on annular gas seals. The present section provides a perspective of this material and a discussion based on specific questions related to the rotordynamic contributions of annular gas seals. We begin with the following brief section, which covers the known and accepted facts, and will then proceed to a discussion of a range of questions related to these seals.

Known General Results

As stated at the outset of this chapter, annular gas seals have a major impact on the *stability* of high-pressure compressors and steam turbines. Many load-dependent instability problems have been attributed to labyrinth seals whose rotordynamic coefficients are proportional to both the seal's ΔP and the average density of the fluid within the seal. Annular gas seals normally have significantly smaller direct stiffness coefficients than annular liquid seals and negligible added-mass coefficients. Labyrinths with more than five cavities will generally have negative direct stiffness coefficients. A long center labyrinth in a back-to-back compressor may have a large enough negative K value to reduce the critical speed and degrade stability. Otherwise, annular-seal *direct stiffness coefficients* are not likely to make a significant impact on either the critical-speed locations or stability characteristics of compressors or turbines.

The cross-coupled stiffness k coefficients of high-pressure centrifugal compressors and steam turbines are extremely important elements in establishing rotordynamic stability or instability. While they are small in comparison to their hydrodynamic-bearing or liquid-pump-seal counterparts, they are located at potentially large modal locations. For units operating on tilting-pad bearings at high speeds (in comparison to the first critical speed), annular-gas-seal k values can easily be the difference between stable or unstable operation.

Direct damping coefficients are small but very important for all annular seals including labyrinths. Rotordynamics analyses for high-pressure compressors which include only the cross-coupled coefficients will yield excessively pessimistic predictions.

Clarifying Questions and Rotordynamic Issues

1 What analytical models are available for labyrinth seals? The three bulk-flow models which have been developed for labyrinth seals are

(a) **One-control-volume model.** Childs and Scharrer (1986) present this type of model which uses a single control volume for the circumferential flow within a labyrinth cavity. Their model uses a leakage equation with constant (eccentricity-independent) flow coefficients. Kurohashi et al. (1980) used an eccentricity-dependent flow coefficient as did Thieleke and Stetter (1990).

Scharrer (1988) extended this analysis to stepped labyrinths. Rajakumar and Sisto (1984) developed eccentricity-dependent rotordynamic motion by considering small motion about an arbitrary eccentric position.

(b) **Two-control-volume model.** Wyssman et al. (1984) initially proposed this type of model, using two control volumes for the circumferential flow in a cavity. One control volume is for the throughflow regime; the other is for the vortex region between labyrinth teeth. Scharrer (1988) extended this model to define the average vortex velocity as a function of the throughflow velocity. He also accounted for the axial component of the vortex velocity in defining the shear stress and used an eccentricity-dependent kinetic-carry-over coefficient.

(c) **Three-control-volume model.** Nordmann and Weiser (1990) have developed a three-control-volume model for labyrinths which is patterned after Florjancic's (1990)* model for liquid labyrinth seals. This model introduces a control volume over the labyrinth tooth which looks like Nelson's plain-seal model (1984, 1985), including a continuity equation, axial- and circumferential-momentum equations, and the energy equation. Within the cavity, an axial-momentum equation replaces Scharrer's leakage equation.

Nordmann et al. (1987), Nordmann and Weiser (1988, 1990), and Weiser and Nordmann (1989) have proposed solutions for the rotordynamic coefficients based on various solutions to the Navier-Stokes (NS) equations for small motion about a centered position.

2 How good are the prediction models for labyrinth seals? Extended comparisons of predictions and measurements have only been made for Scharrer's two-control-volume model by Pelletti (1990). For see-through TOS and TOR seals, Pelletti found that Scharrer's model did well at larger clearances with reduced supply pressures and speeds, but produced systematically diverging results for increased supply pressures and running speeds. Comparisons which are available for either Nordmann and Weiser's three-control-volume model or the (NS) solution approaches of Nordmann and his co-workers at the University of Kaiserslautern are favorable but quite limited. Certainly, the NS solutions are expected to yield much better predictions, but the superiority of their predictions has not yet been confirmed by a systematic comparison to measurements.

3 What models are available for plain annular seals? Nelson's bulk-flow models (1984, 1985) and the NS solution approaches of Nordmann et al. are the two alternatives for plain annular seals. Nelson's model is intuitively attractive and was discussed at length in Section 5.6. For honeycomb seals, Nelson's models correctly predict that honeycomb stators reduce cross-coupled stiffness coefficients and enhance stability, but do a poor job of predicting rotordynamic coefficients. The erratic dependence of the friction factors of honeycomb surfaces on the hole depths, hole widths, clearances, and Reynolds numbers probably precludes accurate predictions for this type of seal for the near future [Ha and Childs (1991a, 1991b), Ha et al. (1991)].

As with labyrinth seals, the NS solution approaches developed by Nordmann and his co-workers at the University of Kaiserslautern probably offers the best future approach for calculating rotordynamic coefficients of smooth annular seals with various geometries; however, the CFD approaches offer no present prospects for prediction of honeycomb-stator seals.

---

*Chapter 4. Florjancic's model is an extension of Scharrer's two-control-volume gas labyrinth model.

4 From a rotordynamics viewpoint, how do the various annular-seal configurations compare? Before considering this question, note that only a limited number of annular-seal configurations have been analyzed or tested completely. The tests at the University of Stuttgart (Benckert and Wachter, Thieleke and Stetter) and at Heriott Watt University (Brown and Leong) considered a wide variety of labyrinths but yielded only direct and cross-coupled stiffness coefficients. No stepped seals have been tested. At operating speed, many TOR labyrinth blade tips grow sufficiently, because of centrifugal stresses, to cut into the stator and operate in an interference mode. This type of seal is referred to generically as an abradable seal, and the abradable stators can be made from honeycomb, powdered metal, or polymer materials. No analysis or tests have been reported for abradable seals. Now, having more clearly in mind the restricted range of available test data, the following known results can be stated:

(a) For labyrinth seals, teeth-on-stator (TOS) seals are marginally more stable than teeth-on-rotor (TOR) seals.

(b) There is a difference of opinion on the relative merits of see-through TOS labyrinths and interlocking labyrinths. Kanemitsu and Ohasawa's (1986) test results show the interlocking seal to have higher damping values. The Childs et al. (1988) results show the opposite.

(c) Absent interference, TOR seals with smooth and honeycomb surfaces have comparable rotordynamic characteristics.

(d) Honeycomb-stator/smooth-rotor seals can be designed to have exceptionally attractive rotordynamic characteristics with much higher damping and (for preswirled flow) much lower whirl-frequency ratios than TOS labyrinth seals or smooth seals.

(e) For smooth floating-ring seals, stepped and convergent-taper geometries can develop significantly higher direct stiffness values than constant-clearance seals.

(f) The data on brush seals are still preliminary, but suggest that replacing a labyrinth seal with a brush seal will not degrade rotordynamics and can improve rotordynamic stability if the labyrinth has a highly prerotated inlet circumferential velocity.

5 What are the biggest holes in our knowledge concerning annular gas seals and rotordynamics? The current "ignorance factor" for the prediction of gas seals is significantly higher than for the liquid seals of Chapter 4. The bulk-flow models which work well for plain annular liquid seals perform poorly for plain annular gas seals.

All of the seal models developed to date assume a perfect-gas-law state equation. More general thermodynamic models are needed for steam and other gases of interest.

The Navier-Stokes (NS) equation solution approaches of Nordmann and his co-workers at Kaiserslautern are clearly the best long-term approach for seal rotordynamic prediction, but they are largely unverified by comparison to available test results. Given the time and expense involved in their execution, this situation is understandable. However, more comparisons are needed both to rotordynamic-coefficient test data and to fundamental flow velocity measurements [Morrison et al. (1990, 1991a, 1991b)].

6 How can annular gas seals be modified to improve rotordynamic response? As demonstrated by Benckert and Wachter (1980a), the most immediate way to improve the rotordynamic characteristics of a labyrinth seal is to introduce a swirl brake. The only new insights concerning swirl brakes since their introduction is that a proper aerodynamic design of these elements yields major benefits. Childs and Ramsey (1991) demonstrate the remarkable effectiveness of a Pratt & Whitney swirl-brake design for the advanced technology development (ATD) high-pressure-fuel turbopump of the Space Shuttle main engine (SSME). The design is illustrated in Figure 5.34 and was exceptionally effective in reducing the cross-coupled stiffness coefficient. Childs and Ramsey (1991) made a direct comparison between the current turbine-interstage swirl brake for the SSME high-pressure oxygen turbopump and a proposed aerodynamically-designed replacement. The aerodynamically designed swirl brake yielded cross-coupled stiffnesses which were no more than half the magnitude developed by the current design.

A second excellent approach for remedying rotordynamic problems caused by long ($L/D \geq \frac{1}{3}$) labyrinth seals is to replace them with honeycomb seals. Extensive tests have clearly demonstrated the marked advantages of properly designed honeycomb-stator/smooth-rotor seals over TOS or TOR labyrinths. Moreover, they leak less and are more robust than labyrinths should contact occur between stationary and rotating seal parts. Benaboud et al. (1984) provide a cautionary note on honeycomb seals. They report a structural failure of the balance-piston honeycomb stator in a high-pressure compressor. Obviously a careful stress analysis is appropriate when honeycomb stators are used with very high pressure differentials. No reports have been published of the use of honeycomb stators in steam-turbine seals.

Figure 5.35 illustrates a "shunt-bypass" approach to remedying or preventing an instability problem in a high-pressure compressor, Kirk (1986). In this approach, flow is diverted from the discharge and injected into an intermediate labyrinth cavity. The recirculated flow is sometimes injected against rotation (as illustrated) with obvious benefits in the reduction of circumferential velocity and cross-coupled stiffness. Flow injection is sufficient to reverse the direction of flow at the seal inlet, and proponents of this approach see some virtues in having this flow proceed up the backside of the last impeller. This approach has been widely and successfully employed to enhance compressor stability. However, the author considers it to be wasteful because of

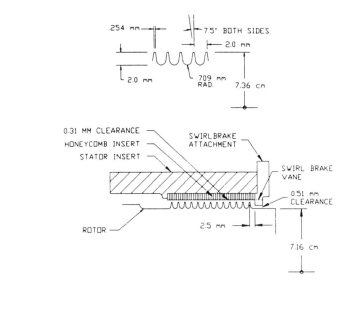

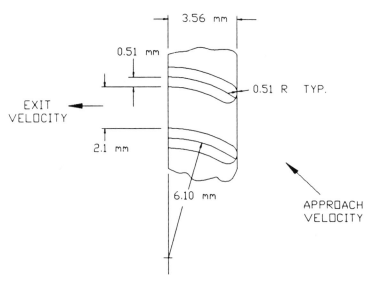

Figure 5.34 Pratt and Whitney swirl-brake design for the ATD-HPFTP SSME turbine interstage seal [Childs and Ramsey (1991)].

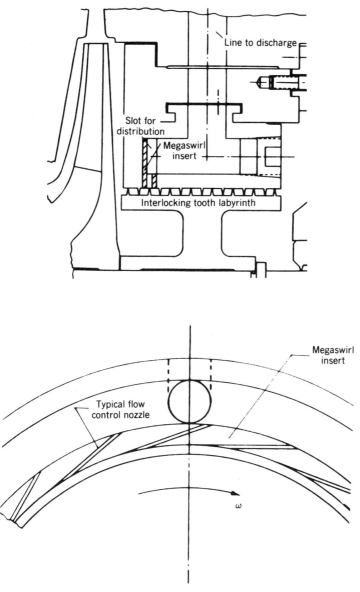

Figure 5.35 Shunt bypass for a balance-piston seal in a throughflow machine [Kirk (1986)].

the immediate and continuing performance loss due to the recirculated flow. Further, the desired objective can be accomplished without a performance penalty (or significant hardware modifications) by using either inlet swirl brakes or replacing the labyrinth with a honeycomb stator.

REFERENCES: CHAPTER 5

Abramovich, G. (1963), *The Theory of Turbulent Jets*, MIT Press, Cambridge, MA.

Benaboud, N., Borchi, M., and Tesei, A. (1984), "Hassi R'mel High Pressure Injection Project with Centrifugal Compression," in *Proceedings, Second European Congress on Fluid Machinery for the Oil, Petrochemical, and Related Industries*, IMechE Conference Publication, 1984-2, pp. 167–176.

Benckert, H., and Wachter, J. (1980a), "Flow Induced Spring Coefficients of Labyrinth Seals for Applications in Turbomachinery," Rotordynamic Instability Problems in High-Performance Turbomachinery, NASA CP No. 2133, proceedings of a workshop held at Texas A&M University, pp. 189–212.

Benckert, H., and Wachter, J. (1980b), "Flow Induced Spring Constants of Labyrinth Seals," IMechE Proceedings of the Second International Conference, *Vibrations in Rotating Machinery*, Cambridge, England, pp. 53–63.

Benckert, H., and Wachter, J. (1979), "Investigations on the Mass Flow and the Flow Induced Forces in Contactless Seals of Turbomachines," in *Proceedings of the Sixth Conference on Fluid Machinery*, Scientific Society of Mechanical Engineers, Akadémiaki Kikado, Budapest, pp. 57–66.

Benckert, H., and Wachter, J. (1978), "Studies on Vibrations Stimulated by Lateral Forces in Sealing Gaps," in *Seal Technology in Gas-Turbine Engines*, AGARD Conference Proceedings No. 237, London, pp. 9.1–9.11.

Brown, D., and Leong, Y. (1984), "Experimental Measurements of Lateral Forces in a Model Labyrinth and the Effect on Rotor Stability," in *Vibrations in Rotating Machinery*, IMechE Proceedings of the Third International Conference, York, England, pp. 215–222.

Childs, D., Elrod, D., and Ramsey, C. (1990), "Annular Honeycomb Seals: Additional Test Results for Leakage and Rotordynamic Coefficients," in *IFToMM, Proceedings of the Third International Conference on Rotordynamics*, Lyon, France, pp. 303–312.

Childs, D., Elrod, D., and Hale, K. (1989), "Annular Honeycomb Seals: Test Results for Leakage and Rotordynamic Coefficients; Comparisons to Labyrinth and Smooth Configurations," *Journal of Tribology*, **111**, 293–301.

Childs, D., Elrod, D., and Hale, K. (1988), "Rotordynamic Coefficient and Leakage Test Results for Interlock and Tooth-on-Stator Labyrinth Seals," ASME Paper No. 88-GT-87, ASME Turbo Expo, Amsterdam, The Netherlands.

Childs, D., Nelson, C., Nicks, C., Scharrer, J., Elrod, D., and Hale, K. (1986), "Theory versus Experiment for the Rotordynamic Coefficients of Annular Gas Seals: Part I—Test Facility and Apparatus," *Journal of Tribology*, **108**, 426–432.

Childs, D. W., and Ramsey, C. (1991) "Seal-Rotordynamic-Coefficient Test Results for a Model SSME ATD-HPFTP Turbine Interstage Seal with and without a Swirl Brake," *Journal of Tribology*, **113**, 113–203.

Childs, D., and Scharrer, J. (1988), "Theory Versus Experiment for the Rotordynamic Coefficients of Labyrinth Gas Seals: Part II—A Comparison to Experiment," *Journal of Vibration Acoustics, Stress, Reliability, and Design*, **110**, 281–287.

Childs, D., and Scharrer, J. (1986a), "An Iwatsubo-Based Solution for Labyrinth Seals: Comparison to Experimental Results," *Journal of Engineering for Gas Turbines and Power*, 325–331.

Childs, D., and Scharrer, J. (1986b), "Experimental Rotordynamic Coefficients Results for Teeth-on-Stator Labyrinth Gas Seals," *Journal of Engineering for Gas Turbines and Power*, **108**, 599–604.

Conner, K., and Childs, D. (1990), "Rotordynamic Coefficient Test Results for a 4-Stage Brush-Seal," AIAA Paper No. 90-2139, AIAA/SAE/ASME/ASEE 26th Joint Propulsion Conference, Orlando, FL.

Deissler, R. (1953), "Analysis of Turbulent Heat Transfer and Flow in the Entrance Region of Smooth Passages," NACA TN No. 3016.

Dunn, M. (1990), "A Comparison of Experimental Results and Theoretical Predictions for the Rotordynamic Coefficients of Stepped Annular Gas Seals," M.S.M.E. thesis, Texas A&M University, and Turbomachinery Laboratory Report No. TL-Seal-3-90.

Elrod, D., Childs, D., and Nelson, C. (1989), "An Annular Gas Seal Using Empirical Entrance and Exit Region Friction Factors," *Journal of Tribology*, **111**, 337–343.

Elrod, D., Nelson, C., and Childs, D. (1990), "An Entrance Region Friction Factor Model Applied to Annular Seals: Theory Versus Experiment for Smooth and Honeycomb Seals," *Journal of Tribology*, **111**, 337–343.

Ferguson, J. (1988), "Brushes as High Performance Gas Turbine Seals," ASME Paper No. 88-GT-182, International Gas Turbine Conference, Amsterdam, The Netherlands.

Fleming, D. (1980), "Damping in Ring Seals for Compressible Fluids," Rotordynamic Instability Problems of High Performance Turbomachinery, NASA CP No. 2133, proceedings of a workshop held at Texas A&M University, pp. 169–188.

Fleming, D. (1979), "Stiffness of Straight and Tapered Annular Gas Seals," *Journal of Lubrication Technology*, **101**, 349–355.

Fleming, D. (1977), "High Stiffness Seals for Rotor Critical Speed Control," ASME Paper No. 77-DET-10, Design Technical Engineering Conference, Chicago, IL.

Fulton, J. (1984), "Full Load Testing on the Platform Module Prior to Tow Out: A Case History of Subsynchronous Instability," Rotordynamic Instability Problems in High-Performance Turbomachinery, NASA CP No. 2338, proceedings of a workshop held at Texas A&M University, pp. 1–16.

Greathead, S., and Bostow, P. (1976), "Investigations into Load Dependent Vibrations of the High Pressure Rotor on Large Turbo-Generators," in *Proceedings of the IMechE Conference on Vibrations in Rotating Machinery*, Cambridge, England, pp. 279–286.

Gurevich, M. I., *The Theory of Jets in an Ideal Fluid*, Pergamon, New York, 1966, pp. 319–323.

Ha, T.-W., and Childs, D. (1991a), "Friction-Factor Data for Flat Plate Tests of Smooth and Honeycomb Surfaces (Including Extended Test Data)," TL-SEAL-1-91, Turbomachinery Laboratory, Texas A&M University.

Ha, T.-W., and Childs, D. (1991b), "Friction-Factor Data for Flat Plate Tests of Smooth and Honeycomb Surfaces," ASME Paper No. 91-Trib-20, ASME/STLE Joint Tribology Conference.

Ha, T.-W., Morrison, G., and Childs, D. (1991), "Friction-Factor Characteristics for Narrow Channels with Honeycomb Surfaces," ASME Paper No. 91-Trib-21, ASME/STLE Joint Tribology Conference.

Hawkins, L., Childs, D., and Hale, K. (1988), "Experimental Results for Labyrinth Gas Seals with Honeycomb Stators: Comparisons to Smooth-Stator Seals and Theoretical Predictions," *Journal of Tribology*, **111**, 161–168.

Hisa, S., Sakakido, H., Asatu, S., and Sakamoto, T. (1986), "Steam Excited Vibrations in Rotor-Bearing Systems," in *Proceedings of the Second IFToMM International Conference on Rotordynamics*, Tokyo, Japan, pp. 635–642.

Iwatsubo, T. (1980), "Evaluation of Instability Forces of Labyrinth Seals in Turbines or Compressors," Rotordynamic Instability Problems in High-Performance Turbomachinery, NASA CP No. 2133, proceedings of a workshop held at Texas A&M University, pp. 139–167.

Iwatsubo, T., Motooka, N., and Kawai, R. (1982), "Flow Induced Force and Flow Pattern of Labyrinth Seal," Rotordynamic Instability Problems in High-Performance Turbomachinery, NASA CP No. 2250, proceedings of a workshop held at Texas A&M University, pp. 205–222.

John, E. (1979), *Gas Dynamics*, Wiley, New York.

Kamotori, K. (1973), *Theory of Non-Contact Seal* (in Japanese), Corona.

Kanemitsu, Y., and Ohasawa, M. (1986), "Experimental Study on Flow-Induced Force of Labyrinth Seal," in *Proceedings, Post IFToMM Conference on Flow Induced Force in Rotating Machinery*, Kobe, Japan, 16–18 September, pp. 106–112.

Kirk, G. (1987), "A Method for Calculating Labyrinth Seal Inlet Swirl Velocity," *Rotating Machinery Dynamics*, ASME, New York, Vol. 2, pp. 345–350.

Kirk, G. (1986), "Labyrinth Seal Analysis for Centrifugal Compressor Design—Theory and Practice," in *Rotordynamics*, Proceedings of the Second IFToMM International Conference on Rotordynamics, Tokyo, Japan, pp. 589–596.

Kirk, G., and Donald, G. (1983), "Design Criteria for Improved Stability of Centrifugal Compressors," in *Rotordynamical Instability*, ASME AMD-Vol. 55.

Kleynhans, G. F. (1991), "A Comparison of Experimental Results and Theoretical Predictions for the Rotordynamic and Leakage Characteristics of Short ($L/D = 1/6$) Honeycomb and Smooth Annular Seals, Texas A&M University Turbomachinery Laboratory Report No. TL-SEAL-91.

Kurohashi, M., Inoue, Y., Abe, T., and Fujikawa, T. (1980), "Spring and Damping Coefficients of the Labyrinth Seals," in *Proceedings IMechE—Second International Conference on Vibrations in Rotating Machinery*, Cambridge, England, pp. 215–222.

Leong, Y., and Brown, D. (1984), "Experimental Investigation of Lateral Forces Induced by Flow Through Model Labyrinth Seals," Rotordynamic Instability Problems in High-Performance Turbomachinery, NASA CP No. 2338, proceedings of a workshop held at Texas A&M University, pp. 187–210.

Millsaps, K., and Martinez-Sanchez, M. (1990), "Static and Dynamic Pressure Distribution in a Short Labyrinth Seal," Rotordynamic Instability Problems in High-Performance Turbomachinery, proceedings of a workshop held at Texas A&M University.

Morrison, G., Johnson, M., and De Otte, R. (1990), "Experimental Investigation of an Eccentric Labyrinth Seal Velocity Field Using 3-D Laser Doppler Anemometry," in D. Rhode and J. Tuzson, Eds., *Fluid Machinery and Components*, ASME, New York, FED-Vol. 101, pp. 61–71.

Morrison, G., Johnson, M., and Tatterson, G. (1991a), "Three-D Laser Measurements in a Labyrinth Seal," *Journal of Engineering for Gas Turbines and Power*, **113**, 119–125.

Morrison, G., Johnson, M., and Tatterson, G. (1991b), "Three-Dimensional Laser Anemometer Measurements in an Annular Seal," *Journal of Tribology*, **113**, 421–427.

Nelson, C. (1985), "Rotordynamic Coefficients for Compressible Flow in Tapered Annular Seals, *Journal of Tribology*, **107**, 318–325.

Nelson, C. (1984), "Analysis for Leakage and Rotordynamic Coefficients of Surface-Roughened Tapered Annular Gas Seals," *Journal of Engineering for Gas Turbines and Power*, **106**, 927–934.

Nelson, C., Childs, D., Nicks, C., and Elrod, D. (1986), "Theory Versus Experiment for the Rotordynamic Coefficients of Annular Gas Seals: Part 2—Constant-Clearance and Convergent-Tapered Geometry," *Journal of Tribology*, 433–438.

Nelson, C., Dunn, M., and Scharrer, J. (1990), "Rotordynamic Coefficients for Stepped Annular Gas Seals," in *Proceedings, The Third International Conference on Transport Phenomena and Dynamics of Rotating Machinery*, Vol. 2 (Dynamics), pp. 259–272.

Neumann, K. (1964), "Zur Frage der Verwendung von Durchblickdichtungen im Dampgturbinebau," *Maschinentechnik*, **13**(4).

Nordmann, R., Dietzen, F., and Weiser, H. (1987), "Calculation of Rotordynamic Coefficients and Leakage for Annular Gas Seals by Means of a Finite-Difference Technique," in *Rotating Machinery Dynamics*, ASME, New York, Vol. Two, DE—Vol. 2.

Nordmann, R., and Weiser, H. (1990), "Evaluation of Rotordynamic Coefficients of Look-Through Labyrinths by Means of a Three Volume Bulk Model," Rotordynamic Instability Problems in High Performance Turbomachinery, proceedings of a workshop held at Texas A&M University, pp. 141–157.

Nordmann, R., and Weiser, H. (1988), "Rotordynamic Coefficients for Labyrinth Seals Calculated by Means of a Finite Difference Technique," Rotordynamic Instability Problems in High Performance Turbomachinery, NASA CP No. 3026, proceedings of a workshop held at Texas A&M University.

Pelletti, J. (1990), "A Comparison of Experimental Results and Theoretical Predictions for the Rotordynamic Coefficients of Short ($L/D = 1/6$) Labyrinth Seals," M.S.M.E. Thesis, Texas A&M University and Turbomachinery Laboratory Report No. TL-Seal-1-90.

Pelletti, J., and Childs, D. (1991), "A Comparison of Theoretical Predictions for the Rotordynamic Coefficients of Short ($L/D = 1/6$) Labyrinth Seals," in *Rotating Machinery and Vehicle Dynamics*, Proceedings of the 13th Vibration Conference, Miami, FL, ASME, New York, DE—Vol. 35, pp. 69–76.

Pollman, E., and Termuehlen, N. (1975), "Turbine Rotor Vibrations Excited by Steam Forces (Steam Whirl)," ASME Paper No. 75-WA/Pwr-11.

Rajakumar, C., and Sisto, F. (1988), "Labyrinth Seal Force Coefficients for Small Motion of the Rotor about an Arbitrary Eccentric Position," ASME Paper No. 88-G7-194, ASME Gas Turbine Conference, Anaheim, CA.

Scharrer, J. (1988), "Rotordynamic Coefficients for Stepped Labyrinth Gas Seals," *Journal of Tribology*.

Scharrer, J. (1987), A Comparison of Experimental and Theoretical Results for Labyrinth Gas Seals, Ph.D. dissertation, Texas A&M University.

Scharrer, J., and Nelson, C. (1991), "Rotordynamic Coefficients for Partially Tapered Annular Seals: Part II—Compressible Seals," *Journal of Tribology*, **113**, 53–57.

Stoff, H. (1980), "Incompressible Flow in a Labyrinth Seal," *Journal of Fluid Mechanics*, **100**, Part 4, 817–829.

Thieleke, G., and Stetter, H. (1990), "Experimental Investigation of Exciting Forces Caused by Flow in Labyrinth Seals," Rotordynamic Instability Problems in High Performance Turbomachinery, NASA CP No. 3122, proceedings of a workshop held at Texas A&M University.

Vermes, G. (1961), "A Fluid Mechanic Approach to the Labyrinth Seal Leakage Problem," *Journal of Engineering for Power*, **83**(2), 161–169.

Weiser, H., and Nordmann, R. (1989), "Calculation of Rotordynamic Coefficients by Means of a Three Dimensional Finite-Difference Method," in *Machinery Dynamics—Applications and Vibration Control Problems*, ASME, New York, DE—Vol. 18-2, pp. 109–113.

Wright, D. V. (1983), "Labyrinth Seal Forces on a Whirling Rotor," in *Rotor Dynamical Stability*, ASME, New York, AMD-Vol. 55, pp. 19–31.

Wyssman, H. (1987), "Rotor Stability of High-Pressure Multistage Centrifugal Compressors," in *Rotating Machinery Dynamics*, ASME, New York, Vol. 2, pp. 561–569.

Wyssman, H., Pham, T., and Jenny, R. (1984), "Prediction of Stiffness and Damping Coefficients for Centrifugal Compressor Labyrinth Seals," *Journal of Engineering for Gas Turbines and Power*, **106**, 920–926.

6

ROTORDYNAMIC MODELS FOR TURBINES AND PUMP IMPELLERS

6.1 INTRODUCTION

Given the large amount of concentrated power transfer at a turbine stage or pump impeller, these elements have always been suspected of making a large contribution to destabilizing forces in high-energy machines. Thomas (1958) initially proposed that an eccentric turbine would generate a destabilizing force due to circumferential variation in efficiency. He called these forces "spälterregnung" (clearance-excitation) forces. Black (1974) suggested that pump impellers could also develop destabilizing forces. Since these initial papers, a considerable volume of literature has been developed involving analysis and tests of turbines and centrifugal impellers. Rotordynamic modeling of turbine stages and impellers for centrifugal pumps is the subject of this chapter.

6.2 MODELS AND MEASUREMENTS FOR AXIAL TURBINE STAGES

As noted above, Thomas (1958) initially suggested that nonsymmetric clearances caused by eccentric operation of a turbine can create destabilizing forces which he called spälterregnung (clearance-excitation) forces. Thomas was concerned with instability problems in steam turbines. Alford (1965) subsequently identified the same mechanism when analyzing stability problems of aircraft gas turbines; hence, in the United States, excitation forces due to clearance changes around the periphery of a turbine are regularly called "Alford forces." In deference to Thomas' seven-year head start, his

355

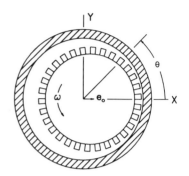

Figure 6.1 Thomas's (1958) model for clearance-excitation due to a turbine stage.

designation of clearance-excitation force is used here. The development which follows is taken from Urlichs (1976).

Figure 6.1 illustrates a turbine which is positioned eccentrically within its stator and housing. In the absence of eccentricity, the force developed by a differential sector of the turbine wheel is

$$dF = F_s \eta_u \frac{d\theta}{2\pi}, \qquad (6.1)$$

where η_u is the tangential efficiency, and F_s is the *ideal* tangential force which would be developed in the absence of losses, defined by

$$F_s = \frac{g\dot{m}\,\Delta H_t}{R_m \omega}. \qquad (6.2)$$

In this equation, $\dot{m}$ is the mass flow rate through the turbine, ΔH_t is the head drop, g is the acceleration of gravity, R_m is the mean radius of the turbine blade, and ω is the angular velocity. Thomas noted that a clearance increase due to eccentricity would increase leakage, and reduce the force. Conversely, a clearance decrease would increase the force. Hence, for an eccentric rotor Eq. (6.1) is stated

$$dF = F_s \left[\eta_u - \zeta_{sp}(\theta) \right] \frac{d\theta}{2\pi}, \qquad (6.3)$$

where ζ_{sp} defines the local efficiency loss due to eccentricity.

The forces, which are developed as a result of the eccentricity e, are defined from Eq. (6.3) by the following integrals

$$F_X = -\int_0^{2\pi} \sin\theta\, dF = \frac{F_s}{2\pi} \int_0^{2\pi} \zeta_{sp}(\theta)\sin\theta\, d\theta,$$

$$F_Y = \int_0^{2\pi} \cos\theta\, dF = -\frac{F_s}{2\pi} \int_0^{2\pi} \zeta_{sp}(\theta)\cos\theta\, d\theta, \qquad (6.4)$$

which must be integrated over the periphery of the turbine.

The efficiency function $\zeta_{sp}(\theta)$ is assumed to be a linear function of the local clearance H defined by

$$H = C_r - e_0 \cos \theta,$$

where C_r is the nominal radial clearance, and e_0 is the eccentricity. As a first approximation, ζ_{sp} is defined as the ratio of the leakage mass flow rate $\dot{m}_{sp}$ to the turbine mass flow rate $\dot{m}_0$

$$\zeta_{sp}(\theta) = \dot{m}_{sp}(\theta)/\dot{m}_0. \tag{6.5}$$

The leakage flow rate $\dot{m}_{sp}(\theta)$ may be stated:

$$\dot{m}_{sp}(\theta) = k_c \rho \overline{W} A_l = k_c \rho \overline{W} [2\pi R_t H(\theta)], \tag{6.6}$$

where k_c is a contraction coefficient, ρ is the fluid density, $\overline{W}$ is the average axial fluid velocity, and $R_t = R_m + L_t/2$ is the blade or stator tip radius. The axial velocity may be defined in terms of the head drop by

$$\Delta H_t = C_d \frac{\overline{W}^2}{2} \Rightarrow \overline{W} = \sqrt{\frac{2 \Delta H_t}{C_d}}, \tag{6.7}$$

where C_d is the discharge coefficient for the labyrinth tip seals. The turbine flow may be stated:

$$\dot{m}_0 = \rho C_1 \sin \alpha_1 A = \rho C_1 \sin \alpha_1 (2\pi R_m L_t), \tag{6.8}$$

where C_1 is the inlet velocity magnitude, α_1 is the inlet flow angle, and L_t is the turbine blade length. Combining Eqs. (6.5)–(6.8) yields

$$\zeta_{sp}(\theta) = \frac{k_c}{C_1 \sin \alpha_1} \left(\frac{R_t}{R_m} \right) \sqrt{\frac{2 \Delta H_t}{C_d}} \cdot \frac{H(\theta)}{L_t}$$

$$= \frac{k_c}{C_1 \sin \alpha_1} \sqrt{\frac{2 \Delta H_t}{C_d}} \left(1 + \frac{L_t}{2 R_m} \right) \frac{H}{L_t} = \frac{2 K_2 H}{L_t}, \tag{6.9a}$$

where

$$K_2 = \frac{k_c}{2 C_1 \sin \alpha_1} \sqrt{\frac{2 \Delta H_t}{C_d}} \left(1 + \frac{L_t}{2 R_m} \right). \tag{6.9b}$$

Substituting from Eq. (6.9a) into Eq. (6.4) gives

$$F_X = \frac{F_s K_2}{\pi L_t} \int_0^{2\pi} (C_r - e_0 \cos \theta) \sin \theta \, d\theta = 0,$$

$$F_Y = -\frac{F_s K_2}{\pi L_t} \int_0^{2\pi} (C_r - e_0 \cos \theta) \cos \theta \, d\theta = e_0 \left(\frac{F_s}{L_t} \right) K_2 = e_0 k_q. \quad (6.10)$$

The zero value for F_X demonstrates that the change in efficiency due to a change in clearance does not develop a direct restoring force due to the displacement e. The positive sign of F_Y demonstrates that the force developed is in the same direction as the shaft rotation, and is accordingly *destabilizing* for forward whirl. The complete force-motion model from Eq. (6.10) may be stated

$$-\begin{Bmatrix} F_X \\ F_Y \end{Bmatrix} = \begin{bmatrix} 0 & k_q \\ -k_q & 0 \end{bmatrix} \begin{Bmatrix} X \\ Y \end{Bmatrix}. \quad (6.11)$$

The clearance-excitation force provides a purely destabilizing force without direct stiffness or damping.

Thomas and Urlichs perform a parallel analysis for the influence of leakage around the stator and add the resulting coefficients for the stator to obtain a combined K_2 coefficient for the stator and turbine combination.

Although the radial restoring force due to clearance excitation is zero according to the first of Eq. (6.10), a coefficient corresponding to the direct restoring force can be defined as follows:

$$F_X = e_0 \left(\frac{F_s}{L_t} \right) K_1. \quad (6.12)$$

This form is used by Urlichs. Note that a positive K_1 indicates a negative spring constant, and vice versa.

Alford's development is similar to the German authors; however, he concentrates on the local change in efficiency due to eccentricity, rather than a change in the force. His analysis yields

$$F_Y = e_0 \frac{T\beta}{D_m L_t} = e_0 k_q, \quad (6.13)$$

where β is defined as "the change in thermodynamic efficiency per unit of rotor displacement, expressed as a fraction of blade height," D_m is the mean blade diameter, and T is the torque. A comparable form for Eq. (6.10) is

$$F_Y = e_0 \frac{F_s R_m \eta_u}{2 R_m L_t} \left(\frac{2 K_2}{\eta_u} \right) = e_0 \left(\frac{T}{D_m L_t} \right) \left(\frac{2 K_2}{\eta_u} \right),$$

and yields the following relationship between Alford's β and Thomas' K_2:

$$K_2 = \beta \left(\frac{\eta_u}{2} \right). \tag{6.14}$$

Available Test Data

A test program at the Technical University of Munich under Professor Thomas provides the only published test results for clearance-excitation forces. Figure 6.2, which is taken from Urlichs (1976) illustrates the turbine, stator, and clearance geometries. The radii for the turbine and stator clearances are, respectively, $R_t = 97$ mm, $R_s = 41$ mm. The tests consist of measuring the forces developed by the stage (turbine and stator) due to an eccentric displacement of the housing with respect to the rotor. Hence, the

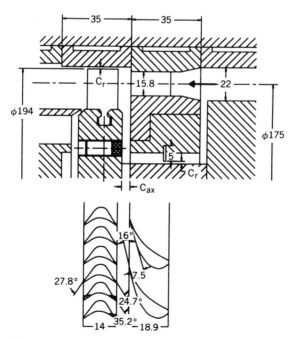

Figure 6.2   Turbine and stator configuration for the T. U. Munich clearance-excitation and turbine-tip-seal test program [Urlichs (1976)].

prescribed eccentricity is the independent variable, and the resultant transverse force is the dependent variable. The parameters which may be varied include the axial clearance C_{ax} between the turbine and the stator and the turbine and stator radial clearances. C_{ax} may be varied continuously between 0.5 and 3 mm. The three sets of turbine and stator clearances (in millimeters) which were tested are

C'_r (stator)	C'_r/R_s	C_r (turbine)	C_r/R_t
0.55	0.0134	0.5	0.005
1.10	0.0268	1.0	0.010
1.55	0.0378	1.5	0.015

The Urlichs dissertation provides very little information on the operating conditions of the turbine stage with respect to flow rate, ΔP, running speed, power, etc. However, the running-speed capacity of the unit varies between 4000 and 8000 rpm. Cited values for pressure differentials across shrouded turbines vary between 57 mba (0.84 psi) and 64 mba (0.94 psi). Hence, stage ΔP's for unshrouded turbines in these tests may be on the order of 136 mba (2 psi).

Figure 6.3 illustrates Urlichs's experimental results for the coefficients K_2 and K_1 of Eqs. (6.10) and (6.12), respectively. The three curves correspond to the radial clearances cited above. The independent variable is the axial clearance. The dashed line for the left figure is the prediction for K_2 from Eq. (6.9b) which is theoretically independent of C_r, the turbine-tip clearance.

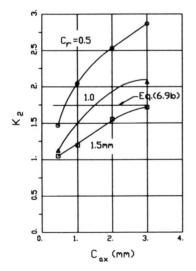 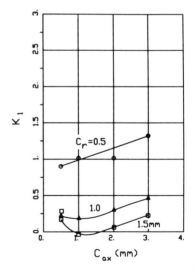

Figure 6.3 Measured clearance excitation and radial stiffness coefficients for unshrouded turbines [Urlichs (1976)].

In addition, the theoretical value for K_1 is zero, while the right figure shows a generally negative restoring force.

Urlichs reasoned that the discrepancy between theory and experiment for K_2 arises because the change in efficiency is a nonlinear function of H rather than the linear function proposed in Eq. (6.6). He suggested the following relationship for ζ_{sp}:

$$\zeta_{sp} = a_1 H + a_2 H^2 + a_3 H^3. \tag{6.15}$$

When substituted into Eqs. (6.4), this yields

$$F_X = 0,$$

$$F_Y = \frac{F_s}{2}\left[(a_1 + 2a_2 C_r + 3a_3 C_r^2)e_0 + \frac{3a_3}{4}e_0^3\right]. \tag{6.16}$$

Figure 6.4 illustrates measured results for $|F_Y/F_s|$ versus e_0 for $C_r = 0.5$, 1.0, and 1.5 mm. Since the results are clearly linear in e_0, the a_3 coefficient of H^3 in Eq. (6.15) is negligible, and may also be dropped in Eq. (6.16). The resultant governing equations are

$$\zeta_{sp} = a_1 H + a_2 H^2,$$

$$F_X = 0, \qquad F_Y = \frac{F_s}{2}(a_1 + 2a_3 C_r)e_0. \tag{6.17}$$

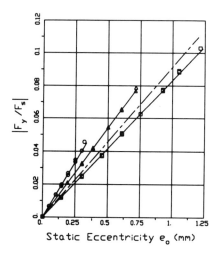

Static Eccentricity e_0 (mm)

O $C_r = 0.5$mm, $C_{ax} = 2$mm
△ $C_r = 1.0$mm, $C_{ax} = 2$mm
□ $C_r = 1.5$mm, $C_{ax} = 2$mm

Figure 6.4   Experimental results for F_Y/F_s versus eccentricity for unshrouded turbines [Urlichs (1976)].

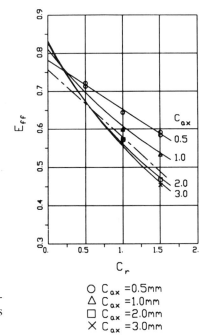

Figure 6.5 Measured efficiencies versus theoretical predictions from Eq. (6.18) [Urlichs
(1976)].

○ C_{ax} =0.5mm
△ C_{ax} =1.0mm
□ C_{ax} =2.0mm
✕ C_{ax} =3.0mm

Urlich curve-fitted the test results of Figure 6.4 to obtain the unknown
coefficients (a_1, a_2) of Eq. (6.17). He then compared the efficiency predicted by

$$\text{Eff} = \eta_u - \bar{\zeta}_p = \eta_u - (a_1 + a_2 C_r)C_r \qquad (6.18)$$

to measured efficiencies for the turbine stage with varying axial and radial
clearances. The results are shown in Figure 6.5. The solid curves in this figure
are quadratic curve-fitted results from Eq. (6.18). The data points are
measured efficiencies. Urlichs concludes from his work that, "the clearance
excitation force can be calculated to a reasonable degree of accuracy from
known efficiencies."

Note that Thomas's and Alford's clearance force could in theory be
obtained by simply testing a turbine at different circumferential clearances to
obtain the change in efficiency caused by a (uniform) change in clearance.
Based on his experience with aircraft gas turbines, Alford suggests a β on the
order of 1–1.5. The Urlichs measurements yield markedly higher values.
Taking the nominal predicted values from Figure 6.3 yields, from Eq. (6.14),

$$\beta = \frac{2K_2}{\eta_u} \simeq \frac{2(1.75)}{0.7} = 5.0. \qquad (6.19)$$

Selecting the point $C_{ax} = C_r = 0.5$ mm yields

$$\beta = \frac{2K_2}{\eta_u} \simeq \frac{2(1.5)}{0.7} = 4.29. \tag{6.20}$$

In the author's experience, these values are simply too high. A rotordynamics analysis of the high-pressure fuel turbopump (HPFTP) of the Space Shuttle main engine (SSME) would yield a prediction of strong instability for this order of β, versus a demonstration of comfortable stability. Values for β on the order of 1.0 have yielded more reasonable predictions for the author.

Shrouded Turbines

The research group at T. U. Munich also tested shrouded turbine stages [Leie (1979), and Leie and Thomas (1980)]. Given that the clearance-excitation phenomenon arises because of the local sensitivity of efficiency to a local change in clearance, one would guess that a shrouded turbine's performance would be less sensitive to changes in clearance than an unshrouded turbine and would accordingly have reduced destabilizing forces. However, measurements showed a contrary result, viz., measured destabilizing forces were on the same order of magnitude for both shrouded and unshrouded turbines. Moreover, for shrouded turbines, the destabilizing forces were much larger than predicted by a clearance-excitation analysis. Pressure measurements demonstrated that the labyrinth* tip seals on the shroud bands accounted for the discrepancy. The discrepancy was largest for turbine stages which introduce a high degree of preswirl upstream of the tip seals. Swirl brakes were demonstrated to sharply reduce the net destabilizing forces.

Hauck (1982) presented additional test results for two shrouded-turbine stages and reached similar conclusions concerning the predominant influence of labyrinth tip seals on the measured destabilizing force coefficients.

Concluding Comments

A NASA-funded research project at the Massachusetts Institute of Technology (MIT) on turbine excitation forces is nearing conclusion and promises to yield additional (excellent) test data, an improved understanding of the detailed flow phenomena which generate clearance-excitation forces, and new and more sophisticated methods for predicting these forces. Unfortunately, these results are not yet completed and/or published. Preliminary results demonstrate that β increases with the degree of reaction of the turbine stage, and that β values on the order of 4 or 5 are possible for high-degrees-of-reaction turbines. A turbine's degree of reaction provides a

---

*See Chapter 5, Sections 5.2–5.5.

measure of the pressure drop through the turbine stage and is an indication of the degree to which the stage acts as a nozzle to the flow. Given that the stage illustrated in Figure 6.2 has a negligible degree of reaction, its associated high β values are particularly puzzling.

Given the strong influence of clearance excitation forces on turbine rotordynamics and the general paucity of available data and analyses, a great deal of additional work is needed on this subject.

6.3 ROTORDYNAMIC-COEFFICIENT TEST RESULTS FOR CENTRIFUGAL-PUMP IMPELLERS

Figure 6.6 illustrates a typical centrifugal-pump impeller. The seals of the impeller provide obvious forces which were discussed in Chapter 4. Initially, the forces developed by the remainder of the impeller were thought to arise primarily due to the interaction of the flow leaving the impeller and entering either a volute or vaned diffuser without any particular concern for the surface forces developed by the shroud. This viewpoint was probably influenced by the measured turbine clearance-excitation forces which were at hand when research was initiated. Test results have since demonstrated that the "shroud forces" are comparable to the "interaction forces" for large clearances between the impeller shroud and the pump housing and can dominate the resultant forces for reduced clearances.

The initial test program for pump impellers was developed by California Institute of Technology (Cal Tech) researchers, using a "shaft-in-a-shaft" test rig [Brennen et al. (1980)]. This apparatus mounts the test shaft eccentrically in another rotatable shaft. Rotation of the outer shaft at the running speed Ω then yields a radius equal to the eccentricity A at a precessional frequency Ω. The impeller forces can generally be modeled by an equation of the form

$$-\begin{Bmatrix} F_X \\ F_Y \end{Bmatrix} = \begin{bmatrix} K & k \\ -k & K \end{bmatrix}\begin{Bmatrix} X \\ Y \end{Bmatrix} + \begin{bmatrix} C & c \\ -c & C \end{bmatrix}\begin{Bmatrix} \dot{X} \\ \dot{Y} \end{Bmatrix} + \begin{bmatrix} M & m_c \\ -m_c & M \end{bmatrix}\begin{Bmatrix} \ddot{X} \\ \ddot{Y} \end{Bmatrix}.$$

(6.21)

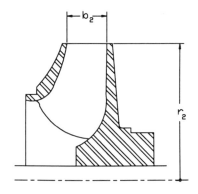

Figure 6.6 Centrifugal pump impeller.

The Cal Tech apparatus can separately identify K_{XX}, K_{YY} or K_{XY}, K_{YX}, etc., and showed that the diagonal elements are equal and that off-diagonal elements were equal in magnitude but opposite in sign; hence, the reduced model of Eq. (6.21). Note in comparison to the liquid-seal model of Eq. (4.1) the presence of a cross-coupled mass coefficient m_c. Tests demonstrate that this term is not negligible.

Returning to Figure 6.6, the radius r_2 and the impeller-discharge width b_2 are combined with the running speed ω and fluid density ρ to develop the reference force

$$F_0 = \rho \pi r_2^3 b_2 \omega^2. \tag{6.22}$$

Nondimensionalizing the reaction-force components of Eq. (6.21) with respect to F_0 and the rotor displacement components with respect to r_2 yields

$$-\begin{Bmatrix} \bar{F}_x \\ \bar{F}_y \end{Bmatrix} = \begin{bmatrix} \bar{K} & \bar{k} \\ -\bar{k} & \bar{K} \end{bmatrix} \begin{Bmatrix} x \\ y \end{Bmatrix} + \begin{bmatrix} \bar{C} & \bar{c} \\ -\bar{c} & \bar{C} \end{bmatrix} \begin{Bmatrix} \dot{x} \\ \dot{y} \end{Bmatrix} + \begin{bmatrix} \bar{M} & \bar{m}_c \\ -\bar{m}_c & \bar{M} \end{bmatrix} \begin{Bmatrix} \ddot{x} \\ \ddot{y} \end{Bmatrix},$$
$$\tag{6.23a}$$

where

$$\bar{F}_x = F_X/F_0, \qquad \bar{F}_y = F_Y/F_0,$$
$$x = X/r_2, \qquad y = Y/r_2. \tag{6.23b}$$

The dimensional and nondimensional rotordynamic coefficients are then related by*

$$K = \bar{K} C_I \omega^2, \qquad k = \bar{k} C_I \omega^2,$$
$$C = \bar{C} C_I \omega, \qquad c = \bar{c} C_I \omega,$$
$$M = \bar{M} C_I, \qquad m_c = \bar{m}_c C_I, \tag{6.24a}$$

where

$$C_I = \pi \rho b_2 r_2^2. \tag{6.24b}$$

Chamieh et al. (1982) presented initial test results for an impeller in a volute giving only the stiffness coefficients

$$\bar{K} = -2.0, \qquad \bar{k} = 0.9. \tag{6.25}$$

The direct stiffness coefficient is negative and could cause a reduction in a

---

*Note that M and m_c are not functions of running speed. As noted in Chapter 7, this characteristic can be used to good advantage in developing rotordynamic models for pumps.

pump's critical speed. For a circular orbit, the positive sign for $\bar{k}$ implies that a tangential force is developed in the direction of rotation which would destabilize forward precession modes.

Full sets of rotordynamic coefficients were subsequently published by Jery et al. (1984, 1985). However, most of their data are given in terms of nondimensional radial (normal) and tangential force coefficients. From Eq. (6.21), a circular orbit of radius A and precessional frequency Ω gives radial and circumferential force components

$$F_r = -(K + c\Omega - M\Omega^2)A,$$
$$F_\theta = (k - C\Omega - m_c\Omega^2)A. \tag{6.26}$$

In terms of the nondimensionalized variables, these equations become

$$\bar{F}_r(f) = \frac{F_r}{F_0(A/r_2)} = -\left(\bar{K} + \bar{c}f - \bar{M}f^2\right), \tag{6.27a}$$

$$\bar{F}_\theta(f) = \frac{F_\theta}{F_0(A/r_2)} = \left(\bar{k} - \bar{C}f - \bar{m}_c f^2\right), \tag{6.27b}$$

where

$$f = \frac{\Omega}{\omega}. \tag{6.28}$$

Jery et al. present $\bar{F}_r$ and $\bar{F}_\theta$ versus f for various flow coefficients,

$$\phi = \dot{Q}/2\pi b_2 \omega r^2, \tag{6.29}$$

and running speeds. Figure 6.7 illustrates typical plots for a five-bladed impeller in a volute with $\phi = 0.092$ (design point). Note that $\bar{F}_r(f)$ is (as assumed) a parabolic function, while $\bar{F}_\theta(f)$ is approximately linear. The $f = 0$ intercepts of $\bar{F}_r(f)$ and $\bar{F}_\theta(f)$ yield $-\bar{K}$ and $\bar{k}$, respectively.

Neglecting (temporarily) m_c in Eq. (6.27b), the zero crossing of $\bar{F}_t$ yields

$$\bar{F}_\theta(f^*) = 0 = \bar{k} - \bar{C}f^*$$

or

$$f^* = \frac{\bar{k}}{\bar{C}} = \frac{k}{C\omega} = \Omega_w, \tag{6.30}$$

where Ω_w is the whirl-frequency ratio. For Figure 6.7, the zero crossing yields $\Omega_w \cong 0.4$. Following the logic developed for liquid seals in Section 4.6, the impeller would become destabilizing at running speeds which are about 2.5 times greater than the first critical speed. Hence, impellers were initially

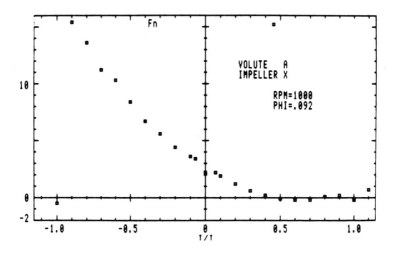

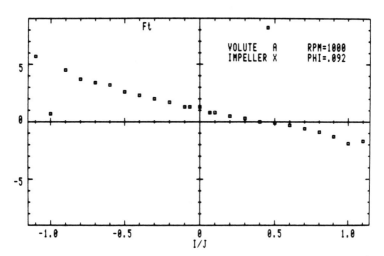

Figure 6.7 $\overline{F}_r$ and $\overline{F}_\theta$ versus $f = \Omega/\omega$ for $\omega = 1000$ rpm and $\phi = 0.092$ [Jery et al. (1985)].

viewed as being less destabilizing than journal bearings or long smooth seals ($\Omega_w = 0.5$).

The rotordynamic coefficients connected with Figure 6.7 are

$$\overline{K} = -2.5, \quad \overline{k} = 1.1,$$
$$\overline{C} = 3.14, \quad \overline{c} = 7.91,$$
$$\overline{M} = 6.51, \quad \overline{m}_c = -0.58. \tag{6.31}$$

These $\overline{K}$ and $\overline{k}$ results compare favorably to the Chamieh et al. values in Eq. (6.25). The nondimensional values for $\overline{c}$ and $\overline{M}$ are large in comparison to the other coefficients and yield large dimensional coefficients as well. From Eq. (6.27b), the negative sign for $\overline{m}_c$ implies that it is destabilizing for forward precession. The Jery et al. results are insensitive to changes in the running speed when ϕ is held constant but do vary modestly when ϕ is changed. Figure 6.8 illustrates $\overline{F}_\theta$ for $\phi = 0.060$ at four running speeds. Note by comparison to Figure 6.7 that $\overline{F}_\theta(f)$ has developed a "hump" around $f \cong 0.3$ and does not cross from positive to negative until $f^* \cong 0.53$. Hence, at reduced ϕ, the impeller tends to be more destabilizing. The Jery et al. results for $\phi = 0.145$ are about the same as at the design point.

Test results for the same basic impeller with five and six blades gave about the same rotordynamic coefficients.

Representative test results for the same impeller in one of several vaned diffusers are

$$\overline{K} = -2.65, \qquad \overline{k} = 1.04,$$
$$\overline{C} = 3.80, \qquad \overline{c} = 8.96,$$
$$\overline{M} = 6.60, \qquad \overline{m}_c = -0.903. \qquad (6.32)$$

These results are comparable to the volute results of Eq. (6.31) but are more benign in terms of the whirl-frequency ratio $\Omega_w = 1.04/3.8 \cong 0.27$. Neglecting $\overline{m}_c$, this value for Ω_w implies that the impeller would not become destabilizing until the running speed was around 3.7 times the rotor's first critical speed.

Shoji and Ohashi (1984, 1987) used a variation on the shaft-in-a-shaft test rig idea, presenting results for a two-dimensional, radial-flow impeller in a vaned and vaneless diffuser. Their results yield extremely low values for $\overline{k}$ and suggest that the tangential impeller coefficient would be stabilizing except at very low flow coefficients.

Test results for one of Shoji and Ohashi's impellers in a vaneless diffuser at design flow is

$$\overline{K} = -0.42, \qquad \overline{k} = -0.09,$$
$$\overline{C} = 1.08, \qquad \overline{c} = 1.88,$$
$$\overline{M} = 1.86, \qquad \overline{m}_c = -0.27.$$

These results strongly suggest that the impeller-diffuser (or volute) *flow* interaction forces in impellers are benign since their radial impeller eliminates any projected axial area for the shroud and thereby eliminates any radial shroud force due to pressure perturbations. The absence of both destabilizing forces and axially extended shroud surfaces suggests that the shroud forces are mainly responsible for destabilizing force coefficients.

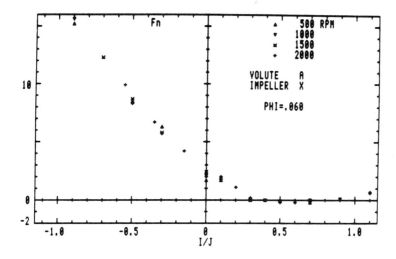

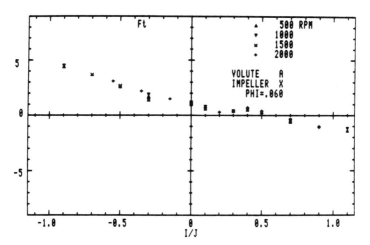

Figure 6.8 $\overline{F}_r$ and $\overline{F}_\theta$ versus $f = \Omega/\omega$ with four running speeds and $\phi = 0.060$ [Jery et al. (1985)].

Additional evidence supporting this view was provided by Adkins (1985) and Adkins and Brennen (1986). Figure 6.9 shows a side view of the Cal Tech impeller tested by both Chamieh et al. and Jery et al. Note that an axial face seal is used to control leakage down the front face of the impeller, and that "flow separation rings" have been installed to isolate the exit-flow region of the volute from the space between the pump impeller shroud and the pump housing. *Before these rings were installed*, Adkins made pressure measurements in the pump housing region to determine the contributions of the

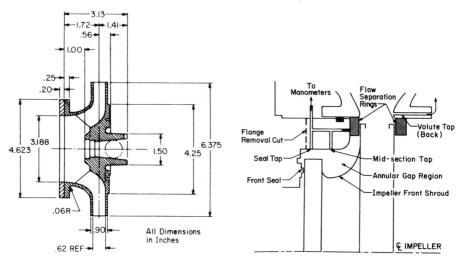

Figure 6.9   Cal Tech test impeller [Adkins and Brennen (1986)].

shroud surfaces. His results were

$$\bar{K} = -1.6, \qquad \bar{k} = 0.3. \tag{6.33}$$

By comparison to Eq. (6.25), note that the shroud region is contributing 80% of $\bar{K}$ and 33% of $\bar{k}$. The large contribution of the annular-gap region is surprising given the quite large clearances illustrated in Figure 6.9. With the separation rings installed, and the front pump housing completely removed, the rotordynamic forces dropped sharply yielding $\Omega_w \cong 0.07$ [Franz and Arndt (1986)], confirming the view that the impeller-volute interaction forces are small and benign.

Bolleter et al. (1987, 1989), presented the first test results for impellers with tighter clearances between the shroud and the housing. These results were developed by Sulzer personnel in a pendulum-type apparatus similar to that developed earlier for gas seals [Section 5.4]. Their pendulum constraint yields vertical oscillating motion of the impeller due to an external shaker. Figure 6.10 illustrates the tested impeller. For BEP conditions, the rotordynamic coefficients for the face-seal* impeller are

$$\bar{K} = -4.2, \qquad \bar{k} = 5.1,$$
$$\bar{C} = 4.6, \qquad \bar{c} = 13.5,$$
$$\bar{M} = 11.0, \qquad \bar{m}_c = 4.0, \tag{6.34a}$$

---

*Cal Tech and Sulzer researchers both initially used face seals to eliminate the influence of the wearing-ring seal forces.

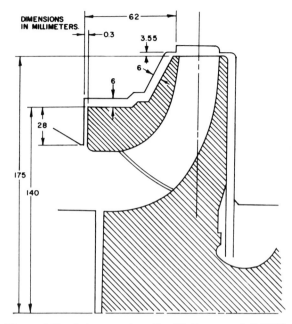

Figure 6.10   Sulzer test impeller [Bolleter et al. (1987)].

yielding

$$\Omega_w = \frac{\bar{k}}{\bar{C}} = 1.1. \tag{6.34b}$$

These coefficients are all larger than the Cal Tech coefficients of Eq. (6.32). The whirl-frequency ratio is also larger and predicts that the impeller becomes destabilizing with the running speed at 90% of the first critical speed. These data demonstrate conclusively that the shroud generates most of the destabilizing forces in a pump impeller.

   Bolleter et al. (1989) provided additional test results for the four impellers of Figure 6.11. Their geometric and performance data are provided in Table 6.1. Figure 6.12 illustrates measured coefficients for these impellers. Impeller D uses a stepped wearing-ring seal and was tested with (DS) and without (D) a swirl brake immediately upstream of the seal. For this impeller, separate measurements were available for the seal rotordynamic coefficients, and Table 6.2 contains the net and separate coefficients for the seal and impeller. Note that the seal values for $\bar{K}$ and $\bar{C}$ are markedly larger than the impeller's, with the opposite result holding for $\bar{c}$, $\bar{M}$, and $\bar{m}_c$. From Table 6.2, the whirl-frequency ratio for impeller D is $\Omega_w = \bar{k}/\bar{C} = 2.26$. This exceptionally high value says that the impeller becomes destabilizing for a running speed that is only 44% of the first critical speed.

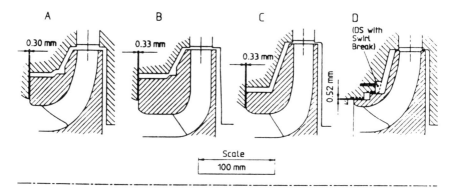

Figure 6.11 Sulzer test impellers [Bolleter et al. (1989)].

TABLE 6.1 Data for Measured Impellers, at 2000 rpm, Cold. Nominal Specific Speed n_q in Metric Units [Bolleter et al. (1989)].

Impeller	$D_2 = 2r_2$ (mm)	b_2 (mm)	H_p (m)	$\dot{Q}$ (l/s)	n_q
A	350	28	67	130	33
B	350	26.9	72	96	26
C	370	25.3	81	87	22
D, DS	350	28	66	130	33

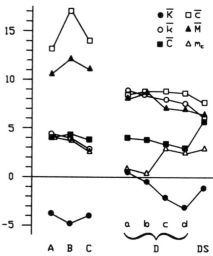

Figure 6.12 Matrix coefficients for various impellers at 100% flow. A, B, C as measured at 2000 rpm, 30°C. D measured at 2000 rpm, 30°C (a) 2000 rpm, 60°C (b) 4000 rpm, 60°C (c) 4000 rpm, 160°C (d) DS measured at 4000 rpm, 160°C [Bolleter et al. (1989)].

TABLE 6.2 Nondimensional Coefficients, Impeller D, at 4000 rpm, 60°C, 100% Flow [Bolleter et al. (1989)].

	$\bar{K}$	$\bar{k}$	$\bar{C}$	$\bar{c}$	$\bar{M}$	$\bar{m}_c$
Measured, total	1.5	12.3	8.8	8.5	6.5	3.3
Measured, seal	3.7	4.6	5.4	−0.01	−0.2	0.2
Difference	−2.2	7.7	3.4	8.6	6.7	3.1

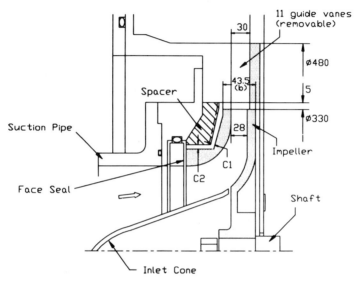

Figure 6.13 Test impeller configuration; $C_1 = 2.7$ mm, $C_2 = 3.0$ mm [Ohashi et al. (1988)].

Ohashi et al. (1988) presented additional results for a 3D impeller in a vaned and vaneless diffuser plus an overview of test results at the University of Tokyo, Cal Tech, Sulzer, and some additional test results at Mitsubishi[*]. They also conclude that the forces developed by an impeller increase sharply with decreasing clearance between the impeller shroud and the housing.

Figure 6.13 illustrates Ohashi's 3D impeller cross-section. Tests were conducted for: (a) no spacer, (b) a smooth spacer, and (c) a spacer with 24 radial grooves (10-mm width, 5-mm depth). Test results are illustrated in Figure 6.14 for these configurations. Ohashi's nondimensionalized force coefficients are

$$\tilde{F}_r = F_r/M'e\omega^2, \qquad \tilde{F}_\theta = F_\theta/M'e\omega^2,$$

---

[*] Yoshida and Tsujimoto (1990) conducted tests in a two-dimensional (radial impeller), obtaining a whirl-frequency ratio of about 0.25.

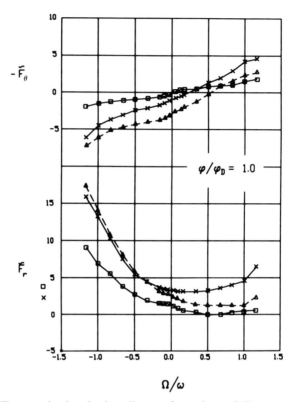

Figure 6.14   Test results for the impeller configurations of Figure 6.13: (a) squares, without spacer, (b) triangles, with spacer, (c) crosses, with grooved spacer (housing) [Ohashi et al. (1988)].

where $M' = \rho \pi r_2^2 b_2$, and e is the eccentricity radius. Without the spacer the whirl-frequency ratio is nearly zero. With the spacer it is about 0.5. Grooving the spacer (housing) reduces Ω_w to about 0.25. These results confirm that the shroud generates most of the destabilizing forces and provides the first suggestion for reducing these destabilizing forces. Presumably, the radial slots reduce the tangential velocity in the leakage annulus and thereby reduce k in the same fashion as the stator roughness in von Pragenau's damper seal of Section 4.5.

6.4   ANALYTICAL MODELS FOR IMPELLER-DIFFUSER (VOLUTE) INTERACTION FORCES

Introduction

Most of the analyses which have been developed for impeller forces have concentrated on the interaction between the flow exiting the impeller and entering either a volute or a vaned diffuser, and most of these analyses are beyond the scope of the present book (and the expertise of its author). Hence, they are only briefly summarized below.

Impeller-Diffuser / Volute Models and Analysis

Colding-Jorgensen (1980) presented the initial analysis for a two-dimensional impeller in a vaneless, logarithmic-spiral volute. A potential-flow solution is developed with the impeller modeled as a vortex which can be displaced eccentrically within the volute or given a velocity relative to the volute. The flow constraint of the volute is enforced by distributed singularities. In Colding-Jorgensen's words, "The diffuser contour is considered as a series of small linear segments. Each segment is covered by a vortex distribution of uniform strength. The vortex strength varies from segment to segment. The normal component of the flow must be zero. If we have N linear segments, we have N vortex strengths, and hence we can satisfy the condition in N points. These points are chosen as the N midpoints of the segments."

Colding-Jorgensen's model yields predictions for general stiffness and damping matrices but no added-mass matrix. His predictions for the stiffness and damping matrices for an $86°$ volute angle at nominal flow is

$$[K_{IJ}] = \begin{bmatrix} \overline{K}_{XX} & \overline{K}_{XY} \\ \overline{K}_{YX} & \overline{K}_{YY} \end{bmatrix} = \begin{bmatrix} -0.42 & 0.02 \\ -0.05 & -0.38 \end{bmatrix},$$

$$[C_{IJ}] = \begin{bmatrix} \overline{C}_{XX} & \overline{C}_{XY} \\ \overline{C}_{YX} & \overline{C}_{YY} \end{bmatrix} = \begin{bmatrix} 0.05 & 2.0 \\ -1.8 & 0.05 \end{bmatrix},$$

where X lies along the volute tongue.

Singularity methods were also used by Shoji and Ohashi (1980) to calculate fluid forces on an impeller in an unbounded volute. Small velocity perturbations in the flow field within the blade passages were introduced to model the effect of an eccentric displacement of the impeller. Shockless entry conditions are assumed for the steady flow, and vortices shed from the blades are carried downstream along the streamlines at constant velocities. Tsujimoto et al. (1984) also used a singularity approach to find reaction forces on a precessing impeller but included the volute in their analysis. No results are presented for precession force coefficients or rotordynamic coefficients. Tsujimoto et al. (1988) present refinements of the (1984) analysis together with predictions of $F_r(\Omega/\omega)$ and $F_\theta(\Omega/\omega)$ for a wide range of operating conditions; however, no comparisons are made to experimental results.

Adkins (1985) presented an analysis for impeller rotordynamic coefficients, also analyzing a 2D model. Adkins analyzed flow through a precessing and spinning impeller. The flow through the impeller is assumed to follow a spiral path. Asymmetry in the impeller discharge flow caused by the volute is accounted for via an imposed circumferential variation. Flow in the volute is defined by a continuity equation, a moment-of-momentum equation and a radial equation of motion. The impeller and volute equations are expanded in terms of constant and harmonic coefficients.

The Adkins model yields stiffness, damping, inertia and "jerk" matrices. At design conditions, his predictions for the Cal Tech test impeller are

$$[\bar{K}_{IJ}] = \begin{bmatrix} -0.40 & 0.08 \\ -0.20 & -0.36 \end{bmatrix},$$

$$[\bar{C}_{IJ}] = \begin{bmatrix} 0.05 & 2.0 \\ -2.5 & 0.05 \end{bmatrix},$$

$$[\bar{M}_{IJ}] = \begin{bmatrix} 3.0 & 0.1 \\ 0.7 & 6.0 \end{bmatrix}.$$

These values are substantially smaller than the Jery et al. measurements of Eq. (6.32).

The singularity analyses and the Adkins analysis all yield the correct form for $\bar{F}_r(f)$ and $\bar{F}_\theta(f)$ curves. They yield reasonable predictions for impellers *which are consistent with their two-dimensional assumptions*. However, because they neglect shroud forces, they consistently underestimate net impeller forces. For reduced clearances between the pump shroud and its housing, the impeller forces are substantially underestimated. The next section reviews an analysis for shroud forces.

6.5 BULK-FLOW ANALYSIS FOR SHROUD FORCES

Introduction

The impeller of Figure 6.10 has a comparatively small clearances across the leakage annulus and suggest that the bulk-flow analysis which works well for liquid seals* could also be used to calculate the rotordynamic reaction forces on the pump shroud. Childs (1989) developed such an analysis and his results and conclusions are the subject of this section; most of this section is taken from this reference.

Geometry and Kinematics

Figure 6.15 illustrates the annular leakage paths along the front and back sides of a typical shrouded impeller of a multistage centrifugal pump. The present discussion concentrates on the flow and pressure fields within the forward annulus; however, the analysis also applies to the rear annulus. As

---

*Section 4.2.

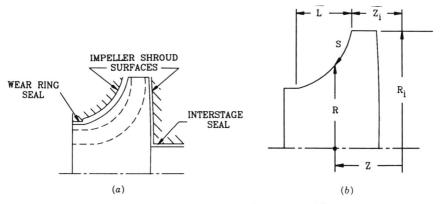

Figure 6.15 Impeller stage [Childs (1989)].

illustrated in Figure 6.16, the outer surface of the impeller is a surface of revolution formed by rotating the curve $R = R(Z)$ about the Z axis. A point on the surface may be located by the coordinates $Z, R(Z), \theta$. The length along the curve $R(Z)$ from the initial point R_i, Z_i to an arbitrary point R, Z is denoted by S and defined by

$$S = \int_{Z_i}^{Z} \sqrt{1 + \left(\frac{dR}{dZ}\right)^2}\, du = \int_{R_i}^{R} \sqrt{1 + \left(\frac{dZ}{dR}\right)^2}\, du. \tag{6.35}$$

In the equations which follow, the path coordinate S and angular coordinate θ are used as independent spatial variables. The coordinates Z, R defining the impeller surface are expressed as parametric functions of S, i.e., $Z(S), R(S)$.

Trigonometric functions of the angle γ, illustrated in Figure 6.17, are defined as follows:

$$\tan \gamma = -\frac{dR}{dZ}, \quad \cos \gamma = \frac{dZ}{dS}, \quad \sin \gamma = -\frac{dR}{dS}. \tag{6.36}$$

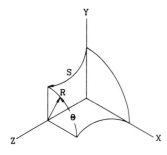

Figure 6.16 Impeller surface geometry [Childs (1989)].

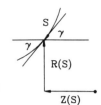

Figure 6.17 Local attitude angle of impeller surfaces [Childs (1989)].

The clearance between the impeller and the housing is denoted as $H(S, \theta, t)$, with the time dependency introduced by impeller motion. In the centered position, the clearance function depends only on S and is denoted by $H_0(S)$. Displacement of the impeller in the X and Y directions obviously causes a change in the clearance function. The Childs analysis also accounts for pitching and yawing of the impeller; however, for brevity this feature is omitted here. For small displacements of the impeller, the clearance function can be stated

$$H(S, \theta, t) = H_0(S) - X \cos \gamma \cos \theta - Y \cos \gamma \sin \theta. \qquad (6.37)$$

Observe in this equation that H_0, R, Z, $\cos \gamma$, and $\sin \gamma$ are solely functions of S, while X and Y are functions only of t.

Governing Equations

Adapting the Hirs bulk-flow equations of Section 4.2 to the present circumstances yields:

Continuity Equation

$$\frac{\partial H}{\partial t} + \frac{\partial}{\partial S}(W_s H) + \frac{1}{R} \frac{\partial}{\partial \theta}(UH) + \left(\frac{H}{R}\right) \frac{\partial R}{\partial S} W_s = 0. \qquad (6.38)$$

*Path-Momentum Equation**

$$-H \frac{\partial P}{\partial S} = \frac{\rho}{2} W_s U_s f_s + \frac{\rho}{2} W_s U_r f_r + \rho H \left(\frac{\partial W_s}{\partial t} + \frac{U}{R} \frac{\partial W_s}{\partial \theta} + W_s \frac{\partial W_s}{\partial S}\right)$$
$$- \rho H \frac{U^2}{R} \frac{dR}{dS}. \qquad (6.39)$$

---

*The continuity equation has been used to simplify the original momentum equations.

Circumferential-Momentum Equation[*]

$$-\frac{H}{R}\frac{\partial P}{\partial \theta} = \frac{\rho}{2}UU_s f_s + \frac{\rho}{2}(U - r\omega)U_r f_r$$
$$+ \rho H \left(\frac{\partial U}{\partial t} + \frac{U}{R}\frac{\partial U}{\partial \theta} + W_s \frac{\partial U}{\partial S} + \frac{UW_s}{R}\frac{dR}{dS} \right). \quad (6.40)$$

U and W_s are the circumferential and path-momentum equations, respectively, and

$$U_s = \left(W_s^2 + U^2 \right)^{1/2},$$
$$U_r = \left[W_s^2 + (U - R\omega)^2 \right]^{1/2} \quad (6.41)$$

are, respectively, the resultant bulk-flow velocities relative to the stator and rotor surfaces. Further, f_s and f_r are the stator and rotor friction factors defined by the Blasius model as

$$f_s = ns \left(\frac{2\rho H U_s}{\mu} \right)^{ms}, \qquad f_r = nr \left(\frac{2\rho H U_r}{\mu} \right)^{mr}. \quad (6.42)$$

The empirical coefficients $(ns, ms), (nr, mr)$ account for different surface roughnesses on the stator and rotor, respectively. Observe in reviewing these equations that they reduce to the seal equations if $dR/dS = 0$, and that a nonzero dR/dS introduces centrifugal and Coriolis acceleration terms.

Impeller-Annulus Geometry

For this analysis, the assumption is made that the impeller is nominally centered in its housing. Hence, *in the centered position*, the clearance function H_0 is only a function of the path coordinate S and does not depend on θ. The inlet clearance $H_0(0)$, the inlet path velocity $W_s(0)$, and the inlet radius $R(0)$ are denoted, respectively, by C_i, V_i, and R_i. In terms of these variables, leakage volumetric flow rate is defined by

$$\dot{Q} = 2R_i C_i V_i. \quad (6.43)$$

The length of the leakage path along the impeller face is defined by

$$L_s = \int_{Z_i}^{Z_i + L} \sqrt{1 + \left(\frac{dR}{dZ} \right)^2} \, dZ. \quad (6.44)$$

[*]The continuity equation has been used to simplify the original momentum equations.

Nondimensionalization and Perturbation Analysis

The governing equations define the bulk-flow velocity components (W_s, U) and the pressure P as a function of the coordinates $(R\theta, S)$ and time t. They are conveniently nondimensionalized by introducing the following variables:

$$
\begin{aligned}
w_s &= W_s/V_i, & u &= U/R_i\omega, & p &= P/\rho V_i^2, \\
h &= H/C_i, & s &= S/L_s, & r &= R/R_i, \\
\tau &= \omega t, & a &= V_i/R_i\omega, & T &= L_s/V_i.
\end{aligned}
\tag{6.45}
$$

The present analysis examines the changes in (w_s, u, p) due to changes in the clearance function $h(\theta, s, t)$ caused by small motion of the impeller within its housing. To this end, the governing equations are expanded in the perturbation variables

$$
\begin{aligned}
w_s &= w_{s0} + \epsilon w_{s1}, & h &= h_0 + \epsilon h_1, \\
u &= u_0 + \epsilon u_1, & p &= p_0 + \epsilon p_1,
\end{aligned}
\tag{6.46}
$$

where $\epsilon = e/C_i$ is the eccentricity ratio. The following equations result:

Zeroth-Order Equations

$$
rh_0 w_{s0} = 1,
$$

$$
\frac{dp_0}{ds} = \frac{1}{r}\frac{dr}{ds}\left(\frac{u_0}{a}\right)^2 + \left(\frac{1}{h_0}\frac{dh_0}{ds} + \frac{1}{r}\frac{dr}{ds}\right)w_{s0}^2 - \frac{1}{2h_0}(u_{s0}\sigma_{s0} + u_{r0}\sigma_{r0}),
$$

$$
\frac{du_0}{dz} = \left[\sigma_{s0}u_0 u_{s0} + \sigma_{r0}(u_0 - r)u_{r0}\right]/2 - \frac{u_0}{r}\frac{dr}{ds},
\tag{6.47}
$$

where

$$
\sigma_{s0} = f_{s0}(L_s/H_0), \qquad \sigma_{r0} = f_{r0}(L_s/H_0).
\tag{6.48}
$$

The continuity equation has been used to eliminate dw_{s0}/ds from the path-momentum equation.

First-Order Equations

$$
\frac{\partial w_{s1}}{\partial s} + \frac{\omega T}{r}\frac{\partial u_1}{\partial \theta} + w_{s1}\left(\frac{1}{r}\frac{dr}{ds} + \frac{1}{h_0}\frac{dh_0}{ds}\right)
$$
$$
= +\frac{h_1 w_{s0}}{h_0^2}\cdot\frac{dh_0}{ds} - \frac{1}{h_0}\left(w_{s0}\frac{\partial h_1}{\partial s} + \omega T\frac{u_0}{r}\frac{\partial h_1}{\partial \theta} + \omega T\frac{\partial h_1}{\partial \tau}\right),
\tag{6.49}
$$

$$
\frac{\partial p_1}{\partial s} + A_{2s}u_1 + A_{3s}w_{s1} + \left[\omega T\frac{\partial w_{s1}}{\partial \tau} + \omega T\frac{u_0}{r}\frac{\partial w_{s1}}{\partial \theta} + w_{s0}\frac{\partial w_{s1}}{\partial s}\right] = A_{1s}h_1,
$$

$$
a\frac{Ls}{R_i}\frac{1}{r}\frac{\partial p_1}{\partial \theta} + A_{2\theta}u_1 + A_{3\theta}w_{s1} + \left[\omega T\frac{\partial u_1}{\partial \tau} + \omega T\frac{u_0}{r}\frac{\partial u_1}{\partial \theta} + u_{s0}\frac{\partial u_1}{\partial s}\right] = A_{1\theta}h_1.
$$

Zeroth-Order-Equation Solutions

Equations (6.47) define the pressure and velocity distributions for a centered impeller position. For a known volumetric flow rate, the continuity equation completely defines w_{s0}. The equations are coupled and nonlinear and must be solved iteratively. The initial condition for $u_0(0)$ is obtained from the exit flow condition of the impeller. The inlet and discharge pressure of the impeller are known and serve, respectively, as the exit (P_e) and supply (P_s) pressures for the leakage flow along the impeller face. The inlet conditions for p_0 are obtained from the inlet relationship

$$P_s - P_0(0, \theta, t) = \rho(1 + \xi)W_{s0}^2(0, \theta, t)/2. \qquad (6.50)$$

From this relationship, the zeroth-order pressure relationship is

$$p_0(0) = P_s/\rho V_i^2 - (1 + \xi)w_{s0}^2(0)/2. \qquad (6.51)$$

The wear-ring seal at the leakage-path exit also provides a restriction, yielding a relationship of the form

$$P(L_s, \theta, t) - P_e = \frac{\rho}{2}C_{de}W_{s0}^2(L_s, \theta, t). \qquad (6.52)$$

The solution to the zeroth-order Eq. (6.47) must be developed iteratively since all of the coefficients depend on the local path velocity W_{s0}. The equations are solved by the following iterative steps:

(a) Guess or estimate V_i which then defines $w_{s0}(s)$.
(b) Calculate $p_0(0)$ from Eq. (6.51) and use a specified $u_0(0)$ as initial conditions to numerically integrate Eqs. (6.47) out to $s = 1$, i.e., the annulus exit.
(c) Based on the difference between a calculated exit pressure and the prescribed exit pressure, calculate a revised V_i and repeat the cycle until convergence is achieved.

Since C_{de} depends on the wear-ring geometry and operating conditions, an additional iteration is required to obtain the same leakage flow rate through the annulus and the seal with the same pressure and tangential velocity at the annulus exit and the seal inlet. A convergent solution defines C_{de} for the wear-ring seal.

First-Order Equations Solutions

The first-order Eqs. (6.49) define the first-order perturbations $w_{s1}(s, \theta, \tau)$, $u_1(s, \theta, \tau)$, and $p_1(s, \theta, \tau)$ resulting from the perturbed clearance function h_1.

From Eqs. (6.37) and (6.45), h_1 can be stated

$$\epsilon h_1 = -x \cos \gamma \cos \theta - y \cos \gamma \sin \theta = h_{1c}(s, \tau) \cos \theta + h_{1s}(s, \tau) \sin \theta. \tag{6.53}$$

The theta dependency of the dependent variables is eliminated by assuming the following, comparable solution format:

$$w_{s1} = w_{s1c} \cos \theta + w_{s1s} \sin \theta,$$
$$u_1 = u_{1c} \cos \theta + u_{1s} \sin \theta,$$
$$p_1 = p_{1c} \cos \theta + p_{1s} \sin \theta.$$

Substituting into Eqs. (6.49) and equating like coefficients of $\cos \theta$ and $\sin \theta$ yields six equations in the independent variables s, τ. By introducing the complex variables,

$$w_{s1} = w_{s1c} + jw_{sis}, \qquad u = u_{1c} + ju_{1s},$$
$$p_1 = p_{p1c} + jp_{1s}, \qquad h_1 = h_{1c} + jh_{1s}, \tag{6.54}$$

these real equations are reduced to three complex equations in s and τ.
From Eqs. (6.36) and (6.53), h_1 can be stated

$$\epsilon h_1 = -q \left(\frac{L}{L_s} \right) \frac{dz}{ds}, \tag{6.55}$$

where

$$q = x + jy. \tag{6.56}$$

The time dependency of the governing complex equations is eliminated by assuming harmonic seal motion of the form

$$q = q_0 e^{jf\tau}, \qquad f = \Omega / \omega, \tag{6.57}$$

where Ω is the impeller precession frequency and q_0 is a real constant. The associated harmonic solution for the dependent variables can then be stated

$$w_{s1} = \bar{w}_{s1} e^{jf\tau}, \qquad u_1 = \bar{u}_1 e^{jf\tau}, \qquad p_1 = \bar{p}_1 e^{jf\tau}. \tag{6.58}$$

Substitution from Eqs. (6.58) and (6.55) into the governing complex partial differential equation yields the following three complex ordinary equations in s

$$\frac{d}{ds}\begin{Bmatrix}\overline{w}_{s1}\\ \overline{u}_1\\ \overline{p}_1\end{Bmatrix} + [A]\begin{Bmatrix}\overline{w}_{s1}\\ \overline{u}_1\\ \overline{p}_1\end{Bmatrix} = \left(\frac{q_0}{\epsilon}\right)\begin{Bmatrix}g_1\\ g_2\\ g_3\end{Bmatrix}, \qquad (6.59)$$

where

$$[A] = \begin{bmatrix} B & -j(\omega T/r) & 0 \\ A_{3\theta}/w_{s0} & (A_{2\theta}+j\Gamma T)/w_{s0} & -j(a/rw_{s0})(L_s/R_i) \\ A_{3s}-w_{s0}B+j\Gamma T & A_{2s}+j(\omega T/r)w_{s0} & 0 \end{bmatrix},$$

$$\begin{Bmatrix}g_1\\ g_2\\ g_3\end{Bmatrix} = \left(\frac{L}{L_s}\right)\begin{Bmatrix} F_1+j(\Gamma T/h_0)(dz/ds) \\ -((A_{1\theta}/w_{s0}))(dz/ds) \\ -A_{1s}(dz/ds)-w_{s0}F_1-jw_{s0}(\Gamma T/h_0)(dz/ds) \end{Bmatrix}, \qquad (6.60)$$

$$B = \frac{1}{r}\frac{dr}{ds} + \frac{1}{h_0}\frac{dh_0}{ds},$$

$$\Gamma = \omega(f - u_{\theta 0}/r),$$

$$F_1 = \frac{u_{s0}}{h_0}\left(\frac{d^2z}{ds^2} - \frac{1}{h_0}\frac{dh_0}{ds}\frac{dz}{ds}\right).$$

The following three boundary conditions are specified for the solution of Eq. (6.59):

(a) The entrance-perturbation, circumferential velocity is zero, i.e.,

$$\overline{u}_1(0) = 0 \qquad (6.61a)$$

(b) The entrance loss at the seal entrance is defined by Eq. (6.51), and the corresponding perturbation-variable relationship is

$$\overline{p}_1(0) = -(1 + \xi)\overline{w}_{s1}(0). \qquad (6.61b)$$

(c) The relationship at the exit is provided by Eq. (6.52) and yields the following perturbation relationship:

$$\overline{p}_1(1) = C_{de}w_{s0}(1)\overline{w}_{s1}(1). \qquad (6.61c)$$

The solution approach to Eq. (6.59) for the boundary conditions of Eqs. (6.61) is covered in Section 4.5. The solution can be stated

$$\left\{\begin{matrix} \overline{w}_{s1} \\ \overline{u}_{\theta 1} \\ \overline{p}_1 \end{matrix}\right\} = \left(\frac{q_0}{\epsilon}\right) \left\{\begin{matrix} f_{1c} + jf_{1s} \\ f_{2c} + jf_{2s} \\ f_{3c} + jf_{3s} \end{matrix}\right\}. \tag{6.62}$$

Reaction Forces

The differential-force components acting on a differential-impeller surface area are

$$dF_X = -P \cos \gamma R \, d\theta \, dS \cos \theta,$$

$$dF_Y = -P \cos \gamma R \, d\theta \, dS \sin \theta.$$

Hence, the perturbed forces are

$$F_{X1} = -\epsilon \int_0^{L_s} \int_0^{2\pi} P_1 \cos \gamma \cos \theta R \, d\theta \, dS,$$

$$F_{Y1} = -\epsilon \int_0^{L_s} \int_0^{2\pi} P_1 \cos \gamma \sin \theta R \, d\theta \, dS. \tag{6.63}$$

Successive substitution from (a) Eqs. (6.45) and (6.36), (b) Eq. (6.54), and (c) Eq. (6.58) into Eqs. (6.63) yields

$$\frac{F_r + jF_\theta}{F_0} = \frac{(F_{X1} + jF_{Y1})e^{-jf\tau}}{F_0} = -\frac{\epsilon \pi L_s}{C_d L} \int_0^1 \overline{p}_1 \left(\frac{L}{L_s}\right) \frac{dz}{ds} r \, ds, \tag{6.64}$$

where

$$F_0 = 2R_i L \, \Delta P. \tag{6.65}$$

Note that

$$\Delta P = P_s - P_e = C_d \frac{\rho V_i^2}{2} \tag{6.66}$$

is the total drop along the leakage path from the impeller discharge to inlet (including the exit wear-ring seal). Substituting the solution for $\overline{p}_1$ from Eq. (6.62) into Eq. (6.64) completes the definition for the reaction forces as a function of f. Equating this result to the comparable definition from Eq.

(6.21) gives

$$f_r(f) = \frac{F_r(f)}{F_0} = -\frac{\pi}{C_d}\int_0^1 f_{3c}\frac{dz}{ds}r\,ds = -\tilde{K} - f\tilde{c} + f^2\tilde{M},$$

$$f_\theta(f) = \frac{F_\theta(f)}{F_0} = -\frac{\pi}{C_d}\int_0^1 f_{3s}\frac{dz}{ds}r\,ds = \tilde{k} - f\tilde{C} - f^2\tilde{m}_c. \quad (6.67)$$

The rotordynamic coefficients are obtained by evaluating the integrals for a range of f values and then performing a least-squares calculation. The dimensional and nondimensional coefficients are related by

$$\tilde{K} = KC_i/F_0, \qquad \tilde{k} = kC_i/F_0,$$

$$\tilde{C} = CC_i\omega/F_0, \qquad \tilde{c} = cC_i\omega/F_0,$$

$$\tilde{M} = MC_i\omega^2/F_0, \qquad \tilde{m}_c = mcC_i\omega^2/F_0. \quad (6.68)$$

Predictions and Comparison to Experimental Results

Predictions and comparisons will be made to the experimental results of Bolleter et al. (1987) for the face-seal impeller of Figure 6.10. Their tests were at best efficiency point (BEP) with the pump running at 2000 rpm, while developing 68 m of head and 130 l/sec of flow rate. The impeller has seven blades and an impeller exit angle of 22.5°. The test fluid is water at 27°C (80°F).

The present analysis requires an estimate of the ΔP across the impeller versus the total head rise of the stage. At Bolleter's suggestion, the impeller ΔP was estimated to be 70 percent of the total ΔP of the pump. An estimate of the inlet tangential velocity is also required. Fortunately, pitot-tube measurements are available, indicating that the inlet tangential velocity is approximately 50 percent of the exit impeller surface velocity; hence, $u_0(0) \cong 0.5$.

Both walls of the impeller were assumed to be smooth and represented by Yamada's (1962) test data; $mr = ms = -0.25$, $nr = ns = 0.079$. The inlet loss for the impeller, ξ, was assumed to be 0.1. The discharge coefficient for the seal was calculated iteratively as discussed above.

Figures 6.18(a) and 6.18(b) illustrate the predicted radial and circumferential force coefficients f_r and f_θ versus the whirl-frequency ratio $f = \Omega/\omega$ for the face-seal impeller. Results are presented for $u_0(0) = 0.5$, 0.6, and 0.7. The $u_0(0) = 0.5$ data of these figures are generally consistent with expectations based on experience with seals except for a slight "dip" in f_r and "bump" in f_θ. However, the peaks exhibited at higher value for $u_0(0)$ are quite unexpected. They arise due to the centrifugal acceleration term in the

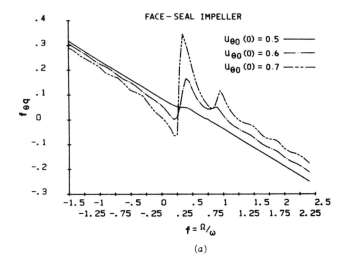

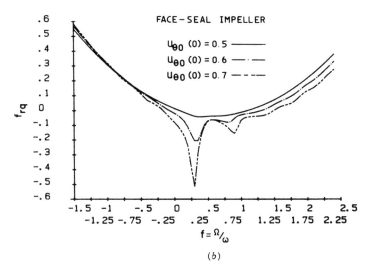

Figure 6.18 Predicted nondimensional force coefficients for the face-seal impeller: (a) tangential-force coefficient, (b) radial-force coefficient [Childs (1989)].

path-momentum equation. If the term

$$\frac{2u_0}{r}\frac{dr}{ds}\bigg/a^2 \qquad (6.69)$$

is dropped from the A_{2s} definition of Appendix D, the "peaks" are substantially eliminated from the force predictions.

TABLE 6.3   Zeroth-Order-Solution Results; C_i = 3.5 mm, C_r = 0.36 mm.

$u_0(0)$	0.500	0.600	0.700
$u_{x0}(0)$	0.880	0.950	1.000
$\mathcal{R}_{s0}$	9380	8910	8426
C_{de}	2.068	2.098	2.130
$\dot{m}$ (kg/sec)	4.448	4.225	3.997

Table 6.3 provides zeroth-order-solution results for the conventional-seal impeller of Figure 6.10(b). Observe that the leakage is reduced by increasing $u_0(0)$. Also observe the relatively high seal-inlet tangential velocity prediction for the exit seals

$$u_{x0}(0) = U_0(1)/R(1)\omega,$$

which will predictably lead to increased cross-coupled stiffness coefficients for the seal and decreased rotor stability. The $\mathcal{R}_{s0} = 2HV_i/\nu$ values on the order of 10,000 are low in comparison to the circumferential Reynolds number $\mathcal{R}_{\theta 0} = 2HU_0/\nu$, which varies along the path but is on the order of 250,000. Given this elevated Reynolds number for a comparatively low-speed pump, impeller-shroud rotordynamic coefficients would be expected to be comparatively insensitive to Reynolds number changes.

Also note from Eq. (6.45) that $a = V_i/R_i\omega = \mathcal{R}_{s0}/\mathcal{R}_{\theta 0} \cong 0.04$. Hence, $a^{-2} = 625$, and the "centrifugal acceleration term" of Eq. (6.69) has a major impact on A_{2s}.

The frequency-dependency of f_r and f_θ exhibited in Figure 6.16 for $u_0(0) = 0.6, 0.7$ cannot be modeled by the rotordynamic-coefficient model of Eq. (6.21). Stated differently, the quadratic dependency of f_r and f_θ on f, which is specified in Eq. (6.67), is simply not true. A significantly more complicated dependency is clearly in order.

The validity of the "resonance" predictions of Figure 6.16 appears to be questionable based on recent measurements by Guinzburg (1992). Guinzburg modified the Cal Tech impeller test rig by replacing the impeller with a solid "dummy" impeller with a straight tapered outside surface to simulate an impeller's surface. An external pump was then used to force "leakage" flow through the annular space between the impeller shroud and its housing. An inlet guide vane was used to prerotate the flow entering the leakage path. The inlet tangential velocity was not measured; however, calculated values up to $u_0(0) = 1.0$ were obtained without any evidence of fluid resonances. Concerning the validity of the model in the absence of resonances, Guinzburg states, "It would appear that the general trends in the magnitudes of the shroud forces are adequately predicted by the numerical model. In particular the inverse proportionality effect of the shroud clearance is predicted. However, as far as describing the particular trends for varying flow coefficient,

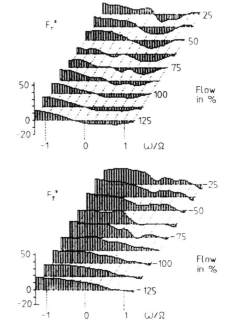

Figure 6.19 Measured $F_r(f)$ and $F_\theta(f)$ functions over a range of flow rates [Bolleter et al. (1989)].

seal clearance or inlet swirl, the numerical results do not agree with the experimental results."

The bulk-flow model is a gross simplification of the true flow field which includes strong recirculation due to flow at the rotor which is moving outwards due to centrifugal forces and flow along the stator which is driven inwards by the pressure gradient. Hence, the true flow field behaves quite differently from the present predictions. Experimental evidence for impellers with inducers [Franz and Arndt (1986)] and for impellers at reduced flow [Bolleter et al. (1989)] show impeller force coefficients which cannot be modeled by the conventional linear model of Eq. (6.21). Some of the Bolleter et al. (1989) results are illustrated in Figure 6.19. Rotordynamic modeling approaches for these nonconventional force coefficients have been demonstrated by Williams and Childs (1991), who use nonlinear whirl-frequency-dependent direct and cross-coupled stiffness coefficients.

The $u_0(0) = 0.5$ results of Figure 6.18 are reasonably modeled by a quadratic dependency of f and can be modeled by rotordynamic coefficients. A comparison of predicted and measured coefficients is provided in Table 6.4. Keeping in mind that the present theory does not account for the momentum flux exiting from the impeller or the pressure forces on the impeller exit, the comparison between theory and experiment is encouraging. The prediction of C is good. The results for k are consistent with Adkins's (1985) statement that the impeller annulus accounts for approximately one-third of the measured stiffness values in Cal Tech test results. Note that the

TABLE 6.4   Theory Versus Experiment for the Face-Seal Impeller.

	Measured	Theory
$K(N/m)$	-0.5×10^6	-0.042×10^6
$k(N/m)$	0.6×10^6	0.288×10^6
C (N sec/m)	2570	2020
c (N sec/m)	7610	2290
M (kg)	29.6	8.96
m_c (kg)	10.8	-0.009
$k/C\omega$	1.11	0.681

measured whirl-frequency ratio is substantially larger than predicted; i.e., measured results are less stabilizing than predicted.

Observe in Table 6.4 that m_c, the cross-coupled mass term, is significant. From Eq. (6.67), this term acts in the same direction as direct damping and would tend to stabilize a pump for forward precession.

Concluding Comments

As noted above, the bulk-flow model discussed here is a gross simplification of the known flow field for the annulus between a pump impeller and its shroud. However, the results confirm that a significant portion of destabilizing impeller forces arise from pressure forces on the shroud. At present, the bulk-flow analysis is the only method available for estimating shroud forces. Various more sophisticated (CFD) approaches are under development to obtain more accurate predictions of the flow field and the rotordynamic coefficients. Baskharone at Texas A & M University is working with finite-element approaches, while Nordmann and his co-workers are pursuing finite-difference methods at the University of Kaiserslautern, Germany.

6.6   HYDRAULIC IMBALANCE FOR PUMP IMPELLERS

Small imperfections in the flow paths within pump impellers can cause rotating loads at running speed which are indistinguishable from mechanical imbalance. These rotating loads are called "hydraulic imbalance" and are characterized in terms of the coefficient

$$K_{HI} = \frac{F_{HI}}{\Delta P D_2 b_2^*}, \tag{6.70}$$

where F_{HI} is the rotating load magnitude, $D_2 = 2r_2$, Δp is the impeller stage rise, and b_2^* is the exit width of the impeller including the side walls. Given that $\Delta P = \alpha \omega^2$ where α is an empirical constant, Eq. (6.70) yields

$$F_{HI} = \alpha K_{HI} D_2 b_2^* \cdot \omega^2. \qquad (6.71)$$

Hence, as with mechanical imbalance, hydraulic imbalance is a synchronous excitation whose magnitude increases with running speed. Measured values for K_{HI} vary dramatically. For precision-cast impellers, Bolleter et al. (1989) found

$$0.005 \leq K_{HI} \leq 0.015. \qquad (6.72)$$

For sand-cast impellers, Verhoeven (1988) reported

$$0.02 \leq K_{HI} \leq 0.12 \qquad (6.73)$$

In practical terms, these results show the impracticality of mechanical balancing for pumps below existing hydraulic imbalance.

6.7 CONCLUSIONS AND RECOMMENDATIONS

Given the exceptional importance of turbine and pump impeller forces on rotordynamics, the present paucity of test data and prediction methods for these elements is discouraging. Thomas and his co-workers at the Technical University of Munich have published measured results for one low-reaction turbine and suggest that the clearance-excitation factor β can be calculated from efficiency considerations. However, their measured β values on the order of 4–5 are much larger than the author's expectation based on rotordynamic modeling of SSME turbopumps. Additional data and analyses will be forthcoming shortly from Manuel Martinez-Sanchez and his co-workers at the Massachusetts Institute of Technology. Perhaps their results will yet provide a basic validated approach for calculating clearance-excitation forces for turbine stages.

With respect to pump impellers, a considerable volume of data is available from the measurements at Cal Tech, Sulzer Brothers, and the University of Tokyo. These studies indicate that the shroud forces contribute most of the destabilizing forces in an impeller, and that these forces go up (briskly) as the clearance between the shroud and its housing are reduced. Nondimensional test results from these studies can only be used confidently for impellers which "reasonably" resemble the test impellers in terms of clearances. Ohashi et al. (1988) have demonstrated that impeller destabilizing forces can be reduced by machining radial slots in the housing.

No comprehensive approach is available for predicting rotordynamic coefficients of pump impellers. Analytical methods have been developed for predicting impeller-diffuser (volute) interaction forces which seem to work

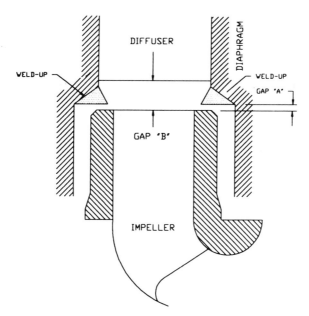

Figure 6.20 Gaps A and B after Makay and Barrett (1984).

for radial diffusers. However, the methods neglect shroud forces and seriously underestimate the destabilizing forces developed by impellers. The bulk-flow method for shroud forces seems to do a better job of predicting shroud destabilizing forces and damping. A combination of force predictions using the bulk-flow model for shroud forces and the singularity method for interaction forces *might* yield improved predictions; however, this combined approach has not been tried.

Concerning pump impellers, the reader may be simply looking for ways to improve the vibration characteristics of a pump at part load without regard to any particular analysis. Makay (1979) pioneered efforts to improve pump reliability by modifying gaps A and B of Figure 6.20. Experience has shown that the intensity of the impeller diffuser or impeller-cutwater interaction forces can be reduced by increasing gap B. Fatigue failures of diffuser blades can be eliminated by increasing this clearance. Increasing the overlap in Figure 6.20 and reducing gap A tends to isolate the impeller sidewalls or shrouds from the interaction flow at part load and eliminate axial vibration and trust-bearing problems. Success stories, based on modifications of these two gaps, are provided by Makay and Nass (1982), Makay and Barrett (1984), and Cooper et al. (1988).

Additional tests data for shroud forces will shortly be available from Cal Tech, and new impeller test programs are underway at Byron Jackson in the Netherlands and the University of Kaiserslautern in Germany. Further, new

CFD results are under development at Texas A & M University (Baskharone) and Kaiserslautern (Nordmann and co-workers).

In summary, there is a large degree of uncertainty involved in modeling turbine and impeller forces, and the reader is advised to include a wide degree of parametric variation in analyzing any rotordynamic unit whose vibration "health" is sensitive to these parameters. Published results suggest that turbine β values can range from 1.0 for any impulse turbine to 5.0 for a turbine with a high degree of reaction. For impellers, the rotordynamic coefficients would seem to lie somewhere between the Cal Tech data (for large clearances) and the Sulzer data (for tight clearances). Perhaps in some future revision of this book, these issues will be resolved more conclusively.

Modeling and analysis of turbine and impeller forces has been the major emphasis of the present chapter. We will make use of this material in Chapter 8 when we model a liquid-rocket-engine turbopump.

REFERENCES: CHAPTER 6

Adkins, D., and Brennen, C. (1986), *Origins of Hydrodynamic Forces on Centrifugal Pump Impellers*, Rotordynamic Instability Problems in High Performance Turbo-machinery, NASA CP No. 2443, proceedings of a workshop held at Texas A & M University, pp. 467–491.

Adkins, D. (1985), "Analysis of Hydrodynamic Forces of Centrifugal Pump Impellers," Ph.D. dissertation, Division of Engineering and Applied Sciences, California Institute of Technology.

Alford, J. (1965), "Protecting Turbomachinery from Self-Excited Rotor Whirl," *Journal of Engineering for Power*, 333–344.

Black, H. (1974), "Lateral Stability and Vibration of High Speed Centrifugal Pumps," in *Proceedings IUTAM Symposium on Dynamics of Rotors*, Lyngby, Denmark.

Bolleter, U., Wyss, A., Welte, I., and Stürchler, R. (1987), "Measurement of Hydrodynamic Interaction Matrices of Boiler Feed Pump Impellers," *Journal of Vibration, Stress, and Reliability in Design*, **109**, 144–151.

Bolleter, U., Leibundgut, E., Stürchler, R., and McCloskey, T. (1989), "Hydraulic Interaction and Excitation Forces of High Head Pump Impellers," in *Pumping Machinery—1989*, Proceedings of the Third Joint ASCE/ASME Mechanics Conference, La Jolla, CA, pp. 187–194.

Brennen, C., Acosta, A., and Caughey, T. (1980), "A Test Program to Measure Cross-Coupling Forces in Centrifugal Pumps and Compressors," Rotordynamic Instability Problems in High Performance Turbomachinery, NASA CP No. 2133, proceedings of a workshop held at Texas A & M University, pp. 229–235.

Chamieh, D., Acosta, A., Brennen, C., Caughey, T., and Franz, R. (1982), "Experimental Measurements of Hydrodynamic Stiffness Matrices for a Centrifugal Pump Impeller," Rotordynamic Instability Problems in High Performance Turbomachinery, NASA CP No. 2250, proceedings of a workshop held at Texas A & M University, pp. 382–398.

Childs, D. (1989), "Fluid-Structure Interaction Forces at Pump-Impeller-Shroud Surfaces for Rotordynamic Calculations," *Journal of Vibrations, Acoustics, Stress, and Reliability in Design*, **111**, 216–225.

Colding-Jørgensen, Jørgen (1980), "Effect of Fluid Forces on Rotor Stability of Centrifugal Pumps and Compressors," Rotordynamic Instability Problems in High Performance Turbomachinery, NASA CP No. 2133, proceedings of a workshop held at Texas A&M University, pp. 249–265.

Cooper, P., Makay, E., and Corsi, L. (1988), "Minimum Continuous Stable Flow in Feed Pumps," in *Proceedings, EPRI Symposium: Power Plant Pumps*, New Orleans, LA, pp. 2-97–2-132.

Guinzburg, A. (1992), "Rotordynamic Forces Generated by Discharge-to-Suction Leakage Flows in Centrifugal Pumps," California Institute of Technology, Division of Engineering and Applied Science, Report No. E249.14.

Franz, R., and Arndt, N. (1986), "Measurements of Hydrodynamic Forces on a Two-Dimensional Impeller and a Modified Centrifugal Pump," Report No. E249.4, Contract NAS 8-33108, California Institute of Technology.

Franz, R., and Arndt, N. (1986), "Measurements of Hydrodynamic Forces on the Impeller of HPOTP of the SSME," Report No. E249.2, California Institute of Technology.

Hauck, L. (1982), "Measurement and Evaluation of Swirl-Type Flow in Labyrinth Seals," Rotordynamic Instability Problems in High Performance Turbomachinery, NASA CP No. 2250, proceedings of a workshop held at Texas A&M University, pp. 242–259.

Jery, B., Brennen, C., Caughey, T., and Acosta, A. (1985), "Forces on Centrifugal Pump Impellers," in *Proceedings of the Second International Pump Symposium*, Houston, TX, pp. 21–32.

Jery, B., Acosta, A., Brennen, C., and Caughey, T. (1984), "Hydrodynamic Impeller Stiffness, Damping, and Inertia in the Rotordynamics of Centrifugal Flow Pumps," Rotordynamic Instability Problems in High Performance Turbomachinery, NASA CP No. 2338, proceedings of a workshop held at Texas A&M University, pp. 137–160.

Leie, B., and Thomas, H. (1980), "Self-Excited Rotor Whirl due to Tip-Seal Leakage Forces," Rotordynamic Instability Problems in High-Performance Turbomachinery, NASA CP No. 2133, proceedings of a workshop held at Texas A&M University.

Leie, B. (1979), "Querkräfte an Turbinenstufen und deren Einfluß auf die Laufstabilität einfacher gleitgelagerter Rotoren," dissertation, TU München.

Makay, E. (1979), "Better Understanding of Sources of Feed Pump Damage Boosts Performance, Reliability," *Power*, 72–74.

Makay, E., and Nass, D. (1982), "Gap Narrowing Rings Make Booster Pumps Quiet at Low Flow," *Power*, 87–88.

Makay, E., and Barrett, J. (1984). "Changes in Hydraulic Component Geometries Greatly Increased Power Plant Availability and Reduced Maintenance Cost: Case Histories," in *Proceedings of the First International Pump Symposium*, Texas A&M University, pp. 85–97.

Ohashi, H., Sakuraki, A., and Nishikoma, J. (1988), "Influence of Impeller and Diffuser Geometries on the Lateral Fluid Forces of Whirling Centrifugal Impeller," Rotordynamic Instability Problems in High Performance Turbomachinery, NASA CP No. 3026, proceedings of a workshop held at Texas A&M University, pp. 285–306.

Shoji, H., and Ohashi, H. (1987), "Lateral Fluid Forces on Whirling Centrifugal Impeller: Experiment in Vaneless Diffuser," Journal of Fluids Engineering, 100–106.

Shoji, H., and Ohashi, H. (1984), "Lateral Fluid Forces Acting on a Whirling Centrifugal Impeller in Vaneless and Vaned Diffusers," Rotordynamic Instability Problems in High Performance Turbomachinery, NASA CP No. 2338, proceedings of a workshop held at Texas A&M University, pp. 109–122.

Shoji, H., and Ohashi, H. (1980), "Fluid Forces on Rotating Centrifugal Impeller with Whirling Motion," Rotordynamic Instability Problems in High Performance Turbomachinery, NASA CP No. 2133, proceedings of a workshop held at Texas A&M University, pp. 317–328.

Thomas, H. (1958), "Instabile Eigenschwingungen von Turbinenläufern angefacht durch die Spaltströmungen Stopfbuschen und Beschauflungen," Bull de L'AIM, **71**, 1039–1063.

Tsujimoto, Y., Acosta, A., and Yoshida, Y. (1988), "A Theoretical Study of Fluid Forces on a Centrifugal Impeller Rotating and Whirling in a Vaned Diffuser," Rotordynamic Instability Problems in High Performance Turbomachinery, NASA CP No. 3026, proceedings of a workshop held at Texas A&M University, pp. 307–327.

Tsujimoto, Y., Acosta, A., and Brennen, C. (1984), "Two-Dimensional Unsteady Analysis of Fluid Forces on A Whirling Centrifugal Impeller in a Volute," Rotordynamic Instability Problems in High Performance Turbomachinery, NASA CP No. 2338, proceedings of a workshop held at Texas A&M University, pp. 161–172.

Urlichs, K. (1976), "Die Spaltstromüng bei Thermischen Turbo-Machinen als Ursache für die Enstehung Schwingungsanfacher Querkrafte," Ingenieur-Archiv, **45**(3), 195–208, also available in English translation, "Leakage Flow in Thermal Turbo-Machines as the Origin of Vibration-Exciting Lateral Forces," NASA TT-17409, March 1977.

Verhoeven, J. (1988), "Rotordynamic Considerations in the Design of High Speed, Multistage Centrifugal Pumps," in Proceedings, International Pump Users Symposium, Texas A&M University, pp. 81–92.

Williams, J., and Childs, D. (1991), "Influence of Impeller Shroud Forces on Pump Rotordynamics," Journal of Vibration and Acoustics, **113**, 508–515.

Yamada, Y., (1962), "Resistance of Flow Through Annulus with an Inner Rotating Cylinder," Bulletin of the JSME, **5**(18), 301–310.

Yoshida, Y., and Tsujimoto, T. (1990), "An Experimental Study of Fluid Forces on a Centrifugal Impeller Rotating and Whirling in a Volute Casing," in Proceedings, the Third International Symposium on Transport Phenomena and Dynamics of Rotating Machinery (ISROMAC-3), Honolulu, HI, **2**, 483–506.

7

DEVELOPING AND ANALYZING A SYSTEM ROTORDYNAMICS MODEL

7.1 INTRODUCTION

A review of Chapters 2–6 shows that Chapter 2 introduces the basic structural dynamic model, with the intervening chapters describing various fluid-structure-interaction forces which act on the rotor's structure to yield a complete rotordynamics model. The present chapter deals with assembly of a system rotordynamics model and techniques for analyzing a system model.

The approach for assembling a general matrix rotordynamics model is explained in Section 7.2 using the simple Stodola-Green model of Section 1.4. Changes in the initial structural-dynamics model due to fluid-structure-interaction forces are demonstrated for both an impeller force and a long seal.

Section 7.3 reviews analysis procedures for an assembled matrix system rotordynamics model, including synchronous response, stability analyses, and transient analysis for nonlinearities. Basic problems caused by the potentially large dimensionality of rotordynamics models are also discussed in this section.

Section 7.4 reviews various approaches for reducing the dimensions of completed matrix rotordynamics models including: (a) model analysis based on real eigenvectors and eigenanalysis, (b) modal analysis based on complex eigenvectors and eigenvalues, (c) component-mode synthesis based on complex eigenanalysis, and (d) direct Guyan reduction of finite-element models.

Section 7.5 reviews transfer-matrix approaches for synchronous-response and stability analysis of a complete rotor including the structure, hydrodynamic bearings, seals, etc.

7.2 COMBINING A STRUCTURAL-DYNAMICS MODEL WITH COMPONENT FORCE MODELS TO OBTAIN A COMPLETE MATRIX ROTORDYNAMICS MODEL

The contents of Chapter 2 showed how to develop a structural-dynamics model for axisymmetric rotor structures. Subsequent chapters have provided definitions for the reaction forces developed by various turbomachinery elements such as bearings and seals. The question addressed by this section is: How do you combine the structural-dynamics model and the component models to obtain a system rotordynamics model?

We will use the Stodola-Green model of Section 1.4 to initially show how a basic structural-dynamics model is modified to account for external fluid-structure-interaction forces. The basic model defined by Eqs. (1.33)–(1.36) is

$$
\begin{bmatrix} m & 0 & 0 & 0 \\ 0 & J & 0 & 0 \\ 0 & 0 & m & 0 \\ 0 & 0 & 0 & J \end{bmatrix} \begin{Bmatrix} \ddot{R}_X \\ \ddot{\beta}_Y \\ \ddot{R}_Y \\ \ddot{\beta}_X \end{Bmatrix} + \begin{bmatrix} 0 & 0 & 0 & 0 \\ 0 & 0 & 0 & -\dot{\phi}J_z \\ 0 & 0 & 0 & 0 \\ 0 & \dot{\phi}J_z & 0 & 0 \end{bmatrix} \begin{Bmatrix} \dot{R}_X \\ \dot{\beta}_Y \\ \dot{R}_Y \\ \dot{\beta}_X \end{Bmatrix}
$$

$$
+ \begin{bmatrix} [K_X] & 0 \\ 0 & [K_Y] \end{bmatrix} \begin{Bmatrix} R_X \\ \beta_Y \\ R_Y \\ \beta_X \end{Bmatrix} = \begin{Bmatrix} F_X \\ M_Y \\ F_Y \\ M_X \end{Bmatrix} + \dot{\phi}^2 \begin{Bmatrix} ma_X \\ J_{XZ} \\ ma_Y \\ J_{YZ} \end{Bmatrix} + \ddot{\phi} \begin{Bmatrix} ma_Y \\ J_{YZ} \\ -ma_X \\ J_{XZ} \end{Bmatrix},
$$

$$
\tag{7.1a}
$$

$$
J_z \ddot{\phi} = M_Z + \ddot{\beta}_X J_{XZ} + \ddot{\beta}_Y J_{YZ} - \ddot{R}_Y ma_X + \ddot{R}_X ma_Y. \tag{7.1b}
$$

The vector of external forces and moments on the right-hand side of Eq. (7.1a) includes the fluid-structure-interaction forces. Suppose, for example, that the cylinder at the end of the Stodola-Green model is a pump impeller. Then, the external forces due to the pump impeller are defined from Eq. (6.21) by

$$
- \begin{Bmatrix} F_{Xp} \\ F_{Yp} \end{Bmatrix} = \begin{bmatrix} K & k \\ -k & K \end{bmatrix} \begin{Bmatrix} R_X \\ R_Y \end{Bmatrix} + \begin{bmatrix} C & c \\ -c & C \end{bmatrix} \begin{Bmatrix} \dot{R}_X \\ \dot{R}_Y \end{Bmatrix} + \begin{bmatrix} M & m_c \\ -m_c & M \end{bmatrix} \begin{Bmatrix} \ddot{R}_X \\ \ddot{R}_Y \end{Bmatrix},
$$

$$
\tag{7.2}
$$

and are readily incorporated into the basic model by modifying the original stiffness, damping, and mass matrices to obtain

$$
[M_s] = \begin{bmatrix} m + M & 0 & m_c & 0 \\ 0 & J & 0 & 0 \\ -m_c & 0 & m + M & 0 \\ 0 & 0 & 0 & J \end{bmatrix}, \tag{7.3a}
$$

$$
[C_s] = \begin{bmatrix} C & 0 & c & 0 \\ 0 & 0 & 0 & -\dot{\phi}J_z \\ -c & 0 & C & 0 \\ 0 & \dot{\phi}J_z & 0 & 0 \end{bmatrix}, \tag{7.3b}
$$

$$
[K_s] = \begin{bmatrix} K_A + K & -K_A L/2 & k & 0 \\ -K_A L/2 & K_A L^3/3 & 0 & 0 \\ -k & 0 & K_A + K & K_A L/2 \\ 0 & 0 & K_A L/2 & K_A L^3/3 \end{bmatrix}, \quad K_A = \frac{12EI}{L^3}. \tag{7.3c}
$$

The system rotordynamic model is then

$$
[M_s]\begin{Bmatrix} \ddot{R}_X \\ \ddot{\beta}_Y \\ \ddot{R}_Y \\ \ddot{\beta}_X \end{Bmatrix} + [C_s]\begin{Bmatrix} \dot{R}_X \\ \dot{\beta}_Y \\ \dot{R}_Y \\ \dot{\beta}_X \end{Bmatrix} + [K_s]\begin{Bmatrix} R_X \\ \beta_Y \\ R_Y \\ \beta_X \end{Bmatrix} = \begin{Bmatrix} F_X \\ M_Y \\ F_Y \\ M_X \end{Bmatrix} + \dot{\phi}^2\begin{Bmatrix} ma_X \\ J_{XZ} \\ ma_Y \\ J_{YZ} \end{Bmatrix} + \ddot{\phi}\begin{Bmatrix} ma_Y \\ J_{YZ} \\ -ma_X \\ J_{XZ} \end{Bmatrix},
$$

$$
J_z\ddot{\phi} = M_Z + \ddot{\beta}_X J_{XZ} + \ddot{\beta}_Y J_{YZ} - \ddot{R}_Y ma_X + \ddot{R}_X ma_Y. \tag{7.4}
$$

Note that the system mass, damping, and stiffness matrices are functions of the running speed $\dot{\phi}$, since the impeller rotordynamic coefficients of Eq. (7.2) are functions of running speed. The vector of external forces and moments on the right of Eq. (7.4) include any additional forces not accounted for by Eq. (7.2).

As a second example, suppose that the end cylinder of the Stodola-Green model is a long ($L/D \geq 0.75$) annular liquid seal instead of an impeller. Then, the model of Eq. (4.63) holds, and the system matrices would be defined by

$$[M_s] = \begin{bmatrix} m+M & M_{\epsilon\alpha} & 0 & 0 \\ M_{\alpha\epsilon} & J+M_\alpha & 0 & 0 \\ 0 & 0 & m+M & -M_{\epsilon\alpha} \\ 0 & 0 & -M_{\alpha\epsilon} & J+M_\alpha \end{bmatrix}, \qquad (7.5a)$$

$$[C_s] = \begin{bmatrix} c & C_{\epsilon\alpha} & c & -c_{\epsilon\alpha} \\ C_{\alpha\epsilon} & C_\alpha & c_{\alpha\epsilon} & -(c_\alpha + \dot{\phi}J_z) \\ -c & -C_{\epsilon\alpha} & C & -C_{\epsilon\alpha} \\ C_{\alpha\epsilon} & c_\alpha + \dot{\phi}J_z & -C_{\alpha\epsilon} & C_\alpha \end{bmatrix}, \qquad (7.5b)$$

$$[K_s] = \begin{bmatrix} K_A+K & K_{\alpha\epsilon}-K_A L/2 & k & -k_{\epsilon\alpha} \\ K_{\alpha\epsilon}-K_A L/2 & K_\alpha+K_A L^3/3 & k_{\alpha\epsilon} & -k_\alpha \\ -k & -k_{\epsilon\alpha} & K_A+K & -K_{\epsilon\alpha}+K_A L/2 \\ k_{\alpha\epsilon} & k_\alpha & -K_{\epsilon\alpha}+K_A L/2 & K_\alpha+K_A L^3/3 \end{bmatrix}. \qquad (7.5c)$$

The long liquid-seal model provides the maximum-complexity, linear, constant-coefficient model for fluid-structure-interaction forces. The turbine model of Eq. (6.11) is an example of an exceptionally simple model which would only change the system stiffness matrix.

Moving forward from the simple Stodola-Green model to a model for a general turbomachinery unit, the typical system matrix *structural-dynamics* model for a flexible rotor can be stated:

$$\begin{bmatrix} [M_X] & 0 \\ 0 & [M_Y] \end{bmatrix} \begin{Bmatrix} (\ddot{R}_X) \\ (\ddot{\beta}_Y) \\ (\ddot{R}_Y) \\ (\ddot{\beta}_X) \end{Bmatrix} + \begin{bmatrix} 0 & 0 & 0 & 0 \\ 0 & 0 & 0 & \dot{\phi}[G] \\ 0 & 0 & 0 & 0 \\ 0 & -\dot{\phi}[G]^T & 0 & 0 \end{bmatrix} \begin{Bmatrix} (\dot{R}_X) \\ (\dot{\beta}_Y) \\ (\dot{R}_Y) \\ (\dot{\beta}_X) \end{Bmatrix}$$

$$+ \begin{bmatrix} [K_X] & 0 \\ 0 & [K_Y] \end{bmatrix} \begin{Bmatrix} (R_X) \\ (\beta_Y) \\ (R_Y) \\ (\beta_X) \end{Bmatrix} = \begin{Bmatrix} (F_X) \\ (M_Y) \\ (F_Y) \\ (M_X) \end{Bmatrix} + \dot{\phi}^2 \begin{Bmatrix} (Ma_X) \\ (J_{XZ}) \\ (Ma_Y) \\ (J_{YZ}) \end{Bmatrix} + \ddot{\phi} \begin{Bmatrix} (Ma_Y) \\ (J_{YZ}) \\ -(Ma_X) \\ (J_{XZ}) \end{Bmatrix}, \qquad (7.6a)$$

$$\bar{J}_z \ddot{\phi} = \sum_{1=1}^n M_{Zi} + \sum_{i=1}^n J_{XZi}\ddot{\beta}_{Xi} + \sum_{1=1}^n J_{YZi}\ddot{\beta}_{Yi} - \sum_{1=1}^n m a_{Xi}\ddot{R}_{Yi} + \sum_{1=1}^n m a_{Yi}\ddot{R}_{Xi}, \qquad (7.6b)$$

where

$$\bar{J}_z = \sum_{1=1}^{n} J_{zi},$$

$$(R_{Xi})^T = (R_{X1}, R_{X2}, \ldots, R_{Xn}),$$

$$(F_{Xi})^T = (F_{X1}, F_{X2}, \ldots, F_{Xn}), \text{ etc.}$$

The stiffness, gyroscopic-coupling, and inertia matrices are defined in Chapter 2 in terms of either lumped-parameter or finite-element models. Defining the fluid-structure-interaction-force contribution to the right-hand-side external force and moment vectors in a manner that is similar to the earlier developments of this section will yield speed-dependent, system stiffness, damping, and inertia matrices.

Most rotordynamics analysis is linear, involving calculation of critical speeds, synchronous response due to imbalance, or complex roots of the homogeneous equations *at constant running speed*. Hence, the $\dot{\phi}$ terms in Eq. (7.6b) are normally zero and only became important for transient simulations involving changes in running speed. In these circumstances, Eqs. (7.6) require the solution (at each time step) for the *coupled* second-order derivatives $(\ddot{R}_{Xi})$, $(\ddot{\beta}_{Yi})$, $(\ddot{R}_{Yi})$, $(\ddot{\beta}_{Xi})$, and $\ddot{\phi}$. This is a numerically expensive requirement which is complicated by the fact that the coefficients of $\ddot{\phi}$ in Eq. (7.6a) are functions of ϕ^* and hence time. In the author's experience, lateral rotor-dynamics can generally be simulated adequately by solving for $\ddot{\phi}$ separately via the following approximation to Eq. (7.6b):

$$\bar{J}_z \ddot{\phi} \cong \sum_{i=1}^{n} M_{zi}. \tag{7.7}$$

This approximation eliminates the $\ddot{\phi}(t)$ oscillation illustrated in Figure 1.46(b) but normally has no appreciable influence on the lateral vibration response of the rotor itself. Obviously, this approximation is not appropriate for coupled lateral-torsional vibrations or for simulation of a slow power-limited critical-speed transition.

7.3 ANALYSIS FOR A COMPLETE MATRIX ROTORDYNAMICS MODEL

Congratulations! Having reached this section of the book and successfully applied its contents, you should now have a completed rotordynamics model which accounts for the rotor's structural-dynamics plus fluid-structure interaction forces, external time-dependent forces, imbalance, etc. This section

---

*See Eqs. (1.2) and (1.35).

briefly reviews procedures which are available for analyzing the completed system matrix rotordynamics model.

Before considering alternative analysis procedures, recall that rotor-dynamic analysis normally seeks to determine the following characteristics:

(a) critical-speed locations,
(b) synchronous response due to imbalance,
(c) stability margins as defined by complex eigenvalue of the system, and
(d) transient response to account for system nonlinearities.

Straightforward Matrix-Based Linear Rotordynamic Calculations

Synchronous-response procedures can be straightforwardly explained in terms of the matrix formulation of the preceding section. More specifically, we wish to calculate the steady-state response of the rotor due to imbalance and product-of-inertia terms on the right-hand side of Eq. (7.6a). From Eqs. (1.1) and (1.35), the excitation terms can be written

$$\dot{\phi}^2 \left\{ \begin{matrix} (Ma_X) \\ (J_{XZ}) \\ (Ma_Y) \\ (J_{YZ}) \end{matrix} \right\} = \dot{\phi}^2 \left\{ \begin{matrix} (Ma_x) \\ (J_{xz}) \\ (Ma_y) \\ (J_{yz}) \end{matrix} \right\} \cos \phi + \dot{\phi}^2 \left\{ \begin{matrix} (-Ma_y) \\ (-J_{yz}) \\ (Ma_y) \\ (J_{xz}) \end{matrix} \right\} \sin \phi$$

$$= \omega^2 (E_c) \cos \omega t + \omega^2 (E_s) \sin \omega t, \qquad (7.8)$$

where $\phi = \omega t$. Note that (E_c) and (E_s) are constant, time-invariant, excitation vectors. The constant-speed rotordynamics model becomes

$$[M_s](\ddot{u}) + [C_s](\dot{u}) + [K_s](u) = \omega^2 (E_c) \cos \omega t + \omega^2 (E_s) \sin \omega t, \quad (7.9)$$

where $[M_s]$, $[C_s]$, and $[K_s]$ are system inertia, damping, and stiffness matrices respectively, and

$$(u)^T = \left\{ (R_X)^T (\beta_Y)^T (R_Y)^T (\beta_X)^T \right\}. \qquad (7.10)$$

Recall that $[M_s]$, $[C_s]$, and $[K_s]$ are functions of the running speed ω. Assuming a solution of the form

$$(u) = (u_c) \cos \omega t + (u_s) \sin \omega t, \qquad (7.11)$$

substituting into Eq. (7.9), and equating coefficients for $\sin \omega t$ and $\cos \omega t$

yields

$$\begin{bmatrix} [K_s] - \omega^2[M_s] & \omega[C_s] \\ -\omega[C_s] & [K_s] - \omega^2[M_s] \end{bmatrix} \begin{Bmatrix} (u_c) \\ (u_s) \end{Bmatrix} = \omega^2 \begin{Bmatrix} (E_c) \\ (E_s) \end{Bmatrix}. \quad (7.12)$$

Solution of these equations for a range of running speeds ω_j yields the desired, synchronous-steady-state response of the rotor due to imbalance and product-of-inertia disturbances.

The model of Eq. (7.9) can also be used to demonstrate an analysis procedure for calculating complex eigenvalue and eigenvectors. The state-variable, homogeneous version of Eq. (7.9) is obtained by introducing

$$(v) = (\dot{u})$$

to yield

$$\begin{bmatrix} [M_s] & 0 \\ 0 & [I] \end{bmatrix} \begin{Bmatrix} (\dot{v}) \\ (\dot{u}) \end{Bmatrix} + \begin{bmatrix} [C_s] & [K_s] \\ -[I] & 0 \end{bmatrix} \begin{Bmatrix} (v) \\ (u) \end{Bmatrix} = 0, \quad (7.13)$$

where $[I]$ is the identity matrix. The eigenvalue problem is obtained by substituting a solution of the form

$$\begin{Bmatrix} (v) \\ (u) \end{Bmatrix} = \begin{Bmatrix} (v_0) \\ (u_0) \end{Bmatrix} e^{st}, \quad (7.14)$$

yielding

$$\{s[I] - [D]\} \begin{Bmatrix} (v_0) \\ (u_0) \end{Bmatrix} = 0, \quad (7.15)$$

where

$$[D] = \begin{bmatrix} [M_s]^{-1}[C_s] & [M_s]^{-1}[K_s] \\ -[I] & 0 \end{bmatrix}. \quad (7.16)$$

The eigenvalues of the dynamic matrix $[D]$ must be calculated [generally using the Q.R. algorithm in Forsythe et al. (1977)] for a range of running speeds. They define both the linear critical speeds and the stability characteristics of the rotor system and are of the form

$$s_i = \sigma_i \pm j\omega_i.$$

A system critical speed occurs when the running speed ω coincides with an imaginary part of the eigenvalue ω_i. The system becomes linearly unstable when the real part of one of the roots σ_i becomes positive.

Analysis Complications—The Curse of Dimensionality

The preceding development glosses over considerable practical difficulties in actually solving the simultaneous Eqs. (7.12) for synchronous response or determining the complex eigenvalues and eigenvectors for the matrix $[D]$ in Eq. (7.16). The potentially large dimensionality of the equations is the major problem. Suppose, for example, that a five-stage pump with a balance-piston seal is to be analyzed. The analysis is to include the influence of forces at the two bearings, five impellers, five wearing-ring seals, four interstage seals, and the balance piston. Hence, each plane of motion will need at least 17 stations. Assuming that displacement and rotation degrees of freedom are provided at all stations, the system model for the X-Z and Y-Z planes will have 68 degrees of freedom. Hence, $[D]$ and the coefficient matrix in Eq. (7.16) will be 136×136. Solving 136 simultaneous equations is not terribly time consuming on current computers; however, repeated solution of large general (nonsymmetric) matrices for eigenvalues and eigenvectors tends to be quite time consuming.

Similar problems arise in developing transient solutions to the governing equations. Suppose that we are concerned with the influence of bearing nonlinearities on the response of our pump at constant running speed. The transient model would than be of the form

$$[M_s](\ddot{u}) + [C_s](\dot{u}) + [K_s](u) = (Q) + \omega^2(E_c)\cos \omega t + \omega^2(E_s)\sin \omega t, \tag{7.17}$$

where

$$(Q)^T = \left\{ (F_X)^T, (M_Y)^T, (F_Y)^T, (M_X)^T \right\}. \tag{7.18}$$

In this formulation, the fluid-structure-interaction forces at seals and impellers are accounted for in the system coefficient matrices, and the nonlinear bearing-reaction forces are defined in terms of the bearing motion and contained in the vector of external forces (Q). Numerical integration of Eq. (7.17) is straightforward in terms of specified initial conditions, or the problem can be started with zero initial conditions and integrated forward until a steady-state solution is obtained. The complication with direct integration of the *original* model is the requirement of an excruciatingly small time step to insure numerical stability.

To illustrate this problem, consider numerical integration of a simple lightly damped harmonic-oscillator equation of the form

$$\ddot{x} + 2\zeta\omega_n\dot{x} + \omega_n^2 x = 0,$$

$$\dot{x}(0) = \dot{x}_0, \qquad x(0) = x_0.$$

A fourth-order Runge-Kutta algorithm requires a time step on the order of $T/20$ to $T/50$ where $T = 2\pi/\omega_n$. For our proposed five-stage pump model with 34 degrees of freedom in each of the X-Z and Y-Z planes, the integration time step has to be selected at $1/20$th to $1/50$th of the period for the largest system natural frequency (or smallest period), which is on the order of $T_{34} = 2\pi/\omega_{34}$, where ω_{34} is the 34th undamped natural frequency of the rotor. To state the obvious, the integration step size required for stability of the numerical integration is generally several orders of magnitude smaller than would be expected for the frequency range of interest, which explains the numerical instability problems which were encountered by early attempts at direct integration of rotordynamic equations, e.g., Kirk and Gunter (1973). In summary, direct stable integration of the original rotordynamic model in Eq. (7.17) can be painfully slow.

7.4 APPROACHES FOR DIMENSIONAL REDUCTION OF MATRIX ROTORDYNAMIC MODELS

The dimensionality difficulty cited above is basically common to all structural-dynamics problems but is exacerbated in rotordynamics by large concentrated damping forces and nonsymmetric stiffness and inertia matrices, which result in complex eigenvalues and eigenvectors. Nonlinear bearing supports are an obvious additional complication which is not generally encountered in conventional structural dynamics. Nonetheless, most of the approaches for reducing rotordynamics-problem dimensionality have been adapted from traditional mechanical vibrations and structural dynamics. Classical modal analysis using real eigenvalues and eigenvectors represents a simple and obvious approach for reducing problem dimensionality [Timoshenko and Young (1954)]. Specifically, real system eigenvectors are used to decouple the original system, and modes whose natural frequencies are substantially above the frequency range of interest are discarded.

Real Eigenvector Modal Reduction

Childs (1972) initially proposed a modal simulation model for flexible-rotating equipment based on a model developed by Likins (1967) for spinning flexible spacecraft. His analysis accounted for gyroscopic-coupling matrices, but did not consider seal or other fluid-structure-interaction forces. Free-free rotor modes were used, with bearing forces treated as external forces and defined in terms of modal coordinates.

Black (1973) presented the first modal analysis for a rotor model including significant fluid-structure-interaction forces. He used an assumed-mode, Rayliegh-Ritz modal approach which *did not* decouple the original system; however, by retaining only the modes having a significant contribution within the frequency range of interest, Black substantially reduced problem dimensionality. Black's approach for modeling fluid-structure-interaction forces will now be reviewed (and expanded to include moment coefficients).

To explain the procedure, Eq. (7.17) is restated in terms of its X-Z and Y-Z plane components as

$$
[M_s] \begin{Bmatrix} (\ddot{u}_X) \\ (\ddot{u}_Y) \end{Bmatrix} + [C_s] \begin{Bmatrix} (\dot{u}_X) \\ (\dot{u}_Y) \end{Bmatrix} + [K_s] \begin{Bmatrix} (u_X) \\ (u_Y) \end{Bmatrix}
$$

$$
= \begin{Bmatrix} (Q_X) \\ (Q_Y) \end{Bmatrix} + \omega^2 \begin{Bmatrix} (E_{cX}) \\ (E_{cY}) \end{Bmatrix} \cos \omega t + \omega^2 \begin{Bmatrix} (E_{sX}) \\ (E_{sY}) \end{Bmatrix} \sin \omega t, \quad (7.19)
$$

where

$$
(u_X)^T = \{(R_X)^T, (\beta_Y)^T\}, \qquad (u_Y)^T = \{(R_Y)^T, (\beta_X)^T\},
$$

$$
(Q_X)^T = \{(F_X)^T, (M_Y)^T\}, \qquad (Q_Y)^T = \{(F_Y)^T, (M_X)^T\}.
$$

The system matrices are defined by

$$
[M_s] = \begin{bmatrix} [M_X] + [MF_{XX}] & [MF_{XY}] \\ [MF_{YX}] & [M_Y] + [MF_{YY}] \end{bmatrix}, \quad (7.20a)
$$

$$
[C_s] = \begin{bmatrix} [CF_{XX}] & \omega[G] + [CF_{XY}] \\ -\omega[G]^T + [CF_{YX}] & [CF_{YY}] \end{bmatrix}, \quad (7.20b)
$$

$$
[K_s] = \begin{bmatrix} [K_X] + [KF_{XX}] & [KF_{XY}] \\ [KF_{YX}] & [K_Y] + [KF_{YY}] \end{bmatrix}, \quad (7.20c)
$$

where $[M_X], [K_X], [M_Y], [K_Y]$ define the rotor's inertia properties and the rotor and its bearings' nominal stiffness properties, $[G]$ defines its gyroscopic-coupling properties, and the remaining matrices arise because of fluid-structure-interaction forces. Stiffness orthotropy of the bearings is possible, implying that $[K_X]$ and $[K_Y]$ may be fundamentally different.

Modal coordinates are introduced to decouple the rotor's *structural-dynamics* definition; viz., the coordinate transformation

$$(u_X) = [A_X](q_X), \qquad (u_Y) = [A_Y](q_Y), \qquad (7.21)$$

where

$$[A_X]^T[M_X][A_X] = [I], \qquad [A_X]^T[K_X][A_X] = [\Lambda_X],$$
$$[A_Y]^T[M_Y][A_Y] = [I], \qquad [A_Y]^T[K_Y][A_Y] = [\Lambda_Y] \qquad (7.22)$$

is introduced. The matrices $[\Lambda_X], [\Lambda_Y]$ are diagonal, containing the eigenvalues $\lambda_{Xi}^2, \lambda_{Yi}^2;\ i = 1, 2 \dots, n$ for the X-Z and Y-Z planes. In Eq. (7.22) the matrices $[A_X], [A_Y]$ decouple the inertia matrices $[M_X], [M_Y]$ and the stiffness matrices $[K_X], [K_Y]$ yielding $[I]$ and $[\Lambda_X], [\Lambda_Y]$, respectively; however, Black's assumed modes did not decouple either the inertia or stiffness matrices. Substitution from Eqs. (7.22) into Eq. (7.19) yields the modal differential equations

$$[M_q] \begin{Bmatrix} (\ddot{q}_X) \\ (\ddot{q}_Y) \end{Bmatrix} + [C_q] \begin{Bmatrix} (\dot{q}_X) \\ (\dot{q}_Y) \end{Bmatrix} + [K_q] \begin{Bmatrix} (q_X) \\ (q_Y) \end{Bmatrix}$$
$$= \begin{Bmatrix} (Q_{qX}) \\ (Q_{qY}) \end{Bmatrix} + \omega^2 \begin{Bmatrix} (E_{qcX}) \\ (E_{qcY}) \end{Bmatrix} \cos \omega t + \omega^2 \begin{Bmatrix} (E_{qsX}) \\ (E_{qsY}) \end{Bmatrix} \sin \omega t, \quad (7.23)$$

where

$$[M_q] = \begin{bmatrix} [I] + [Mq_{XX}] & [Mq_{XY}] \\ [Mq_{YX}] & [I] + [Mq_{YY}] \end{bmatrix}, \qquad (7.24a)$$

$$[C_q] = \begin{bmatrix} [Cq_{XX}] & \omega[CM] + [Cq_{XY}] \\ -\omega[CM]^T + [Cq_{YX}] & [Cq_{YY}] \end{bmatrix}, \qquad (7.24b)$$

$$[K_q] = \begin{bmatrix} [\Lambda_X] + [Kq_{XX}] & [Kq_{XY}] \\ [Kq_{YX}] & [\Lambda_Y] + [Kq_{YY}] \end{bmatrix}, \qquad (7.24c)$$

and

$$[CM] = [A_X]^T[G][A_Y],$$
$$[Mq_{XX}] = [A_X]^T[MF_{XX}][A_X],$$
$$[Mq_{XY}] = [A_X]^T[MF_{XY}][A_Y],$$
$$[Mq_{YX}] = [A_Y]^T[MF_{YX}][A_X],$$
$$[Mq_{YY}] = [A_X]^T[MF_{YY}][A_Y], \text{ etc.} \qquad (7.25)$$

Except for $[CM]$, the matrices in Eq. (7.25) are functions of running speed because of the fluid-structure-interaction coefficient matrices. Returning to Eq. (7.23),

$$(Q_{qX}) = [A_X]^T (Q_X), \qquad (E_{qcX}) = [A_X]^T (E_{cX}), \text{etc.} \qquad (7.26)$$

This modal formulation's appeal is based on the following two points:

(a) Solution for the *real* eigenvalues defined by the *symmetric* inertia and stiffness matrix pairs $[M_X],[K_X]$ and $[M_Y],[K_Y]$ is easily and efficiently accomplished.

(b) Higher-order modes in Eq. (7.21) can be discarded, e.g., modes above approximately twice the system running speed will make a small contribution to the system response. Dropping higher-order modes can *drastically* reduce the dimensionality of the original problem.

To illustrate the modal approach for calculating synchronous-response and stability solutions, assume that n modes are kept in both the X-Z and Y-Z planes. Equation (7.23) is restated as

$$[M_{qn}](\ddot{q}_n) + [C_{qn}](\dot{q}_n) + [K_{qn}](q_n)$$
$$= (Q_{qn}) + \omega^2 (E_{cqn})\cos \omega t + \omega^2 (E_{sqn})\sin \omega t, \qquad (7.27)$$

where $(q_n)^T = \{(q_{nX})^T, (q_{nY})^T\}$ contains the first n modal coordinates, and the remaining matrices and vectors are accordingly truncated. The appropriate steady-state solution is

$$(q_n) = (q_{nc})\cos \omega t + (q_{ns})\sin \omega t.$$

Substitution into Eq. (7.27) yields

$$\begin{bmatrix} [K_{qn}] - \omega^2[M_{qn}] & \omega[C_{qn}] \\ -\omega[C_{qn}] & [K_{qn}] - \omega^2[M_{qn}] \end{bmatrix} \begin{Bmatrix} (q_{nc}) \\ (q_{ns}) \end{Bmatrix} = \omega^2 \begin{Bmatrix} (E_{cqn}) \\ (E_{sqn}) \end{Bmatrix}. \qquad (7.28)$$

Equation (7.28) for the modal-coordinate solution obviously looks like Eq. (7.12) for the *direct* physical-coordinate solution and is only easier to solve because the coefficient matrix is smaller. Equation (7.28) is solved directly for the modal solutions (q_{nc}) and (q_{ns}). The physical-coordinate solution is of the form

$$(u_X) = (u_{Xc})\cos \omega t + (u_{Xs})\sin \omega t,$$
$$(u_Y) = (u_{Yc})\cos \omega t + (u_{Ys})\sin \omega t \qquad (7.29)$$

and is defined from Eq. (7.21) as

$$(u_{Xc}) = [A_{Xn}](q_{Xnc}), \qquad (u_{Xs}) = [A_{Xn}](q_{Xns}),$$
$$(u_{Yc}) = [A_{Yn}](q_{Ync}), \qquad (u_{Ys}) = [A_{Yn}](q_{Yns}). \tag{7.30}$$

The matrices $[A_{Xn}], [A_{Yn}]$ contain the first n (real) eigenvectors.

The reduced system eigenvalues are obtained from the homogeneous eigenvalues of Eq. (7.27) by defining

$$(v_n) = (\dot{q}_n) \tag{7.31}$$

to obtain

$$\begin{bmatrix} [M_{qn}] & 0 \\ 0 & [I_n] \end{bmatrix} \begin{Bmatrix} (\dot{v}_n) \\ (\dot{q}_n) \end{Bmatrix} + \begin{bmatrix} [C_{qn}] & [K_{qn}] \\ -[I_n] & [0] \end{bmatrix} \begin{Bmatrix} (v_n) \\ (q_n) \end{Bmatrix} = 0.$$

Substituting the assumed solution,

$$\begin{Bmatrix} (v_n) \\ (q_n) \end{Bmatrix} = \begin{Bmatrix} (v_{n0}) \\ (q_{n0}) \end{Bmatrix} e^{st},$$

yields the eigenvalue problem

$$\{s[I] - [D_{qn}]\} \begin{Bmatrix} (v_{n0}) \\ (q_{n0}) \end{Bmatrix} = 0,$$

where

$$[D_{qn}] = \begin{bmatrix} [M_{qn}]^{-1}[C_{qn}] & [M_{qn}]^{-1}[K_{qn}] \\ -[I_n] & 0 \end{bmatrix}. \tag{7.32}$$

Again, there is an obvious similarity between $[D_{qn}]$ in Eq. (7.32) for the modal coordinates and $[D]$ in Eq. (7.16) for the original physical coordinates; $[D_{qn}]$ is simply smaller and therefore more speedily and easily solved. The complex eigenvalues of $[D_{qn}]$ are the desired system eigenvalues. The eigenvectors of $[D_{qn}]$ are complex modal eigenvectors. Complex physical eigenvectors are obtained from the modal eigenvectors via Eq. (7.21).

One unpleasant aspect of Eq. (7.31) is the presence of $[M_{qn}]^{-1}$. Inversion of matrices is a computer-intensive activity which should be either avoided or minimized. An inspection of Eqs. (7.24a) and (7.25) shows that added-mass terms cause $[M_{qn}]$ to differ from the identity matrix. These terms are needed to model pump seals and impellers, but are zero for turbines and compressors. Hence, $[M_{qn}]^{-1}$ must be developed for pumps, but this requirement is minimized if the inverse can be performed only once instead of repeatedly at each speed. Fortunately, Section 4.6 shows that the direct added-mass coefficient M, for pump seals *with constant clearances*, is insensitive to changes in running speed. Further, Eq. (6.24a) of Section 6.3 shows that M and the cross-coupled coefficient m_c are speed invariant for pump impellers. Given these results, the direct added-mass terms for pump seals and impellers can be included at the outset in the eigenvalue problem of Eq. (7.22); i.e., Eqs. (7.22) can account for these terms via

$$[A_X]^T\{[M_X] + [MF_{XX}]\}[A_X] = [I], \qquad [A_X]^T[K_X][A_X] = [\Lambda_X],$$

$$[A_Y]^T\{[M_Y] + [MF_{YY}]\}[A_Y] = [I], \qquad [A_Y]^T[K_Y][A_Y] = [\Lambda_Y].$$
$$(7.33)$$

Obviously, a different set of real eigenvalues and eigenvectors result for this mass-matrix definition. The modal-mass matrix is now simply

$$[M_{qn}] = \begin{bmatrix} [I] & [Mq_{XY}] \\ [Mq_{YX}] & [I] \end{bmatrix}. \qquad (7.34)$$

If cross-coupled mass coefficients for pump impellers are included in the model, $[M_{qn}]^{-1}$ must be developed (once); otherwise, the transformation of Eq. (7.33) reduces $[M_{qn}]$ to an identity matrix.

The discussion above has emphasized fluid-structure-interaction forces at seals and impellers and their dependence on running speed. Recall that bearing coefficients, particularly hydrodynamic bearings, are strong functions of running speed. Hence, $[K_X], [K_Y]$ in Eqs. (7.20c) contain speed-dependent terms at the bearing locations. This would imply that the real eigenvalue solution of Eqs. (7.21) and (7.22) would have to be repeated at each running speed with a consequent major increase in computer requirements. This problem is normally circumvented by including nominal bearing stiffness values in $[K_X], [K_Y]$ and speed-dependent *deviations* from these nominal values in $[KF_{XX}], KF_{YY}]$.

Modeling with truncated modes will obviously yield a degree of approximation versus the original system model or a model which retains all of the modes, and investigators have used specific example models to establish a qualitative "feel" for errors involved in modal truncation. Li and Gunter (1981) examined a model for a dual-rotor jet engine. The gas-generator (GG) and power-turbine (PT) rotors were supported to "ground" by four and two bearings, respectively. Two bearings connected the rotors. Component rotor modes were calculated separately for the two rotors and then used in a system modal model which included the intershaft bearings. Undamped critical speeds were calculated for both the system modal model and the complete physical system model for various degrees of truncation. The results were generally excellent with errors less than 6% when component modes were retained, whose natural frequencies were less than 2.5 times the top rotor running speed. To examine stability-analysis-prediction accuracy, an outboard gas-generator bearing was modeled by a stiffness and damping coefficient to simulate a squeeze-film damper. The squeeze-film damper stiffness was 2.15×10^4 N/mm versus the original stiffness used in the component modal calculation of 7.0×10^4 N/mm. Hence, the differential stiffness to be absorbed by the modal model is $\delta K^* = -4.85 \times 10^4$ N/mm. In addition, an intershaft hydrodynamic bearing was added between the two rotors modeled by short-bearing stiffness and damping coefficients. Good correlation was obtained from reduced modal models for damped critical speeds and synchronous response. However, erratic convergence results were obtained for the real parts of damped eigenvalues which would be used for stability analysis. Li and Gunter urge caution when using modal models for stability analysis, suggesting specifically that truncated-model results be verified by recalculation with additional modes.

Lallane and Ferraris (1990) cite very favorable comparisons between a complete model and reduced modal models for damped critical speeds and synchronous response due to imbalance. A single-shaft rotor supported on identical bearings which are modeled with isotropic stiffness and damping coefficients is cited. These authors echo Gunter and Li's suggestion that predictions of truncated modal models be checked by comparison to a model with more modes.

Transient solutions for Eq. (7.27) will now be considered. Recall that transient analysis is normally only undertaken to consider the influence of nonlinearities (normally of the bearings) on rotor response or stability. A nonlinear bearing-reaction force would appear in the (Q_{qn}) vectors of the system model defined by Eq. (7.19). Suppose that there are two bearings and their reaction forces lie at stations i and j. The nonlinear bearing reactions have the form

$$
\begin{aligned}
F_{ibX} &= F_{ibX}\left(R_{Xi}, R_{Yi}, \dot{R}_{Yi}, \dot{R}_{Yi}\right), \\
F_{ibY} &= F_{ibY}\left(R_{Xi}, R_{Yi}, \dot{R}_{Xi}, \dot{R}_{Yi}\right), \\
F_{jbX} &= F_{jbX}\left(R_{Xj}, R_{Yj}, \dot{R}_{Xj}, \dot{R}_{Yj}\right), \\
F_{jbX} &= F_{jbY}\left(R_{Xj}, R_{Yj}, \dot{R}_{Xj}, \dot{R}_{Yj}\right).
\end{aligned} \tag{7.35}
$$

For linear fluid-structure-interaction forces, modal coordinates are used to completely eliminate the physical coordinates by appropriate modification of

---

*The author has experienced a notable lack of success in accounting for negative-stiffness deviations in modal models. Positive deviations are advised; i.e., component modes should be calculated for minimum stiffnesses with positive deviations developed within the modal model.

$[M_q]$, $[C_q]$, and $[K_q]$. For nonlinear bearing reactions, the following steps are required:

(a) The physical coordinates and their derivatives in Eq. (7.35) must be calculated via Eq. (7.21).
(b) Equations (7.35) are used to calculate the physical reaction forces.
(c) Equation (7.26) is used to calculate the modal reaction forces.

Step (a) is accomplished from Eq. (7.21) by

$$R_{Xi} = \sum_{l=1}^{n} A_{Xil} q_{Xl}, \qquad \dot{R}_{Xi} = \sum_{l=1}^{n} A_{Xil} \dot{q}_{Xl},$$

$$R_{Yi} = \sum_{l=1}^{n} A_{Yil} q_{Yl}, \qquad \dot{R}_{Yi} = \sum_{l=1}^{n} A_{Yil} \dot{q}_{Yl},$$

$$R_{Xj} = \sum_{l=1}^{n} A_{Xjl} q_{Xl}, \qquad \dot{R}_{Xj} = \sum_{l=1}^{n} A_{Xjl} \dot{q}_{Xl},$$

$$R_{Yj} = \sum_{l=1}^{n} A_{Yjl} q_{Yl}, \qquad \dot{R}_{Yj} = \sum_{l=1}^{n} A_{Yjl} \dot{q}_{Yl}. \qquad (7.36)$$

From Eq. (7.26), step (c) is

$$Q_{qXm} = A_{Xmi} F_{ibX} + A_{Xmj} F_{jbX}, \qquad m = 1, 2, \ldots, n,$$

$$Q_{qYm} = A_{Ymi} F_{ibY} + A_{Ymj} F_{jbY}, \qquad m = 1, 2, \ldots, n. \qquad (7.37)$$

Physical coordinates from the transient solution are obtained from Eq. (7.21).

A question which arises with transient solutions is: Which and how many modes should be used? The author has traditionally used free-free modes in analyzing transient motion of liquid-rocket-engine turbopumps for which the bearing nonlinearity involves a dead-band clearance comparable to Yamamoto's model of Section 1.9. For these applications, transient motion can involve traversal of the dead band with a complete loss of bearing stiffness when a bearing center of the rotor is within the dead band. Simulations using this approach have generally agreed well with measurements. As a rule, modes are retained whose natural frequencies are 1.5 to 2 times the top running speed.

Transient solutions of Eq. (7.27) involve solving for the highest derivatives, obtaining

$$(\dot{v}_n) = [M_{qn}]^{-1}\{(Q_{qn}) + \omega^2(E_{cqn})\cos \omega t + \omega^2(E_{sqn})\sin \omega t$$

$$- [C_{qn}](v_n) - [K_{qn}](q_n)\},$$

$$(\dot{q}_n) = (v_n). \qquad (7.38)$$

Note the presence of $[M_{qn}]^{-1}$ again, similar to our experience with Eq. (7.32). The same comment applies: $[M_{qn}]$ is the identity matrix for turbines and compressors but not pumps because of added-mass coefficients for pump seals and impellers. For pumps *with constant-clearance seals*, $[M_{qn}]$ is running-speed invariant.

There are many additional publications related to the use of real classical modes for reduction of rotordynamics-problem dimensionality. Childs (1975) considers modal models based on the Jeffcott formulation using rotor-fixed and nonspinning formulations for isotropic rotors. He also presented formulations for a dual-rotor jet-engine system with a flexible housing [Childs (1976a)], and for an orthotropic rotor suitable for a two-pole turbogenerator [Childs (1976b)]. None of these references included significant fluid-structure-interaction forces as analyzed by Black (1973). Childs (1978) used a modal model for a flexible rotor with a flexible housing described by its real eigenvalues and eigenvectors, which does include significant fluid-structure-interaction forces between the rotor and housing. Adams (1980) used a modal formulation for a flexible rotor on nonlinear hydrodynamic bearings. Childs and Bates (1978) examined the influence of residual-flexibility corrections on the response of a modal model showing an improvement in response-prediction accuracy.*

In closing, real classical eigenvalues and eigenvectors are used widely to reduce problem dimensionality. These approaches generally yield good results; however, reduced-dimension-modal-model results should be verified by confirming their insensitivity to the addition of more modes.

Complex-Eigenvector Modal Reduction

Lund (1973) first suggested applying classical linear complex eigenanalysis to the transient rotordynamics problem [Meirovitch (1967)]. The governing equations are stated in first-order form:

$$\begin{bmatrix} [M_s] & 0 \\ 0 & [I] \end{bmatrix} \begin{Bmatrix} (\dot{v}) \\ (\dot{u}) \end{Bmatrix} + \begin{bmatrix} [C_s] & [K_s] \\ -[I] & 0 \end{bmatrix} \begin{Bmatrix} (v) \\ (u) \end{Bmatrix} = \begin{Bmatrix} (Q)_\Sigma \\ (0) \end{Bmatrix}, \quad (7.39)$$

where

$$(Q)_\Sigma = (Q) + \omega^2(E_c)\cos \omega t + \omega^2(E_s)\sin \omega t. \quad (7.40)$$

As noted earlier, the eigenvalue problem is obtained by substituting a

*The rewards in accuracy improvements are generally not worth the programming complexity.

solution of the form

$$\begin{Bmatrix} (v) \\ (u) \end{Bmatrix} = \begin{Bmatrix} (v_0) \\ (u_0) \end{Bmatrix} e^{st} \tag{7.41}$$

to obtain

$$\{s[I] - [D]\} \begin{Bmatrix} (v_0) \\ (u_0) \end{Bmatrix} = 0, \tag{7.42}$$

where

$$[D] = \begin{bmatrix} [M_s]^{-1}[C_s] & [M_s]^{-1}[K_s] \\ -[I] & 0 \end{bmatrix}. \tag{7.43}$$

Eigenanalysis of $[D]$ yields $2n$ complex eigenvalues s_i and their associated eigenvectors,

$$(\phi_{ri}) = \begin{Bmatrix} s_i(a_i) \\ (a_i) \end{Bmatrix}, \tag{7.44}$$

called the right eigenvectors. The eigenvalues normally appear in complex-conjugate pairs, and their associated eigenvectors are also vectors of complex conjugates. Overdamped modes will yield real (negative) eigenvalues and eigenvectors. Eigenanalysis of $[D]^T$ yields the same eigenvalues but a different set of eigenvectors called the left eigenvectors and denoted

$$(\phi_{li}) = \begin{Bmatrix} s_i(b_i) \\ (b_i) \end{Bmatrix}.$$

The matrices of right and left eigenvectors are denoted by $[\mathscr{R}]$ and $[\mathscr{L}]$, and can be normalized to satisfy the orthogonality relationships

$$[\mathscr{L}]^T \begin{bmatrix} [M_s] & 0 \\ 0 & [I] \end{bmatrix} [\mathscr{R}] = [I],$$

$$[\mathscr{L}]^T \begin{bmatrix} [C_s] & [K_s] \\ -[I] & 0 \end{bmatrix} [\mathscr{R}] = -[\mathscr{S}],$$

where $[\mathscr{S}]$ is the diagonal matrix of complex eigenvalues.

Hence, introducing the coordinate transformation

$$\left\{\begin{matrix}(v)\\(u)\end{matrix}\right\} = [\mathscr{R}](q) \qquad (7.45)$$

into Eq. (7.39), and premultiplying the resulting equation by $[\mathscr{L}]^T$ yields

$$(\dot{q}) - [\mathscr{S}](q) = [\mathscr{L}]^T\left\{\begin{matrix}(Q)_\Sigma\\(0)\end{matrix}\right\} = (Q_c). \qquad (7.46)$$

These equations are uncoupled, and for complex-conjugate eigenvalues appear as complex-conjugate equation pairs of the form

$$\dot{q}_i - s_i q_i = (\phi_{li})^T\left\{\begin{matrix}(Q)_\Sigma\\(0)\end{matrix}\right\} = Q_{ci}, \qquad i = 1, 2, \ldots, n, \qquad (7.47a)$$

$$\dot{\bar{q}}_i - \bar{s}_i \bar{q}_i = (-\bar{\phi}_{li})^T\left\{\begin{matrix}(Q)_\Sigma\\(0)\end{matrix}\right\} = \bar{Q}_{ci}, \qquad i = 1, 2, \ldots, n, \qquad (7.47b)$$

where the overbar denotes the complex-conjugate operation. General solutions to these equations can be stated

$$q_i(t) = q_i(0) + \int_0^t e^{s_i(t-\tau)}Q_{ci}(\tau)\,d\tau,$$

$$\bar{q}_i(t) = \bar{q}_i(0) + \int_0^t e^{\bar{s}_i(t-\tau)}\bar{Q}_{ci}(\tau)\,d\tau. \qquad (7.48)$$

Note that the modal-coordinate solutions for complex-conjugate eigenvalues $s_i, \bar{s}_i$ are themselves complex conjugates. Transformation from complex modal coordinates to real physical coordinates is accomplished via Eq. (7.45). Consider just the contribution of the modal pair in Eqs. (7.48) to this transformation, i.e., consider

$$\left\{\begin{matrix}(v)_i\\(u)_i\end{matrix}\right\} = q_i(t)(\phi_{ri}) + \bar{q}_i(t)(\bar{\phi}_{ri}), \qquad (7.49)$$

and observe that the imaginary contribution of $q_i(\phi_{ri})$ is canceled by the imaginary contribution of $\bar{q}_i(\bar{\phi}_{ri})$ and that their real parts are equal. Equation (7.49) can accordingly be replaced by

$$\left\{\begin{matrix}(v)_i\\(u)_i\end{matrix}\right\} = 2\mathscr{R}e\{q_i(t)(\phi_{ri})\}, \qquad (7.50)$$

which implies that only one half of the complex modal equations must be integrated. Assuming for simplicity that all the eigenvalues are complex conjugates, only Eq. (7.47a) must be integrated, and Eq. (7.45) becomes

$$\left\{ \begin{matrix} (v) \\ (u) \end{matrix} \right\} = 2\mathscr{R}e\{[\mathscr{R}](q)\}.$$

Lund adopted this method for the solution of a uniform beam supported by multiple hydrodynamic bearings. However, he used a partial-differential-equation model for the beam rather than the discretized model of Eq. (7.39). Bearing-reaction forces were modeled with stiffness and damping coefficient matrices. Response solutions are presented due to a step increase in foundation velocity.

Since Lund's original presentation, the complex-mode formulation has been rarely used in transient rotordynamic analysis; however, complex modes have been an integral part of the component-mode-synthesis method of the following subsection.

Component-Mode Synthesis

In a series of articles Nelson and his co-workers extended the component-mode synthesis work of Hurty (1965) and Craig and Bampton (1968) for rotordynamics. Glascow and Nelson (1979) first applied the method to stability analysis; Nelson and Meacham (1981) and Nelson et al. (1983) applied the method to transient analysis.

The model of Figure 7.1 consists of four point masses supported by an elastic beam which is itself supported by four springs will be used to explain the component-mode-synthesis approach. The outside supports are assumed to be either nonlinear or linear but subject to variation for purposes of

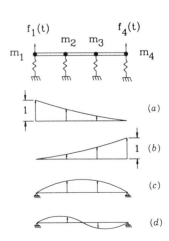

Figure 7.1   Component-mode synthesis example.

analysis. The center springs are linear, and their stiffness coefficients are fixed. The initial physical model is stated

$$\begin{bmatrix} m_1 & 0 & 0 & 0 \\ 0 & m_2 & 0 & 0 \\ 0 & 0 & m_3 & 0 \\ 0 & 0 & 0 & m_4 \end{bmatrix} \begin{Bmatrix} \ddot{x}_1 \\ \ddot{x}_2 \\ \ddot{x}_3 \\ \ddot{x}_4 \end{Bmatrix} + \begin{bmatrix} k_{11} & k_{12} & k_{13} & k_{14} \\ k_{21} & k_{22} & k_{23} & k_{24} \\ k_{31} & k_{32} & k_{33} & k_{34} \\ k_{41} & k_{42} & k_{43} & k_{44} \end{bmatrix} \begin{Bmatrix} x_1 \\ x_2 \\ x_3 \\ x_4 \end{Bmatrix} = \begin{Bmatrix} f_1 \\ 0 \\ 0 \\ f_4 \end{Bmatrix}.$$

$$(7.51)$$

For reasons which will become apparent, the coordinates are rearranged as

$$\begin{bmatrix} m_1 & 0 & 0 & 0 \\ 0 & m_4 & 0 & 0 \\ 0 & 0 & m_2 & 0 \\ 0 & 0 & 0 & m_3 \end{bmatrix} \begin{Bmatrix} \ddot{x}_1 \\ \ddot{x}_4 \\ \ddot{x}_2 \\ \ddot{x}_3 \end{Bmatrix} + \begin{bmatrix} k_{11} & k_{14} & k_{12} & k_{13} \\ k_{41} & k_{44} & k_{42} & k_{43} \\ k_{21} & k_{24} & k_{22} & k_{23} \\ k_{31} & k_{34} & k_{32} & k_{33} \end{bmatrix} \begin{Bmatrix} x_1 \\ x_4 \\ x_2 \\ x_3 \end{Bmatrix} = \begin{Bmatrix} f_1 \\ f_4 \\ 0 \\ 0 \end{Bmatrix}.$$

$$(7.52)$$

A coordinate transformation will be introduced with vectors based on (a) the two static displacement vectors of Figures 7.1(a) and 7.1(b), and (b) the mode shapes of Figures 7.1(c) and 7.1(d). The *static* vectors are obtained from the last two rows of the *stiffness* matrix in Eq. (7.52), via

$$x_1 = 1, \quad x_4 = 0, \quad \begin{Bmatrix} x_2 \\ x_3 \end{Bmatrix} = - \begin{bmatrix} k_{22} & k_{23} \\ k_{32} & k_{33} \end{bmatrix}^{-1} \begin{Bmatrix} k_{21} \\ k_{31} \end{Bmatrix} = \begin{Bmatrix} b_{12} \\ b_{13} \end{Bmatrix},$$

$$x_1 = 0, \quad x_4 = 1, \quad \begin{Bmatrix} x_2 \\ x_3 \end{Bmatrix} = - \begin{bmatrix} k_{22} & k_{23} \\ k_{32} & k_{33} \end{bmatrix}^{-1} \begin{Bmatrix} k_{24} \\ k_{34} \end{Bmatrix} = \begin{Bmatrix} b_{22} \\ b_{23} \end{Bmatrix}.$$

Those results can be used to define the *interior* coordinates x_2, x_3 in terms of the *boundary* coordinates x_1, x_4 as

$$\begin{Bmatrix} x_2 \\ x_3 \end{Bmatrix} = \begin{bmatrix} b_{12} & b_{22} \\ b_{13} & b_{23} \end{bmatrix} \begin{Bmatrix} x_1 \\ x_4 \end{Bmatrix} = [B] \begin{Bmatrix} x_1 \\ x_4 \end{Bmatrix}. \qquad (7.53)$$

The eigenvalues and eigenvectors corresponding to Figures 7.1(c) and 7.1(d) are obtained from the last two rows of Eq. (7.52) with $x_1 = x_4 = 0$; viz., the equation

$$\begin{bmatrix} m_2 & 0 \\ 0 & m_3 \end{bmatrix} \begin{Bmatrix} \ddot{x}_2 \\ \ddot{x}_3 \end{Bmatrix} + \begin{bmatrix} k_{22} & k_{23} \\ k_{32} & k_{33} \end{bmatrix} \begin{Bmatrix} x_1 \\ x_2 \end{Bmatrix} = 0 \qquad (7.54)$$

yields the eigenvalue problem

$$\left\{ -\lambda_i^2 \begin{bmatrix} m_2 & 0 \\ 0 & m_3 \end{bmatrix} + \begin{bmatrix} k_{22} & k_{23} \\ k_{32} & k_{33} \end{bmatrix} \right\} \begin{Bmatrix} a_{i2} \\ a_{i3} \end{Bmatrix} = 0.$$

The eigenvectors can be normalized to satisfy

$$[a]^T \begin{bmatrix} m_2 & 0 \\ 0 & m_3 \end{bmatrix} [a] = [I],$$

$$[a]^T \begin{bmatrix} k_{22} & k_{23} \\ k_{32} & k_{33} \end{bmatrix} [a] = \begin{bmatrix} \lambda_i^2 & 0 \\ 0 & \lambda_3^2 \end{bmatrix} = [\Lambda],$$

where $[a]$ is the matrix of eigenvectors,

$$[a] = \begin{bmatrix} a_{12} & a_{22} \\ a_{13} & a_{23} \end{bmatrix}. \tag{7.55}$$

The desired coordinate transformation defines the interior coordinates as the static deflection due to boundary coordinates defined by Eq. (7.53) plus modal coordinates defined in terms of the eigenvectors in Eq. (7.55); viz.,

$$\begin{Bmatrix} x_2 \\ x_3 \end{Bmatrix} = \begin{bmatrix} b_{12} & b_{22} \\ b_{13} & b_{23} \end{bmatrix} \begin{Bmatrix} x_1 \\ x_4 \end{Bmatrix} + \begin{bmatrix} a_{12} & a_{22} \\ a_{13} & a_{23} \end{bmatrix} \begin{Bmatrix} q_1 \\ q_2 \end{Bmatrix}. \tag{7.56}$$

The proposed transformation does not change the boundary coordinates; hence, the complete transformation is

$$\begin{Bmatrix} x_1 \\ x_4 \\ x_2 \\ x_3 \end{Bmatrix} = \begin{bmatrix} 1 & 0 & 0 & 0 \\ 0 & 1 & 0 & 0 \\ b_{12} & b_{22} & a_{12} & a_{22} \\ b_{13} & b_{23} & a_{13} & a_{23} \end{bmatrix} \begin{Bmatrix} x_1 \\ x_4 \\ q_1 \\ q_2 \end{Bmatrix} = [A] \begin{Bmatrix} x_1 \\ x_4 \\ q_1 \\ q_2 \end{Bmatrix}. \tag{7.57}$$

Substituting into Eq. (7.52) and multiplying the resulting equation by $[A]^T$ yields

$$\begin{bmatrix} [M_{11}] & [M_{12}] \\ [M_{21}] & [I] \end{bmatrix} \begin{Bmatrix} \ddot{x}_1 \\ \ddot{x}_4 \\ \ddot{q}_1 \\ \ddot{q}_2 \end{Bmatrix} + \begin{bmatrix} [K_{11}] & [K_{12}] \\ [K_{21}] & [\Lambda] \end{bmatrix} \begin{Bmatrix} x_1 \\ x_4 \\ q_1 \\ q_2 \end{Bmatrix} = \begin{Bmatrix} f_1 \\ f_4 \end{Bmatrix}. \tag{7.58}$$

Modal reduction possibilities arise through discarding higher-frequency modes in the matrix $[a]$ in Eq. (7.55). Because of the simple nature of the present example, only one mode could reasonably be discarded, replacing two interior coordinates with one modal coordinate. However, a more representative problem would yield a replacement of many interior degrees of freedom by a limited number of modal coordinates. One evident disadvantage of the component-mode-synthesis (CMS) method is that the final equations are not uncoupled. The example problem has a spring support at each coordinate; however, this is obviously not a requirement. The approach is equally valid if the coordinates are unsupported.

Observe in reviewing the current development that the modal-transformation definition has nothing to do with the nature of the supports at the boundary coordinates. Hence, the CMS method can be readily used for analysis of nonlinear supports via transient analyses. To illustrate this statement, suppose that the boundary support reactions are defined by

$$R_1 = -NL_1(x_1, \dot{x}_1), \qquad R_4 = -NL_4(x_4, \dot{x}_4).$$

Hence, in developing the original stiffness matrix, no support would be included at the boundaries, and the final equations would have the form

$$\begin{bmatrix} [M_{11}] & [M_{12}] \\ [M_{21}] & [I] \end{bmatrix} \begin{Bmatrix} \ddot{x}_1 \\ \ddot{x}_4 \\ \ddot{q}_1 \\ \ddot{q}_2 \end{Bmatrix} + \begin{bmatrix} [K_{11}] & [K_{12}] \\ [K_{21}] & [\Lambda] \end{bmatrix} \begin{Bmatrix} x_1 \\ x_4 \\ q_1 \\ q_2 \end{Bmatrix} = \begin{Bmatrix} f_1 - NL_1 \\ f_4 - NL_4 \end{Bmatrix}.$$

The CMS method is applied to the system rotordynamics model of Eq. (7.39) by rearranging the coordinates into boundary and interior variables in the following fashion;

$$(w) = \begin{Bmatrix} (w_B) \\ (w_I) \end{Bmatrix} = \begin{Bmatrix} (v_B) \\ (u_B) \\ (v_I) \\ (u_I) \end{Bmatrix}. \tag{7.59}$$

With this reordering, the original system model can be stated

$$\begin{bmatrix} A_{BB} & A_{BI} \\ A_{IB} & A_{II} \end{bmatrix} \begin{Bmatrix} (\dot{w}_B) \\ (\dot{w}_I) \end{Bmatrix} + \begin{bmatrix} B_{BB} & B_{BI} \\ B_{IB} & B_{II} \end{bmatrix} \begin{Bmatrix} (w_B) \\ (w_I) \end{Bmatrix} = \begin{Bmatrix} (Q_B) \\ (Q_I) \end{Bmatrix}. \tag{7.60}$$

The constrained normal modes are obtained by setting the boundary coordinates to zero, obtaining the following free vibration equations for the interior coordinates:

$$[A_{II}](\dot{w}_I) + [B_{II}](w_I) = 0. \tag{7.61}$$

Eigenanalysis of those equations yields matrices of right and left eigenvectors which satisfy

$$[\mathcal{L}]^T[A_{II}][\mathcal{R}] = [I], \qquad [\mathcal{L}]^T[B_{II}][\mathcal{R}] = -[\mathcal{S}], \tag{7.62}$$

where $[\mathcal{S}]$ is a diagonal matrix of eigenvalues. Complex modal coordinates are introduced via the transformation

$$(w_I) = [\mathcal{R}](q). \tag{7.63}$$

Substitution from Eq. (7.63) into Eq. (7.61) and multiplying the resultant equation by $[\mathcal{L}]^T$ yields the uncoupled modal equations

$$(\dot{q}) - [\mathcal{S}](q) = 0. \tag{7.64}$$

This development uses complex rather than real eigenvalues and eigenvectors but is otherwise identical with the analysis between Eq. (7.54) and (7.55).

The static constraint modes are obtained from the reordered stiffness matrix of Eq. (7.39):

$$\begin{bmatrix} K_{SBB} & K_{SBI} \\ K_{SIB} & K_{SII} \end{bmatrix} \begin{Bmatrix} (u_B) \\ (u_I) \end{Bmatrix} = \begin{Bmatrix} (Q)_\Sigma \\ (0) \end{Bmatrix}. \tag{7.65}$$

The bottom set of equations defines the interior coordinates as a function of the boundary coordinates via

$$(u_I) = -[K_{SII}]^{-1}[K_{SIB}](u_B) = [B](u_B). \tag{7.66}$$

This development coincides with the earlier development of Eq. (7.53). The interior coordinates are defined as the sum of motion due to the boundary coordinates acting through the matrix $[B]$ plus the modal motion due to constraint modes; i.e.,

$$\begin{Bmatrix} (v_I) \\ (u_I) \end{Bmatrix} = \begin{bmatrix} [B] & 0 \\ 0 & [B] \end{bmatrix} \begin{Bmatrix} (v_B) \\ (u_B) \end{Bmatrix} + [\mathcal{R}](q) \tag{7.67}$$

or

$$(w_I) = [\beta](w_B) + [\mathcal{R}](q), \tag{7.68}$$

where

$$[\beta] = \begin{bmatrix} [B] & 0 \\ 0 & [B] \end{bmatrix}. \tag{7.69}$$

The complete transformation is

$$\begin{Bmatrix} (w_B) \\ (w_I) \end{Bmatrix} = \begin{bmatrix} [I] & 0 \\ [\beta] & [\mathscr{R}] \end{bmatrix} \begin{Bmatrix} (w_B) \\ (q) \end{Bmatrix} = [\Psi\mathscr{R}] \begin{Bmatrix} (w_B) \\ (q) \end{Bmatrix} \tag{7.70}$$

or

$$(w) = [\Psi\mathscr{R}](p), \tag{7.71}$$

which can be profitably compared to the earlier simplified result in Eq. (7.57). The related transformation matrix involving the left eigenvectors is defined by

$$[\Psi\mathscr{L}] = \begin{bmatrix} [I] & 0 \\ [\beta] & [\mathscr{L}] \end{bmatrix}. \tag{7.72}$$

Substituting from Eq. (7.71) into Eq. (7.60) and multiplying the resultant equation by $[\Psi\mathscr{L}]^T$ yields transformed system equations of the form

$$[A_p](\dot{p}) + [B_p](\dot{p}) = (P). \tag{7.73}$$

The problem dimensionality is reduced by retaining modes whose natural frequencies are within or moderately above the running speed of the rotor. While $[A_p]$ and $[B_p]$ are coupled, they are normally much smaller than the original system matrices of Eq. (7.60). The reduced modal model can be used for stability, synchronous-response, or transient analysis. Glascow and Nelson (1979) showed a rapid convergence to calculated system eigenvalues using this method with a very restricted number of constrained modes.

As noted earlier, the CMS method is attractive for modeling nonlinear supports at boundary coordinates because the modal development is independent of the nature of the supports at boundary coordinates. Note particularly that boundary coordinates (with their associated constraint modes) are only required at points of nonlinear supports or nonlinear connections. Linear fluid-structure-interaction forces are accounted for in the original mass, damping, and stiffness matrices.

While Nelson et al. used constrained complex modes to decouple the constrained structure, real modes can obviously be used for a CMS analysis. Crandall and Yeh (1987) use independent modes which do not necessarily uncouple anything.

Guyan Reduction

Rouch and Kao (1980) suggested using the Guyan (1965) reduction approach for direct reduction of finite-element-based rotordynamic models. The Guyan reduction approach is similar to the component-mode-synthesis method in using the system stiffness matrix to define interior coordinates in terms of boundary coordinates, but uses this approach to directly reduce problem dimensionality. To illustrate the procedure, assume in Eq. (7.17) that the coordinates have been rearranged to the form

$$(u) = \left\{ \begin{array}{c} (u_r) \\ (u_d) \end{array} \right\}, \tag{7.74}$$

where (u_r) represents coordinates which are to be retained in the model, while (u_d) represents coordinates to be dropped. Generally speaking, retained coordinates would coincide with lumped disks, external-force locations, imbalance locations, bearing locations, and other points within the model which the analyst feels to be important. Discarded-coordinate locations would correspond to points in the model which the analyst views as noncritical or of secondary interest.

The static model from Eq. (7.17) can be partitioned into

$$\begin{bmatrix} K_{srr} & K_{srd} \\ K_{sdr} & K_{sdd} \end{bmatrix} \left\{ \begin{array}{c} (u_r) \\ (u_d) \end{array} \right\} = \left\{ \begin{array}{c} (F_r) \\ (0) \end{array} \right\}. \tag{7.75}$$

The bottom set of equations yields

$$(u_d) = -[K_{sdd}]^{-1}[K_{sdr}](u_r) = [B](u_r), \tag{7.76}$$

and the complete transformation is

$$\left\{ \begin{array}{c} (u_r) \\ (u_d) \end{array} \right\} = \begin{bmatrix} [I] \\ [B] \end{bmatrix} (u_r) = [B^*](u_r). \tag{7.77}$$

Substitution from Eq. (7.77) into Eq. (7.17) and premultiplying the resultant equation by $[B^*]^T$ yields

$$[M_s^*](\ddot{u}_r) + [C_s^*](\dot{u}_r) + [K_s^*](u_r) = (Q_r) + \omega^2(E_c)\cos \omega t + \omega^2(E_s)\sin \omega t. \tag{7.78}$$

The right-hand side is unchanged in format because the discarded coordinates did not include external force or imbalance locations. The reduced mass, damping, and stiffness matrices are

$$[M_s^*] = [B^*]^T[M_s][B^*],$$

$$[C_s^*] = [B^*]^T[C_s][B^*],$$

$$[K_s^*] = [B^*]^T[K_s][B^*]. \tag{7.79}$$

Recall however that $[M_s]$, $[C_s]$, and $[K_s]$ have been rearranged and reordered consistent with the partitioning of retained and discarded coordinates of Eq. (7.73).

The central assumption of the Guyan reduction approach is that the *dynamic* relationship between retained and discarded coordinates is the same as the *static* relationship of Eq. (7.76). The reduced Guyan model of Eq. (7.78) is not uncoupled, and can be used for stability, synchronous-response, and transient analysis. Rouch and Kao provide comparisons between results for reduced models of various dimensions with results for a complete finite-element model for an industrial compressor on hydrodynamic bearings. The complete model had 46 translational degrees of freedom and 5 rotational degrees of freedom. The least-dimension reduced model had 9 translational degrees of freedom and no rotational degrees of freedom but gave quite good correlations with the original complete model out through the fourth mode.

7.5 TRANSFER-MATRIX METHODS FOR SYNCHRONOUS RESPONSE AND STABILITY CALCULATIONS

The Myklestad-Prohl transfer-matrix approach for calculating undamped eigenvalues and eigenvectors was introduced in Section 2.4. The transfer-matrix idea can be readily extended to account for general fluid-structure-interaction forces and used for calculating synchronous response due to imbalance and damped eigenvalues and eigenvectors.

Synchronous Response—Lund and Orcutt (1967)

Lund and Orcutt (1967) initially presented an algorithm for calculating synchronous response of a flexible rotor on hydrodynamic bearings, and the following development is based on their work. New factors to be accounted for, which were absent in Section 2.4, are (a) the presence of excitation, (b) the possibility of a bearing or other fluid-structure-interaction force elements coinciding with a lumped-disc station, and (c) independent motion in the X-Z and Y-Z planes. Using the notation of Section 2.4, the appropriate equations of motion for the rigid body at station i (without fluid-structure-interaction forces) is

$$m_i \ddot{R}_{iX} = V_{iX}^l - V_{iX}^r + \omega^2 m_i a_{iX},$$

$$m_i \ddot{R}_{iY} = V_{iY}^l - V_{iY}^r + \omega^2 m_i a_{iY}, \tag{7.80}$$

$$J_i \ddot{\beta}_{iY} + \omega \bar{J}_i \dot{\beta}_{iX} = M_{iY}^l - M_{iY}^r - \omega^2 J_{iYZ},$$

$$J_i \ddot{\beta}_{iX} - \omega \bar{J}_i \dot{\beta}_{iY} = M_{iX}^l - M_{iX}^r + \omega^2 J_{iXZ}, \tag{7.81}$$

where the imbalance and product-of-inertia excitations are

$$a_{iX} = a_{ix} \cos \omega t - a_{iy} \sin \omega t,$$

$$a_{iY} = a_{ix} \sin \omega t + a_{iy} \cos \omega t, \tag{7.82}$$

$$J_{iXZ} = J_{ixz} \cos \omega t - J_{iyz} \sin \omega t,$$

$$J_{iYZ} = J_{ixz} \sin \omega t + J_{iyz} \cos \omega t. \tag{7.83}$$

Recall that the shear (V_{iX}^r, V_{iX}^l, etc.) and moment (M_{iX}^r, M_{iX}^l, etc.) terms account for the structural forces acting on the rigid body due to relative motion of adjacent rigid bodies. Equations (7.80) and (7.81) are taken from Eqs. (2.1) and (2.8), respectively.

Equations (7.80) and (7.81) vary with the type of forces which act on the rigid body; e.g., if rigid body i coincides with a pump impeller, then the impeller forces are modeled by Eq. (6.21), and Eq. (7.80) becomes

$$\begin{bmatrix} m_i + M & m_c \\ -m_c & m_i + M \end{bmatrix} \begin{Bmatrix} \ddot{R}_{iX} \\ \ddot{R}_{iY} \end{Bmatrix} + \begin{bmatrix} C & c \\ -c & C \end{bmatrix} \begin{Bmatrix} \dot{R}_{iX} \\ \dot{R}_{iY} \end{Bmatrix} + \begin{bmatrix} K & k \\ -k & K \end{bmatrix} \begin{Bmatrix} R_{iX} \\ R_{iY} \end{Bmatrix}$$

$$= \begin{Bmatrix} V_{iX}^l \\ V_{iY}^l \end{Bmatrix} - \begin{Bmatrix} V_{iX}^r \\ V_{iY}^r \end{Bmatrix} + \omega^2 m_i \begin{Bmatrix} a_{iX} \\ a_{iY} \end{Bmatrix}. \tag{7.84}$$

However, if rigid body i coincided with a long seal, the model of Eq. (4.63) defines the reaction forces and would yield

$$
[M_i]\begin{Bmatrix} \ddot{R}_{iX} \\ \ddot{R}_{iY} \\ \ddot{\beta}_{iY} \\ \ddot{\beta}_{iX} \end{Bmatrix} + [C_i]\begin{Bmatrix} \dot{R}_{iX} \\ \dot{R}_{iY} \\ \dot{\beta}_{iY} \\ \dot{\beta}_{iX} \end{Bmatrix} + [K_i]\begin{Bmatrix} R_{iX} \\ R_{iY} \\ \beta_{iY} \\ \beta_{iX} \end{Bmatrix} = \begin{Bmatrix} V^l_{iX} \\ V^l_{iY} \\ M^l_{iY} \\ M^l_{iX} \end{Bmatrix} - \begin{Bmatrix} V^r_{iX} \\ V^r_{iY} \\ M^r_{iY} \\ M^r_{iX} \end{Bmatrix} + \omega^2\begin{Bmatrix} m_i a_{iX} \\ m_i a_{iY} \\ -J_{iYZ} \\ J_{iXZ} \end{Bmatrix},
$$

$$(7.85)$$

$$
[M_i] = \begin{bmatrix} m_i + M & 0 & M_{\epsilon\alpha} & 0 \\ 0 & m_i + M & 0 & -M_{\epsilon\alpha} \\ M_{\alpha\epsilon} & 0 & J_i + M_\alpha & 0 \\ 0 & -M_{\alpha\epsilon} & 0 & J_i + M_\alpha \end{bmatrix}, \qquad (7.86a)
$$

$$
[C_i] = \begin{bmatrix} C & c & C_{\epsilon\alpha} & -c_{\epsilon\alpha} \\ c & C & -c_{\epsilon\alpha} & -C_{\epsilon\alpha} \\ C_{\alpha\epsilon} & c_{\alpha\epsilon} & C_\alpha & -c_\alpha + \bar{\omega}\bar{J}_i \\ c_{\alpha\epsilon} & C_{\alpha\epsilon} & c_\alpha - \bar{\omega}\bar{J}_i & C_\alpha \end{bmatrix}, \qquad (7.86b)
$$

$$
[K_i] = \begin{bmatrix} K & k & K_{\epsilon\alpha} & -k_{\epsilon\alpha} \\ -k & K & -k_{\epsilon\alpha} & -K_{\epsilon\alpha} \\ K_{\alpha\epsilon} & k_{\alpha\epsilon} & K_\alpha & -k_\alpha \\ k_{\alpha\epsilon} & -K_{\alpha\epsilon} & k_\alpha & K_\alpha \end{bmatrix}. \qquad (7.86c)
$$

Hence, in the transfer-matrix approach, component models for fluid-structure interaction simply modify the station equations of motion where they act, and there is no system-matrix development.

To explain the synchronous-response procedure of Lund and Orcutt, we will use their equations for a station acted on by a hydrodynamic bearing with only imbalance excitation. The appropriate governing equations are

$$
m_i\ddot{R}_{iX} = V^l_{iX} - V^r_{iX} - K_{iXX}R_{iX} - K_{iXY}R_{iY} - C_{iXX}\dot{R}_{iX} - C_{iXY}\dot{R}_{iY} + \omega^2 m_i a_{iX},
$$

$$
m_i\ddot{R}_{iY} = V^l_{iY} - V^r_{iY} - K_{iYX}R_{iX} - K_{iYY}R_{iY} - C_{iYX}\dot{R}_{iX} - C_{iYY}\dot{R}_{iY} + \omega^2 m_i a_{iY},
$$

$$
J_i\ddot{\beta}_{iY} = M^l_{iY} - M^r_{iY} - \bar{\omega}\bar{J}_i\dot{\beta}_{iX},
$$

$$
J_i\ddot{\beta}_{iX} = M^l_{iX} - M^r_{iX} + \bar{\omega}\bar{J}_i\dot{\beta}_{iY}. \qquad (7.87)
$$

Synchronous response is obtained by assuming a solution of the form

$$R_{iX} = R_{iXc} \cos \omega t + R_{iXs} \sin \omega t,$$
$$R_{iY} = R_{iYc} \cos \omega t + R_{iYs} \sin \omega t, \text{ etc.,} \tag{7.88}$$

substituting into the governing equations, and equating the coefficients of $\cos \omega t$ and $\sin \omega t$ to zero. This approach yields eight real equations which can be reduced to four complex equations by adding the $\cos \omega t$ equation to ($\sin \omega t$ equation) $\times j$ for each of the original equations. The resulting complex equations are

$$V_{iX}^r = V_{iX}^l + \left(\omega^2 m_i - Z_{iXX}\right)R_{iX} - Z_{iXY}R_{iY} + \omega^2 m_i \bar{a}_i,$$
$$V_{iY}^r = V_{iY}^l - Z_{iYX}R_{iX} + \left(\omega^2 m_i - Z_{iYY}\right)R_{iY} + j\omega^2 m_i \bar{a}_i,$$
$$M_{iY}^r = M_{iY}^l + \omega^2 J_i \beta_{iY} + j\omega \bar{J}_i \beta_{iX},$$
$$M_{iX}^r = M_{iY}^l - j\omega \bar{J}_i \beta_{iY} + \omega^2 J_i \beta_{iX}, \tag{7.89}$$

where

$$V_{ix}^l = V_{ixc}^l + jV_{ixs}^l, \qquad R_{iX} = R_{iXc} + jR_{iXs}, \text{ etc.,}$$
$$Z_{iXX} = K_{iXX} - j\omega C_{iXX}, \qquad Z_{iXY} = K_{iXY} - j\omega C_{iXY}, \text{ etc.} \tag{7.90}$$

Continuity requires that

$$R_{iX}^r = R_{iX}^l = R_{iX},$$
$$R_{iY}^r = R_{iY}^l = R_{iY},$$
$$\beta_{iY}^r = \beta_{iY}^l = \beta_{iY},$$
$$\beta_{iX}^r = \beta_{iX}^l = \beta_{iX}. \tag{7.91}$$

Combining Eqs. (7.89) and (7.91) allows one to write a station transfer matrix of the form

$$(Q_i)^r = [T_{si}(\omega)](Q_i)^l + (E_i), \tag{7.92}$$

where the state vector is

$$(Q_i)^T = (R_{iX}, \beta_{iY}, V_{iX}, M_{iY}, R_{iY}, \beta_{iX}, V_{iY}, M_{iX}), \tag{7.93}$$

and

$$(E_i) = \left(0, 0, \omega^2 m_i \bar{a}_i, 0, 0, 0, j\omega^2 m_i \bar{a}_i, 0\right) \tag{7.94}$$

Absent the imbalance-excitation terms in Eq. (7.94), Eq. (7.92) has the same form as Eq. (2.51). The field transfer matrices can be constructed using the Euler-beam model of Eq. (2.55) or the Timoshenko-beam solution developed by Lund and Orcutt. In either case, the format is

$$(Q_{i+1})^l = [T_{fi}](Q_i)^r. \tag{7.95}$$

Zero shear and moment conditions are assumed to apply to both ends of the rotor; hence,

$$V_{iX}^l = V_{iY}^l = M_{iX}^l = M_{iY}^l = 0, \tag{7.96a}$$

$$V_{nX}^r = V_{nY}^r = M_{nX}^r = M_{nY}^r = 0, \tag{7.96b}$$

The central question in calculating a synchronous-response solution is: What nonzero initial conditions are required to satisfy the terminal end conditions of Eq. (7.96b)? Hence, the unknowns are

$$R_{1X}, R_{1Y}, \beta_{1X}, \beta_{1Y}.$$

Repeated applications of Eqs. (7.92) and (7.93) in moving from left to right across the rotor until the right-hand side is reached will yield an equation of the form

$$
\begin{pmatrix} V_{nX} \\ M_{nY} \\ V_{nY} \\ M_{nX} \end{pmatrix}^r = [d(\omega)] \begin{pmatrix} R_{1X} \\ \beta_{1Y} \\ R_{1Y} \\ \beta_{1X} \end{pmatrix} + \{e\} = 0. \tag{7.97}
$$

Inversion of this equation will yield the required initial conditions; however, the next question is: How are $[d]$ and $\{e\}$ to be calculated? Lund and Orcutt resolve this question by completing the transfer-matrix cycle from left to right through the rotor five times. The first four solutions are obtained without imbalance $[E_i = 0$ in Eq. (7.92)] for the initial-condition vectors

$$(Q_1)_1^{lT} = (1, 0, 0, 0, 0, 0, 0, 0),$$

$$(Q_1)_2^{lT} = (0, 1, 0, 0, 0, 0, 0, 0),$$

$$(Q_1)_3^{lT} = (0, 0, 0, 0, 1, 0, 0, 0),$$

$$(Q_1)_4^{lT} = (0, 0, 0, 0, 0, 1, 0, 0). \tag{7.98}$$

When evaluated at the right-hand side of the nth station, the transfer-matrix solution obtained from these initial conditions will yield the first, second, third, and fourth columns of $[d]$. The fifth solution is obtained for zero state initial conditions but nonzero (E_i). Evaluation of the fifth solution at the right-hand side of the nth station defines $\{e\}$.* Having obtained the required initial conditions from Eq. (7.97), the transfer-matrix algorithm can be used to "crank out" the displacement, slope, shear, and moment results for each station of the model.

Complex Eigenanalysis—Lund (1974)

Lund (1974) developed a transfer-matrix algorithm for calculating the complex eigenvalues and eigenvectors of a rotor supported on fluid-film bearings. With this approach, the station-differential-equation variables of Eq. (7.87) are replaced by their Laplace-domain variables, and the imbalance excitation is dropped, obtaining

$$V_{iX}^r = V_{iX}^l - \left(m_i s^2 + Z_{iXX}\right)R_{iX} - Z_{iXY}R_{iY},$$

$$V_{iY}^r = V_{iY}^l - Z_{iYX}R_{iX} - \left(m_i s^2 + Z_{iYY}\right)R_{iY},$$

$$M_{iY}^r = M_{iY}^l - s^2 J_i \beta_{iY} - s\omega \bar{J}_i \beta_{iX},$$

$$M_{iX}^r = M_{iX}^l - s^2 J_i \beta_{iX} + s\omega \bar{J}_i \beta_{iY}. \tag{7.99}$$

The bold notation in Eq. (7.99) denotes the Laplace-domain variable, s denotes the Laplace variable, and

$$Z_{iXX} = K_{iXX} + sC_{iXX}, \qquad Z_{iXY} = K_{iXY} + sC_{iXY}, \text{ etc.} \tag{7.100}$$

By using the continuity requirements of Eq. (7.91), a complete station transfer matrix is obtained of the form

$$(Q_i)^r = [T_{si}(s)](Q_i)^l. \tag{7.101}$$

Lund developed a field transfer matrix using the Timoshenko-beam model plus internal damping which couples the deflections in the X-Z and Y-Z planes. The field-transfer-matrix definition continues to be independent of s and have the form defined by Eq. (7.95), and the boundary conditions continue to be defined by Eqs. (7.96).

Routine marching from left to right through the rotor using the transfer matrices will yield, at the right-hand side of station n, an equation of the

---

*This approach bears an obvious similarity to the transition-matrix approach used to solve first-order seal equations in Chapter 4.

form

$$\begin{Bmatrix} V_{nX} \\ M_{nY} \\ V_{nY} \\ M_{nX} \end{Bmatrix}^{r} = [d(s)] \begin{Bmatrix} R_{iX} \\ \beta_{iY} \\ R_{iY} \\ \beta_{iX} \end{Bmatrix} = 0. \qquad (7.102)$$

The rotor's complex eigenvalues are defined by

$$\Delta(s) = |d(s)| = 0. \qquad (7.103)$$

Lund suggested a Newton-Raphson algorithm for solving Eq. (7.103). This algorithm assumes that a guess s^i is available for the solution. A Taylor-series expansion of the function $\Delta(s)$ is then performed about the guessed solution, obtaining

$$\Delta(s^i + \delta s^i) = \Delta(s^i) + \frac{d\Delta}{ds}\delta s^i \cong 0,$$

where the derivative $d\Delta/ds$ is evaluated at s^i. The correction to the assumed solution is then

$$\delta s^i = -\Delta(s^i)/\frac{d\Delta}{ds},$$

and the next guess for the root is

$$s^{i+1} = s^i + \delta s^i. \qquad (7.104)$$

The derivative $d\Delta/ds$ is obviously required for this development. The derivative of the determinant can be stated

$$\frac{d\Delta}{ds} = \sum_{k=1}^{4} \Delta_k.$$

Here, Δ_k is the determinant of the matrix $[d_k]$, which is obtained by replacing the elements of the kth column of $[d]$ by their derivatives with respect to s [Sokolnikoff and Redheffer (1966)]. The question of immediate interest is: How are the matrices $[d]$ and $[d_k]$ to be obtained? The matrix $[d]$ is evaluated for a specified value of s by marching from left to right through the rotor using station equation (7.101), field equation (7.95), boundary conditions from Eq. (7.96), and successive use of the initial-condition set in Eq. (7.98). The jth column of $[d]$ is obtained from the jth initial-condition set of Eq. (7.98).

The derivatives of the state variables are needed to calculate the $[d_k]$ matrices. The initial derivatives are set equal to zero; i.e.,

$$R'_{iX} = \beta'_{iY} = V'_{iX} = M'_{iY} = R'_{iY} = \beta'_{iX} = V'_{iY} = M'_{iX} = 0, \quad (7.105)$$

where the prime denoted differentiation with respect to s. Differentiating the station and field equations in Eq. (7.101) and (7.95) yields

$$(Q'_i)^r = [T'_{si}](Q_i)^l + [T_{si}](Q'_i)^l,$$
$$(Q'_{i+1})^l = [T_{fi}](Q'_i)^r. \quad (7.106)$$

With the initial conditions of Eq. (7.105) and (7.98), Eqs. (7.106) can be used to march from left to right, calculating the state-vector-derivative vectors at each station. The result at the right-hand side of station n can be stated

$$\begin{Bmatrix} V'_{nX} \\ M'_{nY} \\ V'_{nY} \\ M'_{nX} \end{Bmatrix} = [D] \begin{Bmatrix} R_{1X} \\ \beta_{1Y} \\ R_{1Y} \\ \beta_{1X} \end{Bmatrix}. \quad (7.107)$$

The jth column of $[D]$ results from the jth initial-condition set of Eq. (7.98) and is the derivative of the jth column of $[d]$. Hence $[d_k]$ is obtained by replacing the kth column of $[d]$ by the kth column of $[D]$. Since the state-variable-derivative vectors (Q'_i) are required for the Newton-Raphson algorithm, they are normally calculated concurrently with the (Q_i)'s.

In review, Lund's algorithm for calculating an eigenvalue starts with an initial guess for a complex eigenvalue s^i. The transfer-matrix definitions are then used to track from left to right through the rotor using the zero *derivative* initial conditions of Eq. (7.105) and the *state-variable* initial conditions of Eq. (7.98). Four passes through the rotor with the four initial conditions of Eq. (7.98) will yield $[d]$, $[d_k]$, Δ, and $d\Delta/ds$, all evaluated at s^i. Equation (7.104) then provides the new (and improved) guess for s^i. The cycle is repeated until eigenvalue convergence is achieved.

The eigenvector convergence is achieved simultaneously with the eigenvalue. The eigenvector is obtained by assigning an arbitrary value to one of the initial state-variable elements of Eq. (7.102) (e.g., $R_{iX} = 1$), solving for the remaining initial station variables from Eq. (7.102), and then marching from left to right through the rotor using the transfer matrices evaluated at the converged eigenvalue.

Lund provides a method to prevent the algorithm from reconverging to a previously calculated eigenvalue.

Lund's algorithm for complex eigenanalysis has the advantage of requiring a minimal amount of computer space. Like most Newton-Raphson-based algorithms, it has the disadvantage of requiring a good guess for the root to achieve rapid conversion. Worse, by failing to make an appropriate initial guess, roots can be entirely missed, which can be embarrassing if the missed root turns out to be unstable.

Characteristic Polynomial Methods—Murphy and Vance (1983)

As noted above, Lund's algorithm can miss complex roots, and its rate of convergence is sensitive to the initial guess. Murphy and Vance (1983) proposed a direct method to obtain the elements of $[d]$ as explicit polynomial functions of s. Calculation of the determinant of $[d]$ then yields the characteristic polynomial

$$\Delta(s) = 0 = a_0 + a_1 s + \cdots + a_m s^m. \tag{7.108}$$

The Murphy and Vance algorithm yields the coefficients $a_0, a_1, \ldots, a_m$. Solution of the roots to Eq. (7.108) yields the system eigenvalues. The Murphy-Vance algorithm will generally outperform the Lund algorithm and does not miss roots [Vance (1988)].

7.6 CLOSING COMMENTS

The results of this chapter show how a system rotordynamics model is assembled, starting with the skeleton of the structural-dynamics model and then adding fluid-structure-interaction forces of bearings, seals, impellers, etc. The general-stiffness-matrix approach yields a large set of system equations, while Lund and Orcutt (1967) and Lund (1974) use an extension of the Myklestad-Prohl method to directly incorporate linear forces of stations into a transfer-matrix formulation. A variety of algorithms are reviewed for analyzing either the system-matrix models or transfer-matrix models.

A list of the comparative advantages of transfer-matrix and system-matrix formulations for rotordynamics would include:

(a) The system-matrix formulation can be readily adapted for transient analysis; the transfer-matrix approach cannot.
(b) The transfer-matrix approach takes less computer space.
(c) The system-matrix formulation is readily adapted to multiple shafts with multiple interconnections; the transfer-matrix approach is not.
(d) The system-matrix approach can be adapted to general finite-element models for the rotor housing; the transfer matrix cannot.

While this list seems to favor the system-matrix approach, for a single rotor in multiple bearings, the Murphy-Vance (1983) polynomial algorithm for calculating complex eigenvalues and eigenvectors is (in the author's experience) the fastest and most effective method available. This advantage disappears as one moves to multiple rotors with multiple interconnections.

Generally speaking, the quality of predictions from a computer code has more to do with the soundness of the basic model and the skill and physical insight of the rotordynamicist than the particular algorithm employed. Good engineers get good results from good models, leading to sound engineering judgments with a variety of algorithms. Superior algorithms or computer codes will not cure bad models or a lack of engineering judgment.

REFERENCES: CHAPTER 7

Adams, M. (1980), "Non-Linear Dynamics of Flexible Multi-Bearing Rotors," *Journal of Sound and Vibration*, **71**(1), 129–144.

Black, H. (1973), "Calculation of Forced Whirling and Stability of Centrifugal Pump Rotor Systems," ASME Paper No. 73-DET-31.

Childs, D. (1978), "The Space Shuttle Main Engine High-Pressure Fuel Turbopump Rotordynamic Instability Problem," *Journal of Engineering for Power*, 48–57.

Childs, D., and Bates, J. (1978), "Residual-Flexibility Corrections for Transient Modal Rotordynamic Analysis," *Journal Mechanical Design*, 251–256.

Childs, D. (1976a), "A Modal Transient Rotordynamic Model for Dual-Rotor Jet Engine System," *Journal of Engineering for Industry*, 876–882.

Childs, D. (1976b), "A Modal Transient Simulation Model for Flexible Asymmetric Rotors," *Journal of Engineering for Industry*, 312–319.

Childs, D. (1975), "Two Jeffcott-Based Simulation Models for Flexible Rotating Equipment," *Journal of Engineering for Industry*, 1000–1014.

Childs, D. (1972), "A Simulation Model for Flexible Rotating Equipment," *Journal of Engineering for Industry*, 201–209.

Craig, R., and Bampton, M. (1968), "Coupling of Substructures for Dynamic Analyses," *AIAA Journal*, **6**(7), 1313–1319.

Crandall, S., and Yeh, N. (1987), "Automatic Generation of Component Modes for Rotordynamic Structures," in *Rotating Machinery Dynamics*, proceedings of the 11th Vibration Conference, Boston, MA, ASME, New York, Vol. 1, pp. 79–84.

Forsythe, G., Malcolm, M., and Moler, C. (1977), *Computer Methods for Mathematical Computations*, Prentice-Hall, Englewood Cliffs, NJ.

Glascow, D., and Nelson, H. (1979), "Stability Analysis of Rotor-Bearing Systems Using Component Mode Synthesis," *Journal of Mechanical Design*, **102**(2), 352–359.

Guyan, R. (1965), "Reduction of Stiffness and Mass Matrices," *AIAA Journal*, **3**(2), 380.

Hurty, W. (1965), "Dynamic Analysis of Structural Systems Using Component Modes," *AIAA Journal*, **3**(4), 678–685.

Kirk, R., and Gunter, E. (1973), "Transient Response of A Rotor-Bearing System," ASME Paper No. 73-DET-102.

Lallane, M., and Ferraris (1990), *Rotordynamic Predictions in Turbomachinery*, Wiley, New York.

Li, D., and Gunter, E. (1981), "A Study of the Modal Truncation Error in the Component Mode Analysis of a Dual-Rotor System," ASME Paper No. 81-GT-144, ASME Gas Turbine Conference, Houston.

Likins, P. (1967), "Modal Method for Analysis of Free Rotations of Spacecraft," *AIAA Journal*, **5**(1).

Lund, J. (1973), "Modal Response of a Flexible Rotor in Fluid-Film Bearings," *Journal of Engineering for Industry*, **96**(2), 525–553.

Lund, J. (1974), "Stability and Damped Critical Speeds of a Flexible Rotor in Fluid-Film Bearings," *Journal of Engineering for Industry*, Ser. B., **96**(2), 509–517.

Lund, J., and Orcutt, F. (1967), "Calculations and Experiments on the Unbalance Response of a Flexible Rotor," *Journal of Engineering for Industry*, 785–796.

Meirovitch, L. (1967), *Analytical Methods in Vibrations*, Macmillan, New York.

Murphy, B., and Vance, J. (1983), "An Improved Method for Calculating Critical Speeds and Rotordynamic Stability of Turbomachinery," *Journal of Engineering for Power*, **105**(3), 591–595.

Nelson, H., and Meacham, W. (1981), "Transient Analysis of Rotor-Bearing System Using Component Mode Synthesis," ASME Paper No. 81-G7-10, Gas Turbine Conference, Houston, TX.

Nelson, H., Meacham, M., Fleming, D., and Kasack, F. (1983), "Nonlinear Analysis of Rotor-Bearing Systems Using Component Mode Synthesis," *Journal of Engineering for Power*, **105**(3), 606–614.

Rouch, K., and Kao, J. (1980), "Dynamic Reduction in Rotor Dynamics by the Finite-Element Method," *Journal of Mechanical Design*, **102**, 360–368.

Sokolnikoff, I., and Redheffer, R. (1966), *Mathematics of Physics and Modern Engineering*, McGraw-Hill, New York, p. 704.

Timoshenko, S., and Young, D. (1954), *Vibration Problems in Engineering* (3rd ed.), Van Nostrand, Princeton, NJ.

Vance, J. (1988), *Rotordynamics of Turbomachinery*, Wiley, New York.

8

EXAMPLE ROTOR ANALYSIS

8.1 INTRODUCTION

In looking back through the contents of this book, recall that the first chapter dealt with rotordynamic phenomena, while the remaining chapters have dealt methodically with modeling. Chapter 2 introduced structural-dynamic models, while Chapters 3–6 covered interaction forces due to bearings, liquid seals, gas seals, and turbines and impellers. Chapter 7 considered assembly and analysis of a complete system rotordynamics model. The present chapter applies the knowledge developed in the preceding chapters by providing a demonstration analysis for the alternate-technology-development (ATD) high-pressure-fuel-turbopump (HPFTP) of the Space Shuttle main engine (SSME).

A factor in choosing the ATD-HPFTP was the availability of data and the helpful assistance of the manufacturing company: Pratt and Whitney (P&W), West Palm Beach, FL. The assistance of P&W personnel and the permission of P&W in using their data is cheerfully acknowledged.

A liquid-rocket-engine turbopump combines a multistage pump and a multistage turbine on a single shaft and is, accordingly, ideal for demonstrating the influences of liquid and gas seals, and turbines and impellers. The ATD-HPFTP is supported by rolling-element bearings with "dead-band" clearances of the outer bearing races which demonstrate the nonlinear effect of bearing clearances discussed in Section 1.9. Both linear (critical-speed and stability-prediction) and nonlinear (transient analyses with dead-band clearances) results are presented.

Figure 8.1 illustrates the ATD-HPFTP. This unit was designed and developed by P&W as an interchangeable replacement for the Rocketdyne-

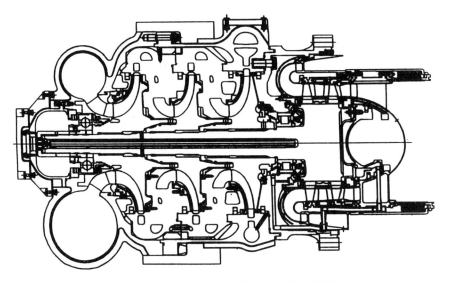

Figure 8.1   The Pratt & Whitney ATD-HPFTP.

manufactured HPFTP. The P&W and Rocketdyne turbopumps are similar in that both use a three-stage hydrogen pump, driven by a two-stage steam turbine. Steam is developed by burning hydrogen in a preburner at the right-hand side of Figure 8.1; hence, the steam flows from right to left through the turbine stages. The first- and second-stage turbines have nominal blade diameters of 23.3 cm (9.17 in.) and 23.6 cm (9.27 in.). The turbine stages develop about 57 MW (76,000 HP) at a top running speed of 37,000 rpm. Pump impeller diameters are about 150 mm (6 in.). Pump discharge pressure is 445 bars (6545 psi) with a flow rate of 72 kg/sec (150 lb/sec) at full power level.

The high-speed roller bearing which supports the turbine end of the ATD-HPFTP is a unique P&W feature in turbopump development. Liquid hydrogen is circulated through the center of the rotor shaft to provide predictable temperatures and predictable radial thermal growth or contraction beneath the turbine bearing. The flow path in the center of the shaft has an inner circular "pipe" and an outer flow annulus between the "pipe" and inside shaft diameter. Hydrogen enters the pipe at the left, flows to the end of the shaft hole at the right, and then flows back from right to left in the annulus, exiting into the main flow between impellers 2 and 3.

The pump end is carried by a single ball bearing. Both bearings are cooled by liquid hydrogen. Axial thrust is absorbed by a balance-piston arrangement at the back side of the third impeller stage. The pump ball bearing would fail under sustained and appreciable axial load; hence, a small dead-band radial clearance is provided between the outer face of the ball bearing and the pump housing to allow relative axial slip between the two surfaces and limit the axial load carried by the bearing. A similar clearance is provided at the outer race of the turbine bearing. These clearances represent an *essential* strong nonlinearity in rotordynamics and can have a dramatic influence on turbopump rotordynamics, as predicted by Yamamoto (1954) in Section 1.9.

The complete rotordynamic model for the present turbopump (TP) or any other turbomachinery unit incorporates models for the rotor's (and housing if required) structural dynamics, bearings, and fluid-structure-interaction forces of liquid and gas seals, impellers, and turbines. A turbopump is attractive as a rotordynamics example because it contains all of these elements. The ATD-HPFTP was selected because the author had recently analyzed this TP in December 1990.

The author originally analyzed the ATD-HPFTP in 1988 prior to any fabrication or tests. The 1988 analysis revealed that the turbine interstage seal represented a major source of destabilizing forces because it combined a high ΔP with a high inlet tangential velocity. The original ATD-HPFTP was predicted to be rotordynamically unstable because of these forces. Based on these predictions, a swirl brake was designed (by P&W) and incorporated into the design. With the swirl brake, the TP was predicted to have a substantial stability margin at its top power level. The initial analysis also predicted a serious critical-speed problem. The author recommended "softening" the turbine-bearing-support stiffness to eliminate this situation; however, this suggestion was not followed.

Initial tests of the ATD-HPFTP in 1990 showed high vibration levels with serious rubbing at the pump interstage seals associated with the predicted critical-speed problem. The author's 1990 study used an updated model for the turbopump with the original (stiff) turbine-bearing support and enlarged clearances at the pump interstage seal (from 0.127 to 0.191 mm). As discussed in Section 4.6, increasing the seal's clearance reduces the direct stiffness and damping significantly; hence, increasing the pump interstage-seal clearances might, or might not, eliminate rubbing. The influence of turbine bearing dead-band clearances on the response of the turbopump with enlarged seal clearances was the central focus of this study.

The following section reviews the elements of the rotordynamic system model and explains how they are defined. Most of the input data for this section were provided by P&W personnel in 1990; however, the data cited here may well have been replaced by more recent and accurate results.

8.2 ELEMENTS OF THE ATD-HPFTP ROTORDYNAMIC MODEL

Structural-Dynamics Models

Figure 8.2 illustrates the position of the ATD-HPFTP rotor within the SSME power head. The HPFTP and HPOTP (high pressure oxygen turbopump) and their preburners are cantilevered from the thrust chamber. Both units

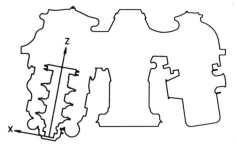

Figure 8.2 Powerhead assembly of the SSME.

have low-frequency "rigid-body" modes defined largely by their rigid-body properties and the flexibility of the cantilever connection. As might be expected from Figure 8.2, rigid-body mode shapes in the X-Z and Y-Z planes differ significantly because of asymmetric stiffness and inertial properties. In addition to low-frequency rigid-body modes, the housings have flexible body modes in the operating-speed range of the TP. Hence, the structural-dynamics model must account for both the rotor and the housing.

The rotor and housing structural-dynamic models were developed by P&W personnel, and their development is not included here. Figure 8.3 illustrates the fine detail used in modeling the rotor structural dynamics.

The system structural-dynamics models for linear and nonlinear analysis differ, but both models start from (a) undamped free-free rotor modes, and (b) undamped housing modes which incorporate connections to "ground" that match measured rigid-body frequencies. Calculated rotor free-free eigenvalues are

$$\lambda_1 = 0, \qquad \lambda_4 = 1088 \text{ Hz},$$
$$\lambda_2 = 0, \qquad \lambda_5 = 1864 \text{ Hz},$$
$$\lambda_3 = 513 \text{ Hz}. \tag{8.1}$$

Zero percent of critical damping is assumed for modes 1 and 2, one-half percent for the remaining modes. The housing eigenvalues in the maximum-

Figure 8.3 P&W ATD-HPFTP rotor structural-dynamics model.

stiffness X-Z plane and minimum-stiffness Y-Z plane are

$$\lambda_{1cx} = 150.0 \text{ Hz,} \qquad \lambda_{1cy} = 88.4 \text{ Hz,}$$

$$\lambda_{2cx} = 429.8 \text{ Hz,} \qquad \lambda_{2cy} = 209.0 \text{ Hz,}$$

$$\lambda_{3cx} = 781.7 \text{ Hz,} \qquad \lambda_{3cy} = 752.4 \text{ Hz,}$$

$$\lambda_{4cx} = 1031.6 \text{ Hz,} \qquad \lambda_{4cy} = 1012.8 \text{ Hz.} \qquad (8.2)$$

One-half percent of critical damping is assumed for all housing modes.

Bearings

The pump and turbine bearings were modeled as linear springs with the following coefficients:

$$K_P = 1.3 \times 10^5 \text{ N/mm} \left(0.75 \times 10^6 \text{ lb/in.}\right),$$

$$K_T = 6.1 \times 10^5 \text{ N/mm} \left(3.5 \times 10^6 \text{ lb/in.}\right). \qquad (8.3)$$

Zero damping is assumed for the bearing. Flexibility of the pump-bearing backup structure is accounted for with the stiffness $K_{PS} = 2.3 \times 10^5$ N/mm $(1.3 \times 10^6$ lb/in.) which is placed in parallel with the pump bearing stiffness.

The bearing dead band is defined to be the *radius* about the centered bearing position for which no bearing reaction is developed. As noted above, these dead bands account for the radial clearances between the outer races of the bearings and the housing. They are zero for linear analysis and constitute the essential rotordynamic nonlinearity. The dead-band clearance ranges used here are

$$\delta_P = 0.0457 \text{ mm } (0.0018 \text{ in.}),$$

$$0.0254 \text{ mm} \leq \delta_T \leq 0.762 \text{ mm,}$$

$$(0.001 \text{ in.} \leq \delta_T \leq 0.003 \text{ in.}). \qquad (8.4)$$

Liquid Seals

The interstage seals between impellers 1 and 2 and impellers 2 and 3 have a significant influence on rotordynamics. By contrast, the wearing-ring seals are deeply grooved with comparatively sharp labyrinth teeth, and their influence is neglected. The interstage seals are "damper seals" using roughened stators and smooth rotors. A round-hole pattern* is used to create stator roughness following the results of Childs et al. (1990) and Childs and Kim (1985). A Moody turbulence model was used with coefficients calculated for the rotor

*See Figure 4.25.

TABLE 8.1   ATD-HPFTP Liquid-Seal Input Data. r = **Radius** = **53.3 mm,** C_r = **Radical Clearance** = **0.19 mm,** L = **Length** = **40.6 mm.**

	ω rpm	ΔP bars	$\mu \times 10^6$ PaS	S.G. —
Seal	27,553	71.8	7.84	0.0437
1-2	34,444	114.0	7.02	0.0556
	36,465	134.0	7.14	0.0554
Seal	27,553	73.2	9.94	0.0666
2-3	34,444	170.0	10.8	0.0687
	36,465	141.0	11.0	0.0693

and stator based on the following assumed relative-roughness parameters

$$\epsilon_r \, (\text{rotor}) = 0.001, \qquad \epsilon_s \, (\text{stator}) = 0.06. \tag{8.5}$$

As noted in Section 4.2, Moody's model underestimates pipe friction factors for relative roughness greater than 0.01. In the present circumstance, the Moody-model-prediction accuracy is on the same order of magnitude as the balance of the input data. The seal inlet-loss coefficient ξ was assumed to be constant at 0.1 for all cases. The exit recovery-factor coefficient ξ_e was held constant at 1.0; i.e., no exit recovery of pressure.

The normalized inlet circumferential velocity for flow entering interstage seal 1-2 (between impellers 1 and 2) is estimated to be 0.5. Most of the flow entering interstage seal 2-3 is recirculating turbine-bearing coolant flow from within the TP shaft. Because the leakage flow is emerging from the rotating shaft, a normalized inlet circumferential velocity of 1.0 is predicted.

The balance of the input data is given in Table 8.1. Note the remarkably low specific gravity (S.G.) for liquid hydrogen.

Table 8.2 provides calculated force coefficients for the model of Eq. (4.1). Note that the damper-seal configuration reduces $k/C\omega$ below 0.5 for seal

TABLE 8.2   Calculated Rotordynamic Coefficients for ATD-HPFTP Interstage Seals.

	ω rpm	$u_0(0)$ —	$\dot{m}$ Kg/sec	K KN/mm	k KN/mm	C NS/mm	c NS/mm	M Kg	$k/C\omega$ —
Seal	27,553	0.5	0.632	16.7	9.75	9.07	0.311	0.210	0.373
1-2	34,444	0.5	0.895	26.6	17.2	12.8	0.492	0.267	0.370
	36,465	0.5	0.973	31.2	19.8	13.9	0.520	0.268	0.371
Seal	27,553	1.0	0.755	15.5	21.2	11.3	0.709	0.322	0.652
2-3	34,444	1.0	1.00	26.1	34.3	14.5	0.849	0.298	0.657
	36,465	1.0	1.10	30.6	39.9	15.9	0.916	0.299	0.659

TABLE 8.3   ATD-HPFTP Turbine-Interstage-Seal Input Data. R = Radius = 115.44 mm, C_r = Radial Clearance = 0.33 mm, L = Pitch = 3.050 mm, b = Tooth Height = 3.30 mm, Tooth-Tip Width = 0.25 mm.

ω rpm	$u_0(0)$ —	P_r (bar)	P_s (bar)	T_r (°K)	$\nu \times 10^6$ m²/sec	Z^* —	R^{**} J/kg °K	GM^{***} —
27,553	0.15	163.9	151.0	463.0	2.71	1.037	3250.	1.393
34,444	0.15	236.8	214.0	559.0	2.13	1.045	3083.	1.432
36,465	0.15	296.4	263.7	639.0	1.76	1.051	2975.	1.535

*Compressibility.
**Gas constant.
***Specific-heat ratio.

1-2. The roughened stator also causes $k/C\omega$ to be lower for seal 2-3 than it would be for a smooth seal. Also, note that the added-mass terms M are small in comparison to the water-seal results of Figure 4.15(c). The very low density of liquid hydrogen explains this result. Table 4.3 shows the results of pitching or yawing of a seal and argues that these seals are not quite long enough (L/D = 0.5) to require the general model of Eq. (4.63), which includes pitch and yaw deflections and reaction moments.

Turbine-Interstage Seal

Labyrinth-seal calculations are based on Scharrer's (1987) analysis. The seal has a honeycomb stator, a five-cavity-labyrinth rotor, and incorporates a swirl brake. Based on measured results for a model of this seal [Childs and Ramsey (1991)], the swirl brake is assumed to reduce the inlet circumferential velocity $u_0(0)$ from about 1.0 to 0.15. The roughness of the rotor and stator is characterized by the following Blasius coefficients:

$$mr = -0.25, \qquad nr = 0.079,$$
$$ms = -0.1083, \qquad ns = 0.282.$$

The honeycomb-stator coefficients are based on measured results from Elrod (1988). Table 8.3 contains the balance of the input data for seal calculations.

Table 8.4 contains the rotordynamic coefficients. K and C from this table are taken directly from Scharrer's calculations. The cross-coupled coefficients

TABLE 8.4   Calculated Rotordynamic Coefficients for the ATD-HPFTP Turbine-Interstage Seal.

ω rpm	$u_0(0)$ —	K KN/mm	k KN/mm	C NS/mm	$k/C\omega$ —
27,553	0.15	0.982	0.162	0.374	0.15
34,444	0.15	1.72	0.298	0.551	0.15
36,465	0.15	2.43	0.396	0.692	0.15

are "back calculated" to yield a whirl frequency of 0.15, based on Pelletti's (1990) measurements for short labyrinth seals. For short seals, Pelletti showed that measured values for $k/C\omega$ were approximately equal to the normalized inlet circumferential velocity $u_0(0)$.

Impeller Rotordynamic Coefficients

Equations (6.23) and (6.24) are used to model the impeller rotordynamic coefficients. Since the clearances between the impeller housing and shroud are comparatively close, the nondimensional data of Bolleter et al. (1987, 1989) of Eq. (6.34a) are used to define the coefficients. Input data for the impellers are given in Table 8.5. Table 8.6 contains the calculated rotordynamic coefficients. The coefficients K, k, and C of Table 8.6 are small in comparison to the pump-interstage-seal results of Table 8.2; however, c and M are much larger. The direct mass coefficient is less than a kilogram, which is small in comparison to the impeller mass. Recall, however, that the impeller coefficients are directly related to the density of the fluid and are small here because of liquid hydrogen's remarkably low density. With water, all of the coefficients would be roughly 14 times larger.

A relative comparison of c and M is obtained from Eq. (6.27a):

$$-\bar{F}_r(f) = \bar{K} + \bar{c}f\left(1 - \frac{\bar{M}}{\bar{c}}f\right), \qquad f = \frac{\Omega}{\omega}, \tag{8.6}$$

where f is the ratio of the precession frequency to the running speed. From

TABLE 8.5 **Impeller Input Data:**
Stage 1: $r_2 = 152$ mm, $b_2 = 13.8$ mm;
Stage 2: $r_2 = 152$ mm, $b_2 = 13.8$ mm;
Stage 3: $r_2 = 144$ mm, $b_2 = 13.8$ mm.

Stage	ω rpm	S.G.* —
	27,553	0.0679
1	34,444	0.0683
	36,465	0.0681
	27,553	0.0705
2	34,444	0.0722
	36,465	0.0728
	27,553	0.0726
3	34,444	0.0750
	36,465	0.0765

*Specific gravity.

TABLE 8.6 Calculated Rotordynamic Coefficients for Impellers 1, 2, and 3.

Impeller	ω rpm	K KN/mm	k KN/mm	C NS/mm	c NS/mm	M Kg	m_c Kg
1	27,520	−2.37	2.88	0.902	2.64	0.748	0.272
	31,800	−3.19	3.86	1.05	3.08	0.752	0.274
	37,115	−4.32	5.25	1.22	3.58	0.750	0.272
2	27,520	−2.46	2.99	0.936	2.75	0.777	0.282
	31,800	−3.37	4.09	1.11	3.25	0.795	0.289
	37,115	−4.62	5.62	1.30	3.83	0.802	0.291
3	27,520	−2.27	2.76	0.865	2.54	0.720	0.261
	31,800	−3.14	3.81	1.03	3.03	0.742	0.270
	37,115	−4.36	5.30	1.22	3.61	0.756	0.275

Eq. (6.34a), $\overline{M} = 11.0$, $\bar{c} = 13.5$, and $\overline{M}/\bar{c} = 0.815$. Hence for synchronous motion with $\Omega/\omega = 1$, the "stiffening" effect of c in increasing $-\overline{F}_r$ is about balanced by the "softening" influence of $\overline{M}$.

Turbine Coefficients

The Alford model of Eq. (6.11) and (6.13) is used for turbine-rotordynamic-coefficient calculations. Alford's cross-coupling coefficient definition is

$$k = \frac{T\beta}{D_m L_t}, \tag{8.7}$$

where β is defined as "the change in thermodynamic efficiency per unit of rotor displacement, expressed as a fraction of blade height," D_m is the mean blade diameter, L_t is the turbine-blade height, and T is the torque. Input data for calculations are given in Table 8.7. Calculated results are given in Table 8.8 for an assumed $\beta = 1.0$.

TABLE 8.7 Turbine-Destabilizing-Force Input Data.
Stage 1: $D = 233$ mm, $L_t = 24.2$ mm;
Stage 2: $D = 235$ mm, $L_t = 26.8$ mm.

Power Level	65%	90%	109%
ω (rpm)	27,520	32,856	37,115
Torque* (Nm)	6,982	10,907	14,574

*Both turbine stages.

**TABLE 8.8 Combined Alford Cross-Coupled Stiffness
Coefficient for Turbines 1 and 2 with $\beta = 1.0$.**

ω rpm	k KN/mm
27,520	1.17
31,800	1.83
37,115	2.44

For rotordynamic calculations, the cross-coupled coefficients of the two turbine stages and the turbine-interstage seal are combined and applied midway between the two turbine stages.

Mechanical and Hydraulic Imbalance

The mechanical-imbalance distribution was estimated by P&W personnel based on their balance specifications and the fact that the balanced rotating assembly must be disassembled and then reassembled into the turbopump. Their results are given in Table 8.9. All of the imbalance is in a single plane of the rotor to yield a worst-case loading.

Hydraulic imbalance can be approximately calculated from Eqs. (6.70)–(6.72). The nominal ΔP per impeller stage at 37,000 rpm is about 145 bars; hence for $\Delta P = \alpha \omega^2$, $\alpha \cong 0.97$. From Table 8.5, $D_2 \cong 300$ mm, and $b_2^* \cong 14.0$ mm. From Eq. (6.71) the hydraulic imbalance is

$$F_{HI} = K_{HI}(4.07 \times 10^{-3})\omega^2.$$

For best-quality cast impellers, Bolleter et al. (1989) provide a lower estimate for K_{HI} of 0.005. The HPFTP-ATD is a machined impeller which should be more precise and yield a lower K_{HI}. Assuming that $K_{HI} = 0.0025$ yields

$$F_{HI} = 10.2 \times 10^{-6}\omega^2.$$

Hence, the equivalent hydraulic imbalance per impeller would be 10.2 g mm. Returning to Table 8.9, the mechanical imbalance of impellers 1, 2, and 3 (locations 2, 3, and 5) are about 200 g mm. Hence, *for the present example* (with K_{HI} assumed to be 0.0025), hydraulic imbalance only accounts for about 5% of mechanical imbalance and is negligible. Precision machined impellers (low K_{HI}) plus liquid hydrogen's low specific gravity (low α) largely account for this outcome.

The reader is cautioned that hydraulic imbalance is frequently quite significant in pump rotordynamics even with precisely manufactured impellers. The author knows of one case where machined impellers were received from two different manufacturers, were balanced in the same

TABLE 8.9   HPFTP-ATD Imbalance Distribution.

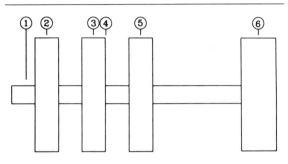

	Z, Axial Location mm	Imbalance g mm
1	85.3	−88.6
2	203	220
3	338	191
4	384	−366
5	461	198
6	635	22

machine prior to installation, and yet consistently ran at markedly different vibration levels when installed in the same pump. Inspection showed that the impellers were manufactured within specified manufacturing tolerances. Hence, quite small differences in impeller flow paths can create significant levels of hydraulic imbalance.

TABLE 8.10   Estimated Minimum Side Loads.

Power Level	65%	90%	109%
ω (rpm)	27,570	32,836	37,115
Stage	Load N	Load N	Load N
First-stage impeller	53	76	93
Second-stage impeller	58	80	98
Third-stage impeller	1700	3700	2870
Turbine	5230	6250	7100

Impeller and Turbine Side Loads

Discharge pressures and flow rates of the impellers and turbine stages are not entirely uniform and hence develop steady-state, nominally fixed-direction side loads. These side loads influence the nonlinear transient solutions for the TP. Minimum and maximum estimated values were calculated by P&W personnel, and the minimum values are given in Table 8.10. All loads are in the X-Z plane of Figure 8.2.

8.3   LINEAR-ROTORDYNAMIC-ANALYSIS PREDICTIONS

The linear analysis used here takes rotor free-free modes and housing modes and calculates coupled, zero-running-speed, rotor-housing modes based on the bearing stiffnesses that connect the two structures [Childs (1978)]. Figures 8.3 illustrate the first five coupled rotor/housing mode shapes in the X-Z plane for $K_p = 1.3 \times 10^8$ N/mm, $K_T = 6.1 \times 10^8$ N/mm. The lowest frequency mode is a "rocking" mode of the rotor and housing together. The housing is significantly stiffer than the rotor and shows very little bending for any of the modes. Comparable modes arise in the Y-Z plane.

The approach for system-modal-model development resembles that of Section 7.4 and incorporates all speed-dependent linear forces arising at bearings, seals, impellers, and turbines. Synchronous response due to imbalance and complex roots can be calculated from the system model. The ATD-HPFTP system model has numerous coupled complex modes, and the complex roots will be reviewed first to help explain the synchronous-response results.

Figure 8.5 illustrates the imaginary parts of the complex system roots versus running speed. At zero running speed, purely imaginary roots are obtained which are associated with the X-Z mode shapes and eigenvalues of Figure 8.4 and their associated (but not illustrated) Y-Z mode shapes. Increasing running speed causes the imaginary parts of some roots to split into modes associated with forward and backward critical speeds. Modes whose roots split involve predominantly rotor motion with large gyroscopic influences, while roots which do not split tend to involve modes with predominantly housing motion. Crossing points for the dashed line in Figure 8.5 with the imaginary part of the complex roots define system critical resonances. Forward critical speeds are approximately defined from this figure at 22,500 and 32,500 rpm and are associated with modes 2 and 3 of Figure 8.4. Backward critical speeds are predicted at about 15,000, 25,000, and 37,500 rpm and are associated with modes 2, 3, and 4. Because the housing is asymmetric, backward critical speeds may well be excited by imbalance.*

*Section 1.6.

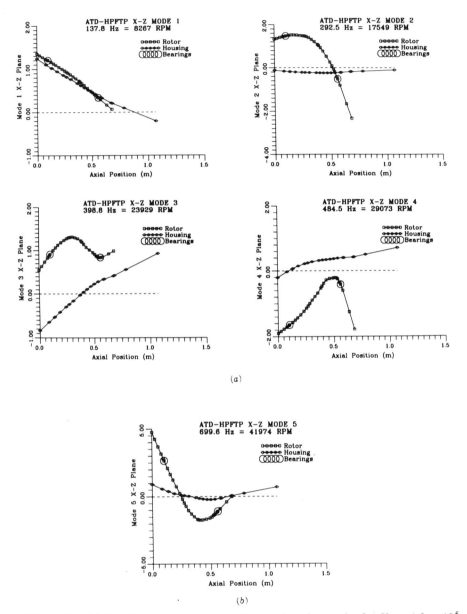

Figure 8.4 (a) Coupled zero-running-speed rotor/housing modes for $K_P = 1.3 \times 10^5$ N/mm, $K_T = 6.1 \times 10^5$ N/mm; first four modes in the X-Z plane. (b) Fifth, coupled zero-running-speed rotor/housing mode in the X-Z plane for $K_P = 1.3 \times 10^5$ N/mm, $K_T = 6.1 \times 10^5$ N/mm.

TABLE 8.11   Complex-Root Results at ω = 37,000 rpm.

Root	ω_{di} (rd/sec)	ω_{di} (rpm)	$\zeta_i \times 100$	Modes
5	2570	24,500	0.545	3 Y-Z, 2 X-Z
6	2700	25,800	12.2	3 X-Z
7	2800	26,700	1.99	4 X-Z, 4 Y-Z (BWD)
8	3570	34,100	0.525	4 X-Z, 4 Y-Z (FWD)
9	3920	37,400	3.48	5 Y-Z (BWD)

The complex roots can be stated

$$\lambda = -\sigma \pm j\omega_{di} = -\zeta_i\omega_{ni} \pm j\omega_{ni}\sqrt{1 - \zeta_i^2}. \tag{8.8}$$

For stability, the real part of the roots are of interest; specifically, a positive real root means that the system is linearly unstable. For stable systems, the damping factor ζ_i is also of interest, since low values for ζ_i suggest a sharp response-excitation peak. Table 8.11 provides ω_{di} and ζ_i at ω = 37,000 rpm.

The "modes" column identifies the coupled rotor-housing modes which are predominantly associated with the root cited. The predicted onset speed of instability (OSI) is around 95,000 rpm; however, extrapolation of the model data to this elevated speed is questionable. The instability is predicted for the forward critical speed associated with the "turbine" mode shapes of the 4 X-Z mode of Figure 8.4 and root 8 of Table 8.11.

Figure 8.6 illustrates the predicted synchronous-response results due to imbalance. The "pump" accelerometers are located at the extreme left end of the pump housing. The "turbine" accelerometers are located at the mounting flange on the pump housing immediately outside of the turbines. The relative deflection plots show relative rotor-housing motion at the pump bearing, the second-stage impeller, the turbine bearing, and the turbine interstage seals.

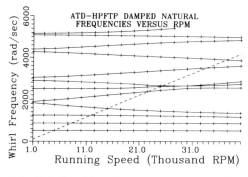

Figure 8.5   Calculated imaginary roots versus running speed.

A sharp "turbine" critical speed around 33,500 rpm dominates the response with two additional critical speeds evident at 22,000 and 25,000 rpm. The large turbine bearing reactions and accelerometer levels predicted by the linear model at the 33,500 rpm critical speed are unacceptable. The possibility of reducing the resonance or eliminating this critical speed via dead-band clearances is the subject of the following section.

8.4 NONLINEAR-ROTORDYNAMIC-ANALYSIS PREDICTIONS

As noted earlier, bearing dead-band clearances constitute the essential nonlinearity for the ATD-HPFTP rotordynamic model. Yamamoto's analysis (1954) for the influence of dead-band clearances on the response of the vertical Jeffcott rotor was introduced in Section 1.9 and, as we shall see, does an excellent qualitative job in predicting the behavior of the current complete rotordynamic model.

The bearing nonlinearities are accounted for here via numerical integration of the governing nonlinear differential equations as discussed in Section 7.4. The linear and nonlinear models are quite similar but differ in the following respects:

(a) The nonlinear model includes the bearing clearances.

(b) The first four rotor modes are used in nonlinear analyses versus five for the linear model. The fifth housing mode has the natural frequency $\lambda_5 = 1864$ Hz $= 111,840$ rpm, versus the top running speed of 36,465 rpm.*

(c) The first three housing modes are used in the nonlinear analysis versus four for the linear analysis. The omitted modes have natural frequencies on the order of 1020 Hz $= 61,350$ rpm.*

(d) The nonlinear model used here does not account for added-mass coefficients at the seals or impellers. To partially compensate for this deficiency, the cross-coupled damping coefficients are also dropped from the nonlinear model.† The comparatively small added-mass terms associated with the pump interstage seals and impellers of the ATP-HPFTP support this approximation; however, for seals with higher densities, these terms must be retained.

(e) The linear model does not account for side loads.

To check the validity of the nonlinear model definition, an initial simulation run was carried out with zero dead-band clearances. Predictions for the

---

*Comparisons between simulations which retained and omitted these modes showed slight differences at speeds around 40,000 rpm, but virtually identical response throughout the operating-speed range.

†Review the discussion following Eq. (4.61) concerning these comments.

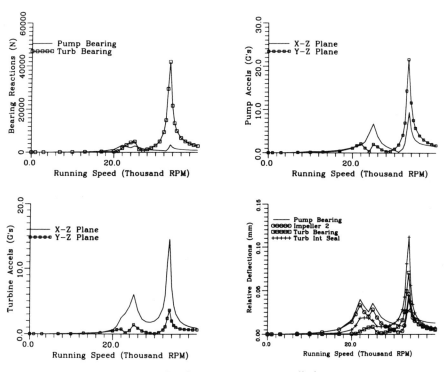

Figure 8.6 Synchronous response predictions.

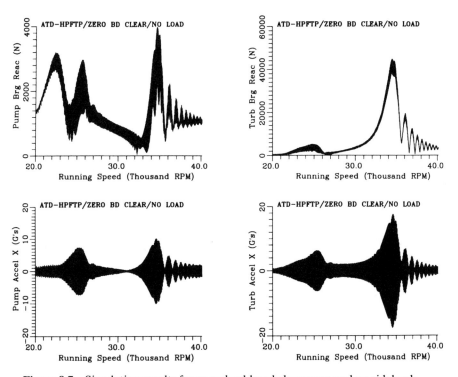

Figure 8.7 Simulation results for zero dead-band clearances and no sideloads.

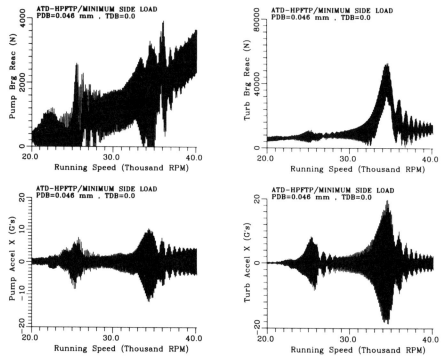

Figure 8.8 Simulation results for $\delta_P = 0.046$ mm, $\delta_T = 0.0$ dead-band clearances with sideloads.

pump and turbine bearing reactions and accelerometers in the X-Z plane of Figure 8.2 are illustrated in Figure 8.7. Initial conditions for the results of Figure 8.7 were obtained by running the simulation model at 20,000 rpm until a steady-state vibration level was achieved. The simulation of Figure 8.7 proceeds from these initial conditions and uses 20,000 integration steps to simulate a steady ramp from 20,000 to 40,000 rpm in 0.6 seconds. The actual turbopump goes from zero to 36,500 rpm in about 2 seconds.

A comparison of Figures 8.6 and 8.7 shows good correlation between predictions of the linear and nonlinear models for critical-speed locations and bearing and accelerometer levels.

Figures 8.8–8.11 provide comparable simulations for ($\delta_P = 0.046$ mm, $\delta_T = 0, 0.0064, 0.0127, 0.0254,$ and 0.0381 mm). Compare in particular the turbine-bearing reaction plots in these figures and observe that the turbine critical-speed location drops progressively as δ_T increases, eventually disappearing for $\delta_T = 0.0254$ and 0.0381 mm. The peak prediction turbine-bearing load drops from around 52 kN at 34,000 rpm to around 14 kN at the top running speed of 36,500 rpm. Predicted peak acceleration levels drop from nearly 20 g's at the turbine to less than 5 g's. Yamamoto's predicted results are just as dramatic in these figures as suggested in Section 1.9. They are

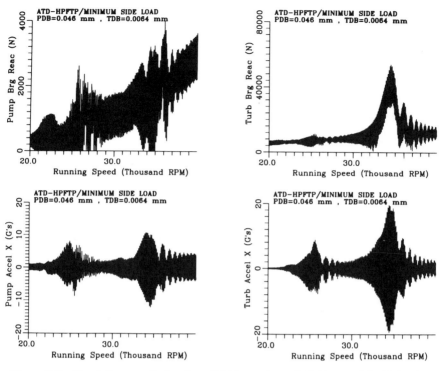

Figure 8.9 Simulation results for $\delta_P = 0.046$ mm, $\delta_T = 0.0064$ mm with sideloads.

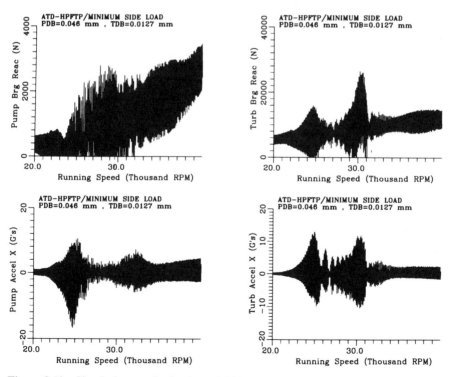

Figure 8.10 Simulation results for $\delta_P = 0.046$ mm, $\delta_T = 0.0127$ mm with sideloads.

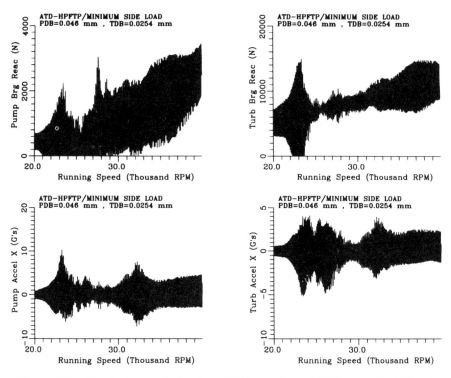

Figure 8.11 Simulation results for $\delta_P = 0.046$ mm, $\delta_T = 0.0254$ mm with sideloads.

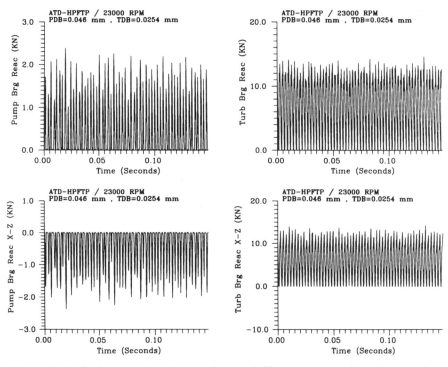

Figure 8.12 Turbopump bearing reactions and X components of bearing reactions versus time at 23,000 rpm for $\delta_P = 0.046$ mm, $\delta_T = 0.0254$ mm.

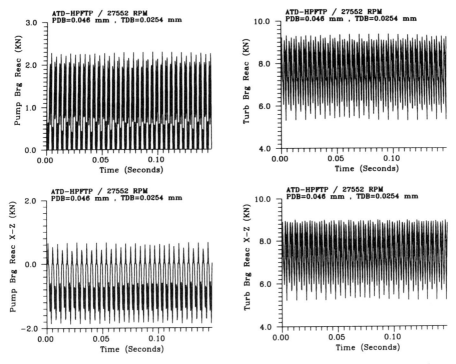

Figure 8.13 Turbopump bearing reactions and X components of bearing reactions versus time at 27,552 rpm for $\delta_P = 0.046$ mm, $\delta_T = 0.0254$ mm.

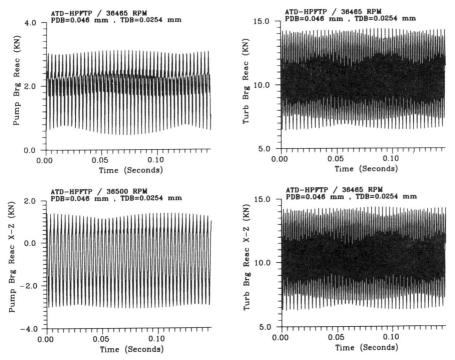

Figure 8.14 Turbopump bearing reactions and X components of bearing reactions versus time at 36,465 rpm for $\delta_P = 0.046$ mm, $\delta_T = 0.0254$ mm.

valid in the present case with a side load, although Yamamoto's model had no side load. Transient results with deadbands can be sensitive to side loads; however the present results held for estimated minimum and maximum side loads.

Figures 8.12–8.14 show the bearing-reaction magnitude and X components versus time for 23,000, 27,752, and 36,465 rpm. Note the "clipping" of bearing-reaction magnitude associated with dead-band traversal at the lower speeds and its absence at the top speed.

Large fractional-frequency motion results due to the dead-band clearances, as illustrated by the Fourier spectra of Figures 8.15 and 8.16. Figure 8.15 provides the Fourier spectra for the X and Y components of the pump and turbine accelerometers. Response at one-half running speed is evident in all of the accelerometer spectra but most pronounced in the pump accelerometers. Note also the sharp twice-frequency response at a running speed of 23,000 rpm. A possible explanation for this nonlinear result is provided by Figure 8.5, which shows a root to be present near 4800 rd/sec (46,000 rpm).

The relative-deflection spectra of Figure 8.16 show the one-half-synchronous response to be dominant at the pump bearing and second-stage impeller. The response is predominantly synchronous at the turbine bearing and about equally synchronous and one-half synchronous at the turbine interstage seal.

The one-half-running-speed response can be eliminated by reducing the clearance at the pump-end bearing. Figure 8.17 shows the predicted bearing reactions versus time at a running speed of 36,465 rpm. Note by comparison to Figure 8.14 that the low-frequency response is eliminated. The associated Fourier spectra of Figure 8.18 shows no one-half-running-speed response, only synchronous and twice-synchronous.

The response at one-half running speed illustrated in the spectra of Figures 8.15 and 8.16 is a *fractional-frequency* phenomena caused by the bearing clearances and is similar to the results of Section 1.8. A stability problem is not indicated by these results.

8.5   SUMMARY AND CONCLUSIONS

The results of the present study supported the conclusion that the ATD-HPFTP could be successfully operated with enlarged clearances at the pump interstage seals and enlarged clearance dead bands at the turbine bearing. Operation of the ATD-HPFTP successfully demonstrated the results of this study with nominal operating accelerometer levels on the order of 1–2 g's at full power levels. The results cited here do not preclude alternative design approaches to achieve a smoothly operating turbopump. Separate simulations predicted that the ATD-HPFTP would also operate successfully with a reduced turbine bearing stiffness ($K_T = 3.1 \times 10^5$ N/mm) and the original HPFTP pump-interstage-seal clearances (0.127 mm).

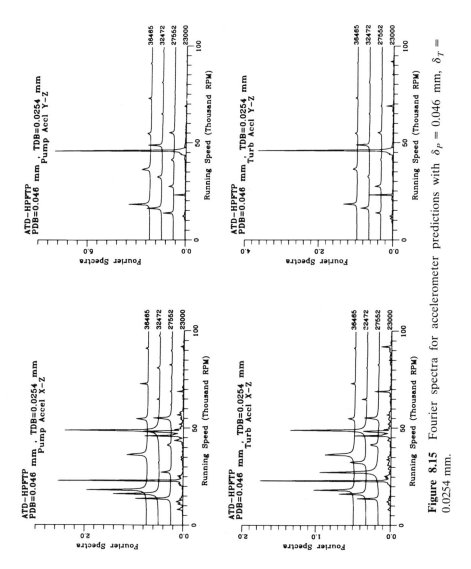

Figure 8.15 Fourier spectra for accelerometer predictions with $\delta_P = 0.046$ mm, $\delta_T = 0.0254$ mm.

453

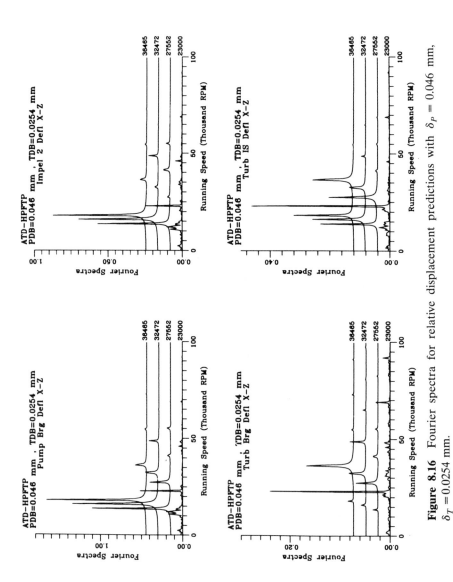

Figure 8.16 Fourier spectra for relative displacement predictions with $\delta_P = 0.046$ mm, $\delta_T = 0.0254$ mm.

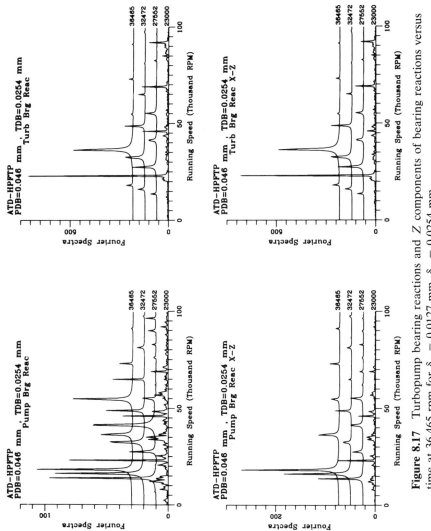

Figure 8.17 Turbopump bearing reactions and Z components of bearing reactions versus time at 36,465 rpm for $\delta_P = 0.0127$ mm, $\delta_T = 0.0254$ mm.

455

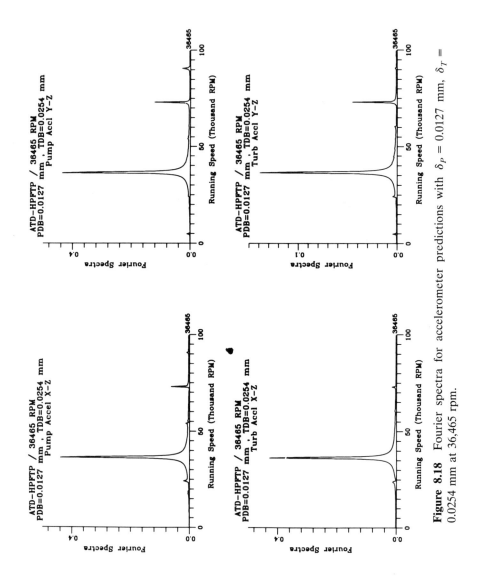

Figure 8.18 Fourier spectra for accelerometer predictions with $\delta_P = 0.0127$ mm, $\delta_T = 0.0254$ mm at 36,465 rpm.

REFERENCES: CHAPTER 8

Bolleter, U., Wyss, A., Welte, I., and Stürchler, R. (1987), "Measurement of Hydrodynamic Interaction Matrices of Boiler Feed Pump Impellers," *Journal of Vibration, Stress, and Reliability in Design*, **109**, 144–151.

Bolleter, U., Leibundgut, E., Stürchler, R., and McCloskey, T. (1989), "Hydraulic Interaction and Excitation Forces of High Head Pump Impellers," in *Pumping Machinery—1989*, Proceedings of the Third Joint ASCE/ASME Mechanics Conference, LaJolla, CA, pp. 187–194.

Childs, D. W., and Ramsey, C. (1991), "Seal-Rotordynamic-Coefficient Test Results for a Model SSME ATD-HPFTP Turbine Interstage Seal with and without a Swirl Brake," *Journal of Tribology*, **113**, 113–203.

Childs, D., Nolan, S., and Kilgore, J. (1990), "Additional Test Results for Round-Hole-Pattern Damper Seals: Leakage, Friction Factors, and Rotordynamic Force Coefficients," *Journal of Tribology*, **112**, 365–371.

Childs, D., and Kim, C.-H. (1985), "Analysis and Testing for Rotordynamic Coefficients of Turbulent Annular Seals with Different, Directionally-Homogeneous Surface-Roughness Treatment for Rotor and Stator Elements," *Journal of Tribology*, **107**, 296–306.

Childs, D. (1978), "The Space Shuttle Main Engine High Pressure Fuel Turbopump Rotordynamic Instability Problem," *Journal of Engineering for Power*, 48–57.

Elrod, D. (1988), "Entrance and Exit Region Friction Factor Models for Annular Seal Analysis," Ph.D. dissertation, TL-Seal-5-88, Turbomachinery Laboratory, Texas A&M University.

Pelletti, J. (1990), "A Comparison of Experimental Results and Theoretical Predictions for the Rotordynamic Coefficients of Short ($L/D = 1/6$) Labyrinth Seals," M.S.M.E. Thesis, Texas A&M University and Turbomachinery Laboratory Report No. TL-Seal-1-90.

Scharrer, J. (1987), "A Comparison of Experimental and Theoretical Results for Labyrinth Gas Seals," Ph.D. dissertation, Texas A&M University.

Yamamoto, T. (1954), "On the Critical Speeds of a Shaft," *Memoirs of the Faculty of Engineering, Nagoya University*, **6**(2).

APPENDIX A

SOLUTION FOR ELLIPTIC-ORBIT PARAMETERS*

The steady-state response of a rotor at a given station is generally of the form

$$x(t) = x_c \cos \omega t + x_s \sin \omega t,$$
$$y(t) = y_c \cos \omega t + y_s \sin \omega t, \tag{A.1}$$

and a synchronous rotor-response solution will yield the coefficients x_c, x_s, y_c, y_s. The orbit associated with the solution of Eq. (A.1) can be obtained by first solving for $\cos \omega t$ and $\sin \omega t$ as

$$\cos \omega t = (xy_s - yx_s)/(x_c y_s - y_c x_s),$$
$$\sin \omega t = (yx_c - xy_c)/(x_c y_s - y_c x_s),$$

and then summing the squares of these two terms to obtain

$$x^2(y_s^2 + y_c^2) - 2xy(x_s y_s + x_c y_c) + y^2(x_s^2 + x_c^2) = (x_s y_c - x_c y_s)^2. \tag{A.2}$$

This equation defines the ellipse of Figure A.1, and the major and minor semiaxes a and b and orbital inclination angle α are to be defined in terms of x_c, x_s, y_c, y_s. Equation (A.2) can be expressed in the quadratic form

$$\{x, y\} \begin{bmatrix} (y_s^2 + y_c^2) & -(x_s y_s + x_c y_c) \\ -(x_s y_s + x_c y_c) & (x_s^2 + x_c^2) \end{bmatrix} \begin{Bmatrix} x \\ y \end{Bmatrix} = (x_s y_c - x_c y_s)^2, \tag{A.3}$$

*The results of this Appendix were initially stated by Lund (1966) (References—Chapter 1).

458

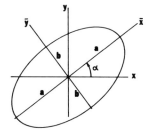

Figure A.1

or symbolically as

$$(u)^{T}[D](u) = g^{2}. \tag{A.4}$$

The eigenvalues of $[D]$ in this quadratic form are a^2 and b^2 defined by

$$a^2 = A + (B^2 + C^2)^{1/2},$$
$$b^2 = A - (B^2 + C^2)^{1/2},$$
$$A = (x_c^2 + y_c^2 + x_s^2 + y_s^2)/2,$$
$$B = (x_c^2 + y_c^2 - x_s^2 - y_s^2)/2,$$
$$C = x_c x_s + y_c y_s. \tag{A.5}$$

The desired eigenvector coordinate transformation to the principal $\bar{x}, \bar{y}$ system of Figure A.1 is stated

$$\begin{Bmatrix} x \\ y \end{Bmatrix} = \begin{bmatrix} \cos \alpha & -\sin \alpha \\ \sin \alpha & \cos \alpha \end{bmatrix} \begin{Bmatrix} \bar{x} \\ \bar{y} \end{Bmatrix}, \tag{A.6}$$

or symbolically as

$$(u) = [R(\alpha)](\bar{u}). \tag{A.7}$$

The angle α defines the eigenvectors which uncouple $[D]$, and we want to define it in terms of the parameters x_c, x_s, y_c, y_s. To this end, substitution from Eq. (A.7) into Eq. (A.4) yields

$$[R]^{T}[D][R] = \begin{bmatrix} a^2 & 0 \\ 0 & b^2 \end{bmatrix}. \tag{A.8}$$

Equation (A.8) provides four equations in terms of $\sin \alpha$ and $\cos \alpha$. Selecting and solving two of these equations defines α as follows:

$$\tan 2\alpha = \frac{2(x_s y_s + x_c y_c)}{x_s^2 + x_c^2 - y_s^2 - y_c^2}.$$

If one uses the first of Eq. (A.5) to initially calculate the semimajor axis, a, a useful expression for b follows from the principal-axis definition of the quadratic form

$$(b\bar{x})^2 + (a\bar{y})^2 = (ab)^2,$$

which yields, by comparison to Eq. (A.3),

$$b = \frac{x_c y_s - x_s y_c}{a}. \tag{A.9}$$

The time solution for the principal-axis coordinates may be stated

$$\bar{x} = a \cos(\omega t + \phi_a - \alpha_e),$$
$$\bar{y} = b \sin(\omega t + \phi_a - \alpha_e).$$

Hence, positive and negative solutions for b from Eq. (A.9) define forward- and backward-precessional motion, respectively.

APPENDIX B

FINITE-ELEMENT MODELS FOR SLENDER BEAMS

The developments of Section 2.6 yield the finite-element model

$$\{[M_T] + [M_R]\}(\ddot{u}) - \omega[G](\dot{u}) + [K](u) = (F). \quad (2.103)$$

The matrices in these expressions are

$$[M_T] = \frac{mL}{420}\begin{bmatrix} 156 \\ 0 & 156 & & & & sym \\ 0 & -22L & 4L^2 \\ 22L & 0 & 0 & 4L^2 \\ 54 & 0 & 0 & 13L & 156 \\ 0 & 54 & -13L & 0 & 0 & 156 \\ 0 & 13L & -3L^2 & 0 & 0 & 22L & 4L^2 \\ -13L & 0 & 0 & -3L^2 & -22L & 0 & 0 & 4L^2 \end{bmatrix},$$

$$\tag{B.1}$$

$$[M_R] = \frac{j}{30L}\begin{bmatrix} 36 \\ 0 & 36 & & & & sym \\ 0 & -3L & 4L^2 \\ 3L & 0 & 0 & 4L^2 \\ -36 & 0 & 0 & -3L & 36 \\ 0 & -36 & 3L & 0 & 0 & 36 \\ 0 & -3L & -L^2 & 0 & 0 & 3L & 4L^2 \\ 3L & 0 & 0 & L^2 & -3L & 0 & 0 & 4L^2 \end{bmatrix},$$

$$\tag{B.2}$$

$$[G] = \frac{\ddot{j}}{30L} \begin{bmatrix} 0 \\ 36 & 0 & & & \text{skew} & \text{sym} \\ -3L & 0 & 0 \\ 0 & -3L & 4L^2 & 0 \\ 0 & 36 & -3L & 0 & 0 \\ -36 & 0 & 0 & -3L & 36L & 0 \\ -3L & 0 & 0 & L^2 & 3L & 0 & 0 \\ 0 & -3L & -L^2 & 0 & 0 & 3L & 4L^2 & 0 \end{bmatrix}.$$

$$\tag{B.3}$$

If r is the cross-sectional radius, $j = mr^2/4$; $\ddot{j} = mr^2/2$ in Eqs. (B.1) and (B.2).

$$[K] = \frac{EI}{L^3} \begin{bmatrix} 12 \\ 0 & 12 & & & & \text{sym} \\ 0 & -6L & 4L^2 \\ 6L & 0 & 0 & 4L^2 \\ -12 & 0 & 0 & -6L & 12 \\ 0 & -12 & 6L & 0 & 0 & 12 \\ 0 & -6L & 2L^2 & 0 & 0 & 6L & 4L^2 \\ 6L & 0 & 0 & 2L^2 & -6L & 0 & 0 & 4L^2 \end{bmatrix}.$$

$$\tag{B.4}$$

For the linear imbalance distribution

$$a_x(s) = a_{x0}\left(1 - \frac{s}{L}\right) + a_{xL}\left(\frac{s}{L}\right),$$

$$a_y(s) = a_{y0}\left(1 - \frac{s}{L}\right) + a_{yL}\left(\frac{s}{L}\right), \tag{2.109}$$

the vectors $(ma_c), (ma_s)$ of the joint force definition

$$(F_a) = -\omega^2\{(ma_c)\cos \omega t + (ma_s)\sin \omega t\} \tag{2.108}$$

are defined by

$$(ma_c) = \frac{mL}{20} \left\{ \begin{array}{c} 7a_{x0} + 3a_{xL} \\ 7a_{y0} + 3a_{yL} \\ -La_{y0} - 2a_{yL}/3 \\ La_{x0} + 2a_{xL}/3 \\ 3a_{x0} + 7a_{xL} \\ 3a_{y0} + 7a_{yL} \\ 2La_{y0}/3 + La_{yL} \\ 2La_{x0}/3 - La_{xL} \end{array} \right\}, \tag{B.5}$$

$$(ma_s) = \frac{mL}{20} \left\{ \begin{array}{c} -7a_{y0} - 3a_{yL} \\ 7a_{x0} + 3a_{xL} \\ -La_{x0} - 2La_{xL}/3 \\ -La_{y0} - 2La_{yL}/3 \\ -3a_{y0} - 7a_{yL} \\ 3a_{x0} + 7a_{xL} \\ 2La_{x0}/3 + La_{xL} \\ 2La_{y0}/3 + La_{yL} \end{array} \right\}. \tag{B.6}$$

APPENDIX C

FLUIDITY MATRIX FOR A TWO-DIMENSIONAL THREE-NODE TRIANGULAR ELEMENT

The fluidity matrices of Eq. (3.122) reduce [Booker and Heubner (1972)], see page 203 to the following terms for the linear interpolation functions of Eq. (3.134):

$$K_{pij} = -\frac{\rho}{480\mu A} \left(\sum_{k=1}^{3} H_k^2 \sum_{k=1}^{3} H_k + \prod_{k=1}^{3} H_k \right)(b_i b_j + c_i c_j),$$

$$KUx_{ij} = \frac{\rho}{24} f_j b_i, \qquad KUz_{ij} = \frac{\rho}{24} f_j c_i,$$

$$K\dot{H}_{ij} = -\frac{\rho A}{12}(1 + \delta_{ij}),$$

where δ_{ij} is the Kronecker-delta function, and

$$f_j = \sum_{k=1}^{3} H_k(1 + \delta_{kj}).$$

APPENDIX D

SEAL-PERTURBATION COEFFICIENTS

This appendix defines the linear coefficents of Eq. (4.27). For a convergent-tapered seal, the nondimensional clearance function is

$$h_0 = 1 + q(1 - 2z), \qquad u_{s0} = 1/h_0,$$

where

$$q = \frac{C_0 - C_1}{C_0 + C_1}, \qquad \overline{C}_r = (C_0 + C_1)/2,$$

and C_0 and C_1 are, respectively, the inlet and exit clearances. The following functions are used for both the Blasius and Moody models:

$$u_{s0} = \left(w_0^2 + b^2 u_0^2 \right)^{1/2},$$

$$u_{r0} = \left[w_0^2 + b^2 (u_0 - 1)^2 \right]^{1/2},$$

$$b = R\omega/W_0,$$

$$\sigma_{s0} = f_{s0}(L/\overline{C}), \qquad \sigma_{r0} = f_{r0}(L/\overline{C})$$

$$\mathscr{R}_{z0} = 2W_0\overline{C}/\nu.$$

D.1 BLASIUS-EQUATION MODEL, $f_B = n\mathcal{R}^m$

$$A_{1z} = w_0\big[\sigma_{s0}u_{s0}(1 - ms) + \sigma_{r0}u_{r0}(1 - mr)\big]\big/2h_0^2,$$

$$A_{2z} = b^2 w_0\big[\sigma_{s0}u_0(1 + ms)/u_{s0} + \sigma_{r0}(u_0 - 1)(1 + mr)/u_{r0}\big]\big/2h_0,$$

$$A_{3z} = \left\{\sigma_{s0}\big[u_{s0}^2 + w_0^2(1 + ms)\big]\big/u_{s0}\right.$$

$$\left. +\sigma_{r0}\big[u_{r0}^2 + w_0^2(1 + m_r)\big]\big/u_{r0} - \frac{1}{h_0}\frac{dh_0}{dz}\right\}\bigg/2h_0,$$

$$A_{1\theta} = \big[u_0u_{s0}\sigma_{s0}(1 - ms) + (u_0 - 1)u_{r0}\sigma_{r0}(1 - mr)\big]\big/2h_0^2,$$

$$A_{2\theta} = \left\{\sigma_{s0}\big[u_{s0}^2 + b^2u_0^2(1 + ms)\big]\big/u_{s0}\right.$$

$$\left. +\sigma_{r0}\big[u_{r0}^2 + b^2(u_0 - 1)^2(1 + mr)\big]\big/u_{r0}\right\}\bigg/2h_0,$$

$$A_{3\theta} = -\left\{\sigma_{s0}u_0\big(b^2u_0^2 - m_sw_0^2\big)\big/u_{s0}\right.$$

$$\left. +\sigma_{r0}(u_0 - 1)\big[b^2(u_0 - 1)^2 - m_rw_0^2\big]\big/u_{r0}\right\}\bigg/2,$$

$$f_{s0} = ns\mathcal{R}_{z0}^{ms}(h_0u_{s0})^{ms}, \qquad f_{r0} = nr\mathcal{R}_{z0}^{mr}(h_0u_{r0})^{mr}.$$

D.2 MOODY-EQUATION MODEL, $f_m = a_1[1 + (a_2e/2\overline{C} + a_3/\mathcal{R})^{1/3}]$

$$A_{1z} = \frac{w_0}{2h_0^2}\left\{\sigma_{s0}u_{s0}\left[1 + \frac{1}{3}\left(1 - \frac{a_1}{f_{s0}}\right)\right] + \sigma_{r0}u_{r0}\left[1 + \frac{1}{3}\left(1 - \frac{a_1}{f_{r0}}\right)\right]\right\},$$

$$A_{2z} = \frac{w_0b^2}{2h_0}\left[\frac{\sigma_{s0}}{u_{s0}}u_0(1 + bs) + \frac{\sigma_{r0}}{u_{r0}}(u_0 - 1)(1 + b_r)\right],$$

$$A_{3z} = \frac{1}{2h_0}\left\{\frac{\sigma_{s0}}{u_{s0}}\big[u_{s0}^2 + w_0^2(1 + bs)\big] + \frac{\sigma_{r0}}{u_{r0}}\big[u_{r0}^2 + w_0^2(1 + br)\big] - \frac{1}{h_0}\frac{dh_0}{dz}\right\},$$

$$A_{1\theta} = \frac{1}{2h_0^2}\left\{\sigma_{s0}u_0u_{s0}\left[1 + \frac{1}{3}\left(1 - \frac{a_1}{f_{s0}}\right)\right] + \sigma_{r0}(u_0 - 1)u_{r0}\left[1 + \frac{1}{3}\left(1 - \frac{a_1}{f_{r0}}\right)\right]\right\},$$

$$A_{2\theta} = \frac{1}{2h_0}\left\{\frac{\sigma_{s0}}{u_{s0}}\big[u_{s0}^2 + b^2u_0^2(1 + bs)\big] + \frac{\sigma_{r0}}{u_{r0}}\big[u_{r0}^2 + b^2(u_0 - 1)^2(1 + br)\big]\right\},$$

$$A_{3\theta} = -\frac{1}{2}\left\{\frac{\sigma_{s0}}{u_{s0}}u_0\big(b^2u_0^2 - bsw_0^2\big) + \frac{\sigma_{r0}}{u_{r0}}(u_0 - 1)\big[b^2(u_0 - 1)^2 - brw_0^2\big]\right\},$$

$$f_{s0} = a_1\left(1 + B_{s0}^{1/3}\right), \qquad B_{s0} = \frac{a_{2s}}{h_0} + \frac{a_3}{h_0 u_{s0}},$$

$$f_{r0} = a_1\left(1 + B_{r0}^{1/3}\right), \qquad B_{r0} = \frac{a_{2r}}{h_0} + \frac{a_3}{h_0 u_{s0}},$$

$$bs = -\frac{1}{3B_{s0}}\left(1 - \frac{a_1}{f_{s0}}\right)\left(\frac{a_3}{h_0 u_{s0}}\right), \qquad br = -\frac{1}{3B_{r0}}\left(1 - \frac{a_1}{f_{r0}}\right)\left(\frac{a_3}{h_0 u_{s0}}\right),$$

$$a_1 = 1.375 \times 10^{-3}, \qquad a_3 = 10^6/\mathscr{R}_{z0},$$

$$a_{2s} = 2 \times 10^4\left(e_s/2\overline{C}_r\right), \qquad a_{2r} = 2 \times 10^4\left(e_r/2\overline{C}_r\right).$$

APPENDIX E

IMPELLER PERTURBATION COEFFICIENTS

This appendix defines the linear coefficients of Eq. (6.49).

PERTURBATION COEFFICIENTS

$$A_{1s} = [\sigma_s(1 - ms) + \sigma_r(1 - mr)]w_{s0}^2/2h_0,$$

$$A_{2s} = -\frac{2u_0}{r}\frac{dr}{ds}\bigg/b^2 + [\sigma_r(mr + 1)\beta_0 + \sigma_s(ms + 1)\beta_1]w_{s0}/2,$$

$$A_{3s} = \frac{dw_{s0}}{ds} + [(2 + mr)\sigma_r + (2 + ms)\sigma_s]w_{s0}/2$$
$$- [(1 + mr)\sigma_r\beta_0(u_0 - r) + (1 + ms)\sigma_s\beta_1 u_0]/2,$$

$$2A_{1\theta} = w_{s0}[(1 - mr)(u_0 - r)\sigma_r + (1 - ms)u_0\sigma_s]/h_0,$$

$$2A_{2\theta} = w_{s0}(\sigma_r + \sigma_s) + \sigma_r(mr + 1)(u_0 - r)\beta_0$$
$$+ \sigma_s(ms + 1)u_0\beta_1 + 2\frac{w_{s0}}{r}\frac{dr}{ds},$$

$$2A_{3\theta} = \sigma_r(u_0 - r)[mr - (1 + mr)\beta_0(u_0 - r)/w_{s0}]$$
$$+ \sigma_s u_0[ms - (1 + ms)\beta_1 u_0/w_{s0}],$$

$$\beta_0 = (u_0 - r)/b^2 w_{s0}\left\{1 + [(u_0 - r)/bw_{s0}]^2\right\},$$

$$\beta_1 = u_0/b^2 w_{s0}\left[1 + (u_0/bw_{s0})^2\right].$$

$\tau_{r\theta}$ PERTURBATION COEFFICIENTS

$$B_{\theta 1} = f_{r0}(1 + mr)(u_0 - r)\left[1 - \beta_0(u_0 - r)/w_{s0}\right]\big/2b,$$

$$B_{\theta 2} = f_{r0}\left[w_{s0} + (1 + mr)(u_0 - r)\beta_0\right]\big/2b,$$

$$B_{\theta 3} = f_{r0}mr(u_0 - r)w_{s0}/2bh_0.$$

τ_{rs} PERTURBATION COEFFICIENTS

$$B_{s1} = f_{r0}\left[(2 + mr)w_{s0} - (1 + mr)\beta_0(u_0 - r)\right]\big/2,$$

$$B_{s2} = f_{r0}(1 + mr)\beta_0 w_{s0}/2,$$

$$B_{s3} = f_{r0}mrw_{s0}^2/2h_0.$$

INDEX